Electric Vehicle Efficient Power and Propulsion Systems

Electric Vehicle Efficient Power and Propulsion Systems

Editors

João Pedro F. Trovão
Minh Cao Ta

MDPI • Basel • Beijing • Wuhan • Barcelona • Belgrade • Manchester • Tokyo • Cluj • Tianjin

Editors

João Pedro F. Trovão
University of Sherbrooke
Canada
Polytechnic of Coimbra
(IPC-ISEC) and INESC
Coimbra
Portugal

Minh Cao Ta
Hanoi University of Science
and Technology
Vietnam
University of Sherbrooke
Canada

Editorial Office
MDPI
St. Alban-Anlage 66
4052 Basel, Switzerland

This is a reprint of articles from the Special Issue published online in the open access journal *Energies* (ISSN 1996-1073) (available at: https://www.mdpi.com/journal/energies/special_issues/Electric_Vehicle_Efficient_Power_Propulsion_Systems).

For citation purposes, cite each article independently as indicated on the article page online and as indicated below:

LastName, A.A.; LastName, B.B.; LastName, C.C. Article Title. *Journal Name* **Year**, *Volume Number*, Page Range.

ISBN 978-3-0365-5291-0 (Hbk)
ISBN 978-3-0365-5292-7 (PDF)

Cover image courtesy of João Pedro F. Trovão

Contents

About the Editors . . . . **vii**

Preface to "Electric Vehicle Efficient Power and Propulsion Systems" . . . . **ix**

João Pedro F. Trovão and Minh Cao Ta
Electric Vehicle Efficient Power and Propulsion Systems
Reprinted from: *Energies* **2022**, *15*, 3863, doi:10.3390/en15113863 . . . . **1**

Jorge Varela Barreras, Ricardo de Castro, Yihao Wan and Tomislav Dragicevic
A Consensus Algorithm for Multi-Objective Battery Balancing
Reprinted from: *Energies* **2021**, *14*, 4279, doi:10.3390/en14144279 . . . . **5**

Quan Ouyang, Rui Ma, Zhaoxiang Wu, Guotuan Xu and Zhisheng Wang
Adaptive Square-Root Unscented Kalman Filter-Based State-of-Charge Estimation for Lithium-Ion Batteries with Model Parameter Online Identification
Reprinted from: *Energies* **2020**, *13*, 4968, doi:10.3390/en13184968 . . . . **31**

Joon-Hyoung Ryu, June-Hee Lee and June-Seok Lee
Switching Frequency Determination of SiC-Inverter for High Efficiency Propulsion System of Railway Vehicle
Reprinted from: *Energies* **2020**, *13*, 5035, doi:10.3390/en13195035 . . . . **45**

Thanh Vo-Duy, Bao-Huy Nguyen, Joao P. Trovao and Minh C. Ta
Experimental Platform for Evaluation of On-Board Real-Time Motion Controllers for Electric Vehicles
Reprinted from: *Energies* **2020**, *13*, 6448, doi:10.3390/en13236448 . . . . **59**

Binh-Minh Nguyen, Hung Van Nguyen, Minh Ta-Cao and Michihiro Kawanishi
Longitudinal Modelling and Control of In-Wheel-Motor Electric Vehicles as Multi-Agent Systems
Reprinted from: *Energies* **2020**, *13*, 5437, doi:10.3390/en13205437 . . . . **87**

Stefano De Pinto, Pablo Camocardi, Christoforos Chatzikomis, Aldo Sorniotti, Francesco Bottiglione, Giacomo Mantriota and Pietro Perlo
On the Comparison of 2- and 4-Wheel-Drive Electric Vehicle Layouts with Central Motors and Single- and 2-Speed Transmission Systems
Reprinted from: *Energies* **2020**, *13*, 3328, doi:10.3390/en13133328 . . . . **115**

Caiyang Wei, Theo Hofman, Esin Ilhan Caarls
Co-Design of CVT-Based Electric Vehicles
Reprinted from: *Energies* **2020**, *14*, 1825, doi:10.3390/en14071825 . . . . **137**

Waruna Maddumage, Malika Perera, Rahula Attalage and Patrick Kelly
Power Management Strategy of a Parallel Hybrid Three-Wheeler for Fuel and Emission Reduction
Reprinted from: *Energies* **2021**, *14*, 1833, doi:10.3390/en14071833 . . . . **171**

Bao-Huy Nguyen, João Pedro F. Trovão, Ronan German and Alain Bouscayrol
Real-Time Energy Management of Parallel Hybrid Electric Vehicles Using Linear Quadratic Regulation
Reprinted from: *Energies* **2020**, *13*, 5538, doi:10.3390/en13215538 . . . . **201**

Amin Ghobadpour, Ali Amamou, Sousso Kelouwani, Nadjet Zioui and Lotfi Zeghmi
Impact of Powertrain Components Size and Degradation Level on the Energy Management of a Hybrid Industrial Self-Guided Vehicle
Reprinted from: *Energies* **2020**, *13*, 5041, doi:10.3390/en13195041 . . . . **221**

Lixing Wang, Zhenning Wu and Changyong Cao
Integrated Optimization of Routing and Energy Management for Electric Vehicles in Delivery Scheduling
Reprinted from: *Energies* **2021**, *14*, 1762, doi:10.3390/en14061762 . . . . **241**

About the Editors

João Pedro F. Trovão

João Pedro F. Trovão (Senior Member, IEEE) received M.Sc. degree and the Ph.D. degree in Electrical Engineering from the University of Coimbra, Coimbra, Portugal, in 2004 and 2013, respectively. Dr. Trovão was a Teaching Assistant, and he serves as Professor of Polytechnic of Coimbra–Coimbra Institute of Engineering (IPC–ISEC), Portugal. Since 2014, he has been a Professor with the Department of Electrical Engineering and Computer Engineering, University of Sherbrooke, Sherbrooke, QC, Canada, where he holds the Canadian Research Chair position in Efficient Electric Vehicles with Hybridized Energy Storage Systems. He is an author/coauthor of over 170 journal and conference papers. His research interests cover the areas of electric vehicles, hybridized energy storage systems, energy management and rotating electrical machines. Dr. Trovão was the General Chair of the 2018 IEEE Vehicle Power and Propulsion Conference, Chicago, US. He was a Guest Editor for the Special Issue of IET Electrical Systems in Transportation on Energy Storage and Electric Power Sub-Systems for Advanced Vehicles. He was a Guest Editor for the Special Issues of IEEE Transactions on Vehicular Technology on Electric Powertrains for Future Vehicles and on Advanced Vehicle Power Propulsion Systems. He is a founding member and the director of the electric-Transport, Energy Storage and Conversion (e-TESC) Lab of the University of Sherbrooke. He is a Senior Editor for the Automotive Electronics topic of the IEEE Vehicular Technology Magazine.

Minh Cao Ta

Minh Cao Ta (Senior Member, IEEE) received the B.S. (Hons.) degree from the Institute of Technology (now the University of West Bohemia), Pilsen, Czech Republic, in 1986, and the Ph.D. degree from Laval University, QC, Canada, in 1998, both in electrical engineering. From 1998 to 2004, he was with Kyushu University; The University of Tokyo; and NSK Steering Systems Ltd. Co., Japan. In 2009, he became an Associate Professor with the Hanoi University of Science and Technology (HUST), Vietnam, and was the Founding Director of the Center for Technology Innovation (CTI) 2009–2018, and Head of the Department of Industrial Automation, HUST, 2018–2020. He is currently with the Department of Electrical and Computer Engineering, University of Sherbrooke, Sherbrooke, QC, Canada. He is the author/coauthor of 14 Japanese/US/European patents and over 100 journal and conference papers. His main research interests include motor drives and advanced control techniques and their applications to electric vehicles and energy conversion systems. Dr. Ta was a Recipient the 2017 Nagamori Awards. He was the General Chair of the 2019 IEEE Vehicle Power and Propulsion Conference, Hanoi, Vietnam. He was a Guest Editor for several Special Issues of IEEE Transactions. He severed as Vice-President and General Secretary of the Vietnam Automation Association (VAA) 2014–2020.

Preface to "Electric Vehicle Efficient Power and Propulsion Systems"

Nowadays, vehicle electrification is an important field of research since it is a large umbrella of almost all engineering activities and systems. The challenges are much more than the development of lithium-ion batteries that were proposed in the beginning of the nineties and have been the most used vehicle energy storages to this day, mostly in commercial electric vehicles. Technological advances and affordability have enabled the deployment of new energy storage systems, power electronics semiconductors, electric machines, energy management systems, and in particular, efficient systems and intelligent systems. These new-generation systems cover a broad range of implementation for which traditional approaches have practical limitations. As a result, new development in component and system level related to power efficiency and computational intelligence has been expanding at a high speed over the last two decades. Electric vehicle itself is still a current research trend. Indeed, there are hundreds of publications on this topic that may cause even a knowledgeable person in the subject to lose track of all novelties.

The main purpose of the book is therefore to present some recent and important developments related to vehicle electrification, with consciousness to the efficiency of the powertrain in EVs through state-of-the-art surveys. Future directions and research perspectives on the subject are clearly presented. This book is designed to give to the readers a better understanding of different challenges faced in electrified vehicles, from the power to the propulsion systems, to provide insights that enable the development of new approaches. The reading of this book is not intended to be linear, i.e., each chapter stands on its own. However, the chapter order follows a logical flow of ideas that may help a novice in vehicle electrification to better understand the presented concepts. Chapter 1 is an overview of this book that shows how it should be read to make important links between the contributions and positions readers to the next challenges in this important field of engineering knowledge.

The following two chapters of the book introduce recent development in the sub-field of "Energy Storage Systems". The contributions are focused on the enhancement of the battery's performance based on multi-agent consensus and on effectiveness of the estimation technique by adaptative square-root unscented Kalman filter. The fourth chapter presents the recent development on wide-bandgap (WBG) power semiconductors. Recent research on WBG semiconductors such as silicon carbide (SiC) and gallium nitride (GaN) have attracted significant attention, since the increase in converter switching frequency contributes to the decrease in global power losses. The next four chapters are related to the evolution of traction and motion control techniques and co-design technique for vehicle transmission. The last four chapters are the demonstrations of the effective usage of on-board stored energy. In this sub-field "Energy Management Systems", readers can experience methodologies such as rule-based supervisory controller, novel real-time optimization-based torque distribution strategy, energy management strategy for an industrial hybrid self-guided vehicle, considering the size of a fuel cell stack and the degradation of a battery pack. To replace fossil-fueled vehicles by EVs in logistics applications, a system to manage energy and to schedule the vehicle route has been proposed using genetic algorithm. Most of the powertrain improvement "from energy storage to wheels" presented in this book deal with trends, implementation difficulty and their limitations. In summary, the contributed book covers theoretical aspects of vehicle electrification engineering along with practical applications. This book empowers the readers to establish the appropriate research planning and methods for devoted specific application over the next generation of efficient powertrains.

The book would not be possible without the contributions of various authors. The advanced control and optimization techniques, the valuable findings and practical implementation described in the high-quality manuscripts have been very much appreciated.

João Pedro F. Trovão and Minh Cao Ta
Editors

energies MDPI

Editorial

Electric Vehicle Efficient Power and Propulsion Systems

João Pedro F. Trovão [1,2,*] and Minh Cao Ta [1,3]

1 Department of Electrical Engineering and Computer Engineering, University of Sherbrooke, Sherbrooke, QC J1K 2R1, Canada; cao.minh.ta@usherbrooke.ca or minh.tacao@hust.edu.vn
2 Polytechnic of Coimbra, IPC-ISEC and INESC Coimbra, 3030-199 Coimbra, Portugal
3 Control Technique and Innovation Lab. for EVs, School of Electrical and Electronic Engineering, Hanoi University of Science and Technology, Hanoi 11615, Vietnam
* Correspondence: joao.trovao@usherbrooke.ca or jtrovao@isec.pt; Tel.: +1-(819)-821-8000 (ext. 65401)

Citation: Trovão, J.P.F.; Ta, M.C. Electric Vehicle Efficient Power and Propulsion Systems. *Energies* **2022**, *15*, 3863. https://doi.org/10.3390/en15113863

Received: 16 May 2022
Accepted: 16 May 2022
Published: 24 May 2022

Publisher's Note: MDPI stays neutral with regard to jurisdictional claims in published maps and institutional affiliations.

1. Introduction

Vehicle electrification is one of the main growing trends with an identified growth capacity of 15% until 2030. In the coming years, many internal combustion engine vehicles, hybrid vehicles and all-electric vehicles will be on the road as consumers switch to more efficient and environmentally friendly propulsion systems. To remain competitive in this electrically powered future, carmakers and researchers are investing in a wide range of propulsion technologies to increase efficiency and power capacity, developing the next generation of powertrains. The development of more efficient pure EVs, HEVs and fuel cell electric vehicles (FCEV) presents both a challenge and a definite solution to current mobility issues. A reliable EV solution should therefore harness the advantages of more efficient and powerful energy storage systems from multiple sources through their effective management and new and improved power converters, including the new generation of switching devices, and explore advanced configurations for electric motors, reducing the use of rare-earth materials.

This Special Issue entitled "Electric Vehicle Efficient Power and Propulsion Systems" was established to encourage researchers working in this field to share the latest developments on electric vehicle efficient power and propulsion systems for road, rail and air vehicles, both manned and unmanned. Therefore, this Special Issue aims to provide recent developments in any field of electric vehicle efficiency powertrains, covering a wide range of scientific topics, including, but not limited to, the following:

- Energy storage systems;
- Power Electronics;
- Electric machines;
- Energy management systems.

The Guest Editors are Prof. João Pedro F. Trovão and Minh C. Ta (University of Sherbrooke, Sherbrooke, QC, Canada). Each paper was reviewed by three independent reviewers from countries differing from each author's country of origin. In order to ensure the highest quality, only 11 submissions were accepted for inclusion in this Special Issue of *Energies*.

2. Special Issue Content

In total, 11 papers relating to electric vehicle efficiency powertrains have been accepted and published in this Special Issue. This subject encompasses a wide range of topics, as illustrated by the diversity of the accepted papers: four papers were selected on energy management algorithms and topologies, four on wheel-drives control development, transmission, and validation, two on Lithium-ion batteries and one on SiC-inverter efficiency. All four of these topics are addressed in this Special Issue.

In the first sub-topic "Energy storage Systems" we received two papers. First, Barreras, J.V. et al. [1] present novel research on the distributed multi-objective control algorithm, which is based on a multi-agent consensus algorithm. The authors implemented and validated the control in simulations, considering an electro-thermal lithium-ion battery model and an electric vehicle model parameterized with experimental data. The results show that the proposed multi-functional balancing can enhance the performance of batteries with substantial cell-to-cell differences under the most demanding operating conditions. The second paper in this sub-topic focuses on state-of-charge (SOC) as a fundamental indicator of lithium-ion batteries, which has an important role in the battery's optimized operation. Ouyang, Q. et al. [2] propose an adaptive square-root unscented Kalman filter (SRUKF) to estimate the battery's SOC. The effectiveness of the estimation technique is demonstrated by their extensive experimental results, which show higher precision compared with other commonly used Kalman filter-based methods.

For the second subtopic "Power Electronics", one contribution was received focusing on wide-bandgap (WBG) power semiconductors. Recent research on WBG power semiconductor devices such as silicon carbide (SiC) and gallium nitride (GaN) have attracted significant attention, since the increase in inverter switching frequency contributes to the decrease in switching power loss. In paper [3], Ryu, J.-H. et al. investigate a method for switching the frequency determination of a SiC-inverter to improve the efficiency of the railway propulsion system. The hybrid switching method combined with the synchronous PWM and asynchronous PWM was considered in the analysis of the SiC inverter that feeds a PMSM. The efficiency curve of the propulsion system as a function of the switching frequency demonstrated the potential to serve as a guideline for switching frequency selection.

For the third subtopic "Electric machines", three papers focus on electric machines for traction/motion control, and one paper on the co-design technique of a continuously variable transmission (CVT) for EVs. Vo-Duy, T. et al. present an experimental platform for the evaluation of on-board real-time motion controllers for EVs [4], providing a good example of the hardware-in-the-loop (HiL) system, on which the most important vehicle dynamics, including longitudinal, lateral, yaw, kinematic, and position models, have been built and validated in various testing scenarios. The proposed HiL system can be employed for research on various topics of EV control. Paper [5] demonstrates an alternative point of view, using multi-agent-system theory. Nguyen, B.-M. et al. propose three different ways of modeling in-wheel-motor (IWMs) vehicles: a nonlinear model with a hierarchical structure for passivity-based motion control, a linearized model with a rank-1 interconnection matrix for stability analysis, and a time-varying state-space model for optimal control, using a linear quadratic regulator (LQR). The effectiveness of the three models and their design approaches are discussed in several examples with a Matlab/Carsim co-simulator. Additionally, paper [6] addresses the multi-motor configuration and control of EVs. De Pinto, S. et al. provide a comparison between 2-wheel-drive (2WD) and 4WD configurations for the same EV from the viewpoint of drivability and energy consumption. An optimization routine is used to calculate the energy-efficient gear state and/or torque distribution for each considered configuration. The results highlight that the single-speed 4WD layout can reduce the energy consumption during driving cycles by approximately 9% compared to the conventional 2WD layout with single-speed transmission. Finally, Wei, C. et al. consider the co-design of a continuously variable transmission (CVT) for EVs. In paper [7], a novel convex programming (CP)-based co-design method is proposed to minimize the total-cost-of-ownership (TCO). The strength of the co-design is highlighted in comparison with a sequential design, and insights into the design of a low-power EV that is energy-efficient and cost-effective for urban driving are provided. A highly integrated EM-CVT system, which is efficient, low-cost, and lightweight, can be expected for future EV applications.

The effective utilization of energy storage systems is still a critical issue in electrified vehicles. The subtopic of "Energy management systems" attracted four papers in this Special Issue [8–11]. First, Maddumage, W. et al. propose a design methodology for a rule-based supervisory controller of a pre-transmission parallel hybrid three-wheeler based on the optimal control strategy, i.e., dynamic programming (DP). The developed rule-based strategy shows performance within 10% of the DP results on WLTC and UDC-NEDC drive cycles and has the advantage of being near-optimal, easy-to-implement and computationally less demanding. Nguyễn, B.-H. et al. propose a novel real-time optimization-based torque distribution strategy for a parallel hybrid truck in [9]. The strategy aims to minimize the engine fuel consumption while ensuring battery charge-sustaining by using linear quadratic regulation in a closed-loop control scheme. By reformulating the problem, the obtained strategy does not require the information of the engine efficiency map like the previous studies in the literature. The proposed method has been evaluated via simulation in comparison with dynamic programming as a benchmark, and experimentally validated by using a power hardware-in-the-loop simulator. Additionally, Ghobadpour, A. et al. present the design of an energy management strategy (EMS) for an industrial hybrid self-guided vehicle (SGV), considering the size of a fuel cell (FC) stack and the degradation of a battery pack [10]. In the paper, a realistic energy model of the SGV was first proposed and validated, based on experiments. The performance of the developed FC/battery SGV powertrain was then validated under three EMS modes. Each mode was studied by considering four different FC sizes and three battery degradation levels. The results showed several important remarks, such that a small FC as a range extender is recommended to reduce system costs. Battery SOC must be kept at a high level during SGV operation to support the FC during SGV acceleration. Finally, EVs to replace fossil-fueled vehicles in logistics applications are studied in [11] by Wang, L. et al. The authors propose a system to manage energy and to schedule the vehicle route, which comprises two parts: (i) a case-based reasoning subsystem to forecast the energy consumption and travel time for each route section, and (ii) a genetic algorithm to optimize vehicle routing with an energy consumption situation as a new constraint. A dynamic adjustment algorithm is also adopted to achieve a rapid response to accidents in which the vehicles might be involved. Simulation study, validations on Solomon benchmarks, and result analysis have been performed to show that the proposed vehicle management system is more economical than the traditional method.

3. Closing Remarks and Future Challenges

The articles presented in this Special Issue cover important aspects of electric vehicle efficiency powertrains. With the wide range of topics covered in the selected papers, the contributions to this Special Issue can stimulate the EV community to undertake further research concerning more efficient EVs. Therefore, we believe that the presented papers will have practical importance for forthcoming developments in the automotive industries. The contributions included in this Special Issue offer new data, information, and findings to continue the R&D effort in the field, with the aim of stimulating the research community to further contribute to the development of the field.

Author Contributions: Conceptualization and methodology J.P.F.T. and M.C.T.; formal analysis, J.P.F.T. and M.C.T.; resources, J.P.F.T. and M.C.T.; writing—original draft preparation, J.P.F.T. and M.C.T.; writing—review and editing, J.P.F.T. and M.C.T.; visualization, J.P.F.T. and M.C.T.; supervision, J.P.F.T. and M.C.T.; project administration, J.P.F.T. and M.C.T. All authors have read and agreed to the published version of the manuscript.

Funding: This research received no external funding.

Acknowledgments: Each paper was reviewed by at least three reviewers; therefore, the Guest Editors would like to thank them for their work, which helped the authors to improve their manuscripts. In addition, the Guest Editors would like to thank Sherwin Chen for his help and support in the Editorial process of the Electric Vehicle Efficient Power and Propulsion Systems Special Issue of the *Energies* journal.

Conflicts of Interest: The authors declare no conflict of interest.

References

1. Barreras, J.V.; de Castro, R.; Wan, Y.; Dragicevic, T. A consensus algorithm for multi-objective battery balancing. *Energies* **2021**, *14*, 4279. [CrossRef]
2. Ouyang, Q.; Ma, R.; Wu, Z.; Xu, G.; Wang, Z. Adaptive square-root unscented kalman filter-based state-of-charge estimation for lithium-ion batteries with model parameter online identification. *Energies* **2020**, *13*, 4968. [CrossRef]
3. Ryu, J.-H.; Lee, J.-H.; Lee, J.-S. Switching Frequency Determination of SiC-inverter for high efficiency propulsion system of railway vehicle. *Energies* **2020**, *13*, 5035. [CrossRef]
4. Vo-Duy, T.; Ta, M.C.; Nguyễn, B.-H.; Trovão, J.P.F. Experimental platform for evaluation of on-board real-time motion controllers for electric vehicles. *Energies* **2020**, *13*, 6448. [CrossRef]
5. Nguyen, B.-M.; Nguyen, H.V.; Ta-Cao, M.; Kawanishi, M. Longitudinal modelling and control of in-wheel-motor electric vehicles as multi-agent systems. *Energies* **2020**, *13*, 5437. [CrossRef]
6. De Pinto, S.; Camocardi, P.; Chatzikomis, C.; Sorniotti, A.; Bottiglione, F.; Mantriota, G.; Perlo, P. On the comparison of 2- and 4-wheel-drive electric vehicle layouts with central motors and single- and 2-speed transmission systems. *Energies* **2020**, *13*, 3328. [CrossRef]
7. Wei, C.; Hofman, T.; Ilhan Caarls, E. Co-design of CVT-based electric vehicles. *Energies* **2021**, *14*, 1825. [CrossRef]
8. Maddumage, W.; Perera, M.; Attalage, R.; Kelly, P. Power management strategy of a parallel hybrid three-wheeler for fuel and emission reduction. *Energies* **2021**, *14*, 1833. [CrossRef]
9. Nguyễn, B.-H.; Trovão, J.P.F.; German, R.; Bouscayrol, A. Real-time energy management of parallel hybrid electric vehicles using linear quadratic regulation. *Energies* **2020**, *13*, 5538. [CrossRef]
10. Ghobadpour, A.; Amamou, A.; Kelouwani, S.; Zioui, N.; Zeghmi, L. Impact of powertrain components size and degradation level on the energy management of a hybrid industrial self-guided vehicle. *Energies* **2020**, *13*, 5041. [CrossRef]
11. Wang, L.; Wu, Z.; Cao, C. Integrated optimization of routing and energy management for electric vehicles in delivery scheduling. *Energies* **2021**, *14*, 1762. [CrossRef]

energies

Article

A Consensus Algorithm for Multi-Objective Battery Balancing

Jorge Varela Barreras [1,2,*], Ricardo de Castro [3], Yihao Wan [4] and Tomislav Dragicevic [4]

1 Department of Mechanical Engineering, Imperial College London, London SW7 1AY, UK
2 The Faraday Institution, Didcot OX11 0RA, UK
3 Department of Mechanical Engineering, University of California, Merced, CA 95343, USA; rpintodecastro@ucmerced.edu
4 Department of Electrical Engineering, Technical University of Denmark, 2800 Kongens Lyngby, Denmark; wanyh@elektro.dtu.dk (Y.W.); tomdr@elektro.dtu.dk (T.D.)
* Correspondence: jvarelab@ic.ac.uk

Abstract: Batteries stacks are made of cells in certain series-parallel arrangements. Unfortunately, cell performance degrades over time in terms of capacity, internal resistance, or self-discharge rate. In addition, degradation rates are heterogeneous, leading to cell-to-cell variations. Balancing systems can be used to equalize those differences. Dissipative or non-dissipative systems, so-called passive or active balancing, can be used to equalize either voltage at end-of-charge, or state-of-charge (SOC) at all times. While passive balancing is broadly adopted by industry, active balancing has been mostly studied in academia. Beyond that, an emerging research field is multi-functional balancing, i.e., active balancing systems that pursue additional goals on top of SOC equalization, such as equalization of temperature, power capability, degradation rates, or losses minimization. Regardless of their functionality, balancing circuits are based either on centralized or decentralized control systems. Centralized control entails difficult expandability and single point of failure issues, while decentralized control has severe controllability limitations. As a shift in this paradigm, here we present for the first time a distributed multi-objective control algorithm, based on a multi-agent consensus algorithm. We implement and validate the control in simulations, considering an electro-thermal lithium-ion battery model and an electric vehicle model parameterized with experimental data. Our results show that our novel multi-functional balancing can enhance the performance of batteries with substantial cell-to-cell differences under the most demanding operating conditions, i.e., aggressive driving and DC fast charging (2C). Driving times are extended (>10%), charging times are reduced (>20%), maximum cell temperatures are decreased (>10 °C), temperature differences are lowered (~3 °C rms), and the occurrence of low voltage violations during driving is reduced (>5×), minimizing the need for power derating and enhancing the user experience. The algorithm is effective, scalable, flexible, and requires low implementation and tuning effort, resulting in an ideal candidate for industry adoption.

Keywords: lithium-ion battery; balancing systems; electric vehicles; consensus algorithm; distributed control; state-of-charge equalization; temperature equalization

Citation: Barreras, J.V.; de Castro, R.; Wan, Y.; Dragicevic, T. A Consensus Algorithm for Multi-Objective Battery Balancing. *Energies* **2021**, *14*, 4279. https://doi.org/10.3390/en14144279

Academic Editors: João Pedro Trovao and Ta Cao Minh

Received: 17 May 2021
Accepted: 11 July 2021
Published: 15 July 2021

1. Introduction

The 2015 Paris agreement has established global guidelines to tackle climate change, including electrification of transportation [1]. Supported by policies, technological improvements, and falling costs, the global fleet of light-duty passenger vehicles (LDPVs) with a higher degree of electrification, such as battery electric vehicles (BEVs) and plug-in hybrid electric vehicles (PHEVs), have expanded from 2014–2019 by an annual average of 60%, reaching about 1% of the global fleet by 2019 [2].

In 2020 the COVID-19 pandemic has brought great uncertainty for the global electric vehicle (EV) market and the auto market. Although, recent studies confirm that the future remains bright for passenger EVs in the following decades, which will gain market share supported by battery cost reductions and technological advances [3].

Virtually all EVs in today's market store electrical energy in batteries, an essential technology to electrify LDPVs in the next 20 years [2,3]. EVs with a higher degree of electrification, i.e., BEVs and PHEVs, are equipped with larger batteries that can be directly recharged from the grid, in addition to regenerative braking, a common feature in EVs with lower degrees of electrification. Nowadays, a BEV or PHEV battery is formed of a combination of hundreds to thousands of lithium-ion cells connected in a certain series-parallel arrangement.

Lithium-ion battery technology is improving, showing increased energy storage capability, and their cost is falling, mainly thanks to economies of scale. However, there are still some significant technical barriers to overcome for EV adoption [2,3], particularly regarding range anxiety, charging infrastructure, battery supply chain, battery safety [4], and battery management [5].

Regarding the latter, for safety reasons, lithium-ion cell manufacturers limit their operating window by certain voltages, currents, and temperatures. Outside that so-called safety operating window (SOA), the degradation rate accelerates and there is a risk of catastrophic failure [5]. In addition, original equipment manufacturers (OEMs) typically define more restrictive operating windows to increase safety, ensure reduced degradation rates, and/or enhance the user experience. That is implemented in the form of conventional current or power derating algorithms, as reviewed by Barreras et al. [6] and Sun et al. [6]. Derating can be improved through degradation-aware algorithms, as recently shown by Sowe et al. [7] and Schimpe et al. [8,9].

Indeed, battery degradation mechanisms are complex and enhanced by certain operating and storage conditions that are fundamentally related to SOC, temperature, or current levels, as discussed by Schimpe et al. [9] orPellettier et al. [10] or For instance, extreme temperatures are a major concern. On one hand, the degradation rates are accelerated at high temperatures, due to rapid growth of the solid electrolyte interface (SEI) layer, existing even the risk of thermal runaway over certain thresholds (Figure 1). On the other hand, at low temperatures, there is accelerated aging due to lithium-plating and stripping, being more intense at higher C-rates. There is also a risk of internal short due to the formation of dendritic lithium (Figure 1). Even within a moderate temperature range (~10–50 °C), power and energy capability are influenced by battery temperature, since lithium-ions diffusivity in the electrodes increases with temperature. Indeed, the temperature window within which the battery lifetime (calendar and cycle life) and power and energy performance are balanced, is relatively small (~20–40 °C).

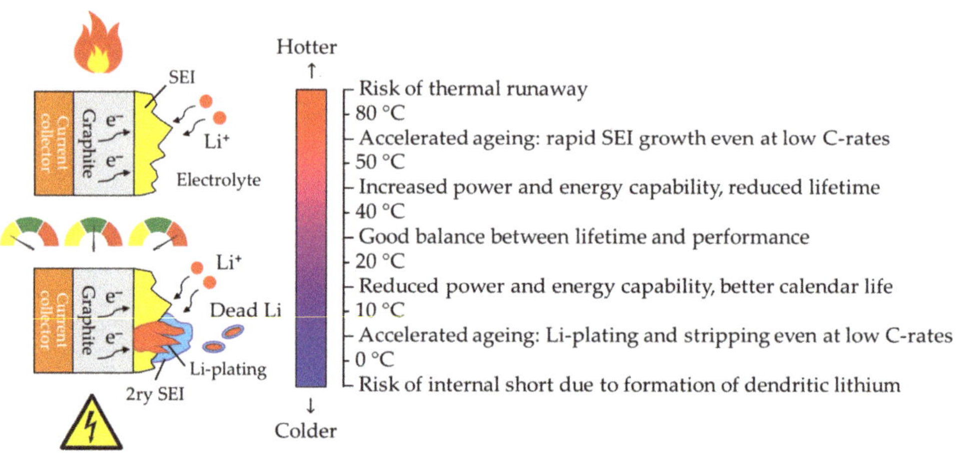

Figure 1. Overview of major thermal issues in lithium-ion batteries.

Regarding SOC, for example, in BEVs and PHEVs the minimum allowable SOC is typically over 5–10%, since the battery power capability, and thus the vehicle acceleration is lower at low SOC levels, and also to avoid the risk of overdischarge. Battery degradation rate is also higher at extreme SOCs, and therefore the maximum SOC window in EV batteries is set typically below 90–95%—although the user may occasionally select operating modes that allow larger SOC windows for extended range operation.

To alleviate those problems, active heating and cooling strategies are implemented in EVs, and the battery power is derated under eventual extreme SOC or temperature conditions [11]. However, the former translates into higher auxiliary loads (i.e., lower efficiency), more complex integration, extra costs, volume, weight, and reliability and maintenance issues, while the latter typically means longer charging times at low temperatures and reduced power capability at high temperatures.

To make everything more complex, the battery pack should also be managed at a cell level, due to the existence of cell-to-cell variations in SOC, temperatures, or power and energy capability (Figure 2). While cell-to-cell variations may not be substantial at the beginning-of-life (BOL), they increase cumulatively over time due to intrinsic and system-level induced issues, such as uneven power due to parallelization, non-uniform heating/cooling, heterogeneous mechanical stress, uneven presence of heat sources nearby, or battery management issues. For example, in series-connected packs, lower capacity cells charge and/or discharge faster, leading to a reduced battery operating window. In other words, less energy is charged and discharged, since the battery pack performance is limited by the cell(s) with the minimum voltage/SOC during discharge and the maximum voltage/SOC during charge.

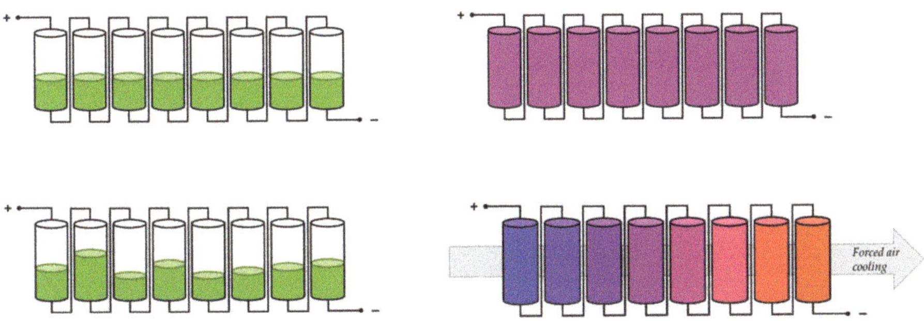

Figure 2. Schematic representation of cell-to-cell differences on SOC (**left**) and temperature (**right**), showing balanced (**top**) and imbalanced modules (**bottom**) of 8 cells in series. As suggested on the bottom-left schematic, non-uniform cooling can lead to temperature imbalances, but other reasons are possible.

There are two main approaches to balance these cell-to-cell differences. The first relies on dissipative mechanisms, and it is so-called passive balancing. The idea is to dissipate the energy of the highest-charged cell in a passive element, such as a resistor, during the charging phase [12]. The main advantages of this approach are simplicity and low cost but at the expense of higher energy losses. This dissipative approach also does not fully address the balancing problem during discharge: the effective capacity and power of the battery pack are still defined by the weakest cell(s), i.e., the cell(s) with the highest/lowest SOC, lowest power capability and/or maximum/minimum temperature [13].

The second approach is non-dissipative, a so-called active balancing, which aims to continuously re-distribute the energy within the battery pack using power conversion. In the literature, there is a wide range of power conversion topologies that can be used [5,12,14,15], enabling the transfer of energy from one cell to another, a common energy buffer (e.g., a capacitance of inductance) or from the cell to the pack. This re-distribution of energy brings several advantages. Firstly, it provides higher energy efficiency than pure dissipative approaches and can be much faster. But more importantly, it opens new possibilities from the point of view of the control algorithm beyond the conventional SOC-based, OCV-based, and terminal voltage-based balancing algorithms, which are broadly used both in passive and active balancing, as described by Fleischer et al. in [12].

As shown in pioneering works of Altaf et al. [16,17], Barreras et al. [5,18], Pinto et al. [14,19], and de Castro et al. [20], active balancing opens the way for novel balancing functions, including not only SOC and voltage balancing, but also thermal control, power capability equalization, or distributed hybridization. For instance, it is well known the connection between battery current and temperature, due to self-heating effects, as illustrated in Figure 3. This means that increase in battery temperature can be controlled, up to some extent, through current limitation [5].

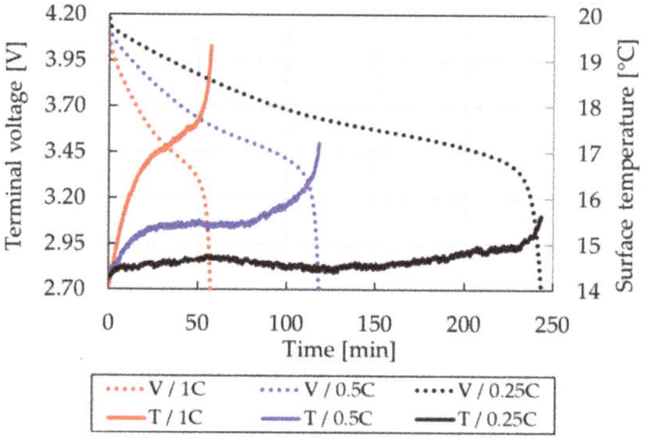

Figure 3. Temperature and voltage in time for CC discharge at various C-rates for a pristine Kokam 53 Ah cell.

All in all, active balancing can enable improved battery performance, prolonged lifetime, and improved or additional features, as already stated by Barreras et al. in 2014 [18]. Regarding battery performance, the usable capacity of the entire pack over its lifetime can be increased, since both the charge and discharge performance of the battery pack can be dictated by the average characteristics of the cells, instead of the weakest cell, as is the case with dissipative approaches [13]. On the other hand, higher costs, complexity, and volume represent the main obstacles for the deployment of this technology in practice, especially in automotive applications [5,12,15].

From a control architecture perspective, battery balancing functions can be performed in the following ways: centralized; de-centralized, or distributed architectures. That can be related to the battery management system (BMS) topology [5], as illustrated in Figure 4.

In centralized networks, control functions are concentrated in a single unit (aka agent), which receives state information from all battery cells and centrally decides the control actions for the balancing hardware. In decentralized networks, each cell (or module) decides the control action locally, so the biggest reliability problem of centralized networks, i.e., single point of failure, is solved. However, the controllability is limited because there is no cell-to-cell communication. In distributed paradigms, the decentralization is only partial, i.e., the cells can communicate variables of interest with their neighboring cells. Therefore, the control intelligence is divided among a full network of control agents with partial information about the state of the battery.

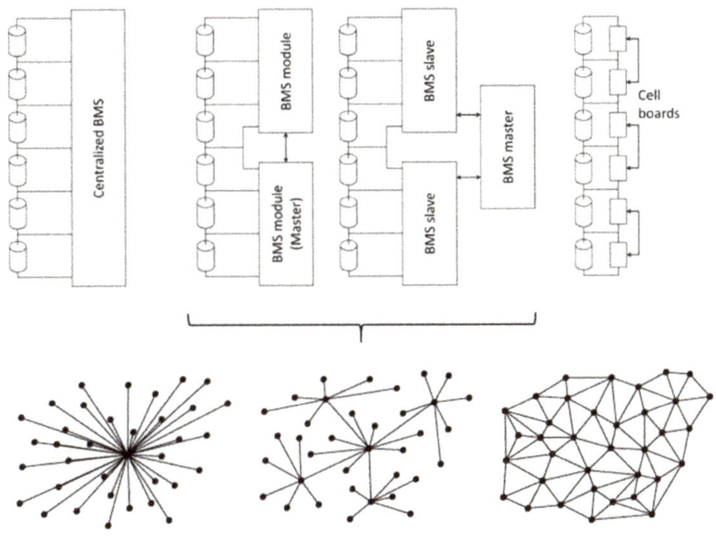

(a) (b) (c)

Figure 4. Basic types of BMS topologies (**top**) and their corresponding control networks (**bottom**): (**a**) centralized; (**b**) decentralized (modular and master-slave topologies), and (**c**) distributed.

In general, centralized and distributed control paradigms have the potential to achieve optimal performance, since they access global state information; however, they are also more complex to solve and implement, especially for large-scale systems with dozens or hundreds of actuators. On the other hand, decentralized architectures execute the control effort only locally, thereby requiring no information exchange and lower computational power. However, since each agent uses local information of the battery pack, it is challenging to achieve optimal control performance.

From an algorithmic perspective, three classes of control methods have been proposed for advanced active battery balancing, either in centralized or de-centralized architectures:

- Model-based: they rely on the mathematical model of the battery modules, often based on equivalent electric circuits. Battery models can be just used to estimate the current battery state, such in conventional OCV- and SOC-based balancing methods [12], but also to predict the future behavior of the battery pack, aiming to synthesize a control policy that fulfills the balancing goals, such as minimization of SOC or thermal unbalances, enforcing actuation and safety constraints. Model-predictive control [21], or linear state feedback [22] represent paradigmatic examples of this approach.
- Machine learning: in this case, the control policy is derived based on interactions with the real battery pack or with a simulation model. It usually decreases modeling efforts and domain knowledge expertise but at the expense of higher data needs and computational effort (especially during training). Reinforcement learning [23] is a good example of this approach, which has been gaining increased attention over the last few years.
- Fuzzy logic: in contrast with previous approaches, it relies mainly on expert knowledge to derive control algorithms [24], but this also means that there is no single systematic approach or implementation framework.

Beyond that, a novel method for battery balancing control are consensus algorithms, a distributed architecture based on graph theory in the area of communication. Consensus algorithms only require communication links between adjacent units, i.e., previous and next cells in a series string in our case, featuring advantages of less information requirement, scalability (i.e., modularity), and robustness, while keeping the potential for optimal performance due to global information discovery [25].

So far, only a couple of papers in the literature have implemented consensus algorithms, and for a single function or objective, i.e., either cell voltage [26] or SOC equalization [27]. In contrast, in this study, we propose for the first time in the literature a multi-functional/multi-objective consensus algorithm for distributed battery balancing. Here we show how this algorithm can simultaneously perform SOC, temperature, and voltage balancing, but it could be easily extended to equalize other variables. For example, the surface concentration of electrode particles, or the rates of degradation, being able to perform derating actions, such as the ones presented in [9], at cell or module level.

The overall framework is presented in Figure 5. The concept is evaluated in simulations in the context of an e-mobility application, considering a cell-to-pack-to-cell balancing system configuration and a series-connected battery pack.

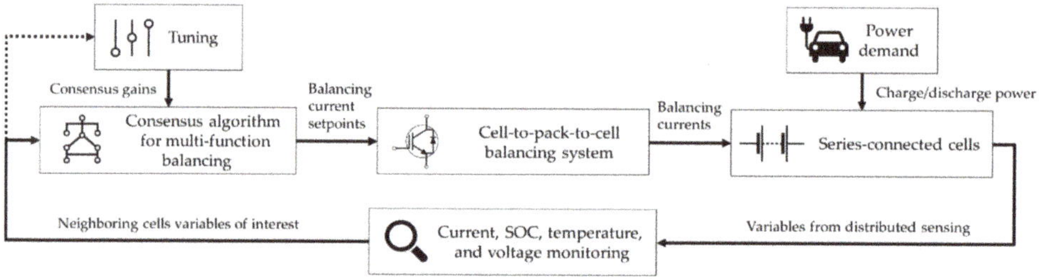

Figure 5. Block diagram of the distributed consensus control framework for multi-function EV battery balancing.

2. System Modeling and Control Algorithm

A single module of a battery system in an e-mobility application is modeled in this work, including 8-cells connected in series and the cell-to-pack-to-cell (CPC) balancing system. The battery power profiles are generated considering the full-pack and different driving conditions. The battery module electrical equivalent circuit model (EECM) is summarized in Figure 6.

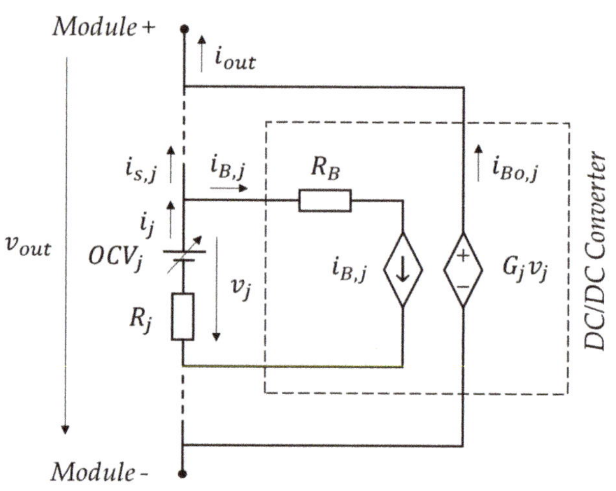

Figure 6. EECM of the 8S battery module and the DC/DC converter of the CPC balancing system.

2.1. Electro-Thermal Battery Model

A battery module made of 8 large format Li-ion pouch cells (NMC cathode, 53 Ah) connected in series is simulated by means of a multi-cell model. The electro-thermal model of each cell is composed of a simple lumped EECM coupled with a simple lumped thermal equivalent circuit model (TECM), which considers heat transfer between neighboring cells. Apart from that, different sets of parameters are generated to create different scenarios for cell-to-cell variations. Experimental data from single-cell characterization tests are combined with statistical data from screening tests on more than 200 pristine cells from Barreras et al. [28] and data from the literature to parameterize the models.

2.1.1. Electrical Model

Regarding the EECM of each cell j, the circuit implemented consists of a variable DC voltage source, OCV_j [V], in series with a constant resistor, R_j [Ω] (Figure 6). The former represents the cell equilibrium voltage, usually referred to in this context as the open-circuit voltage (OCV), while the latter accounts for all the cell internal resistance effects. This is mathematically represented as:

$$v_j(t) = \mathrm{OCV}_j(\mathrm{SOC}_j) - R_j i_j(t), \quad j \in \{1, \ldots, N_s\} \tag{1}$$

$$\dot{\mathrm{SOC}}_j(T) = -\frac{1}{\overline{Q}_j} i_j(t) \tag{2}$$

where v_j [V] is the terminal voltage of the cell j, N_s [–] is the number of cells in series in each module, SOC_j [–] is the cell SOC, $\overline{Q}_j$ [As] is the cell actual capacity, and i_j [A] is the cell current.

The average non-linear OCV vs. SOC characteristic is obtained from step-response tests conducted on an automated battery tester on a single cell (Figure 7). For implementation purposes, the average OCV vs. SOC relationship is linearized between 95% and 5% SOC, as illustrated in Figure 8. That is the typical SOC range of a lithium-ion battery in a full-electric vehicle. For the interest of the reader, the electrical test system setup is shown in Figure 9. This linear model is described as:

$$\mathrm{OCV}_j(t) = a_j + b_j \mathrm{SOC}_j(t) \tag{3}$$

where (a_j, b_j) are constant parameters derived for the model, and equal to 3.406 V and 0.673 V, respectively.

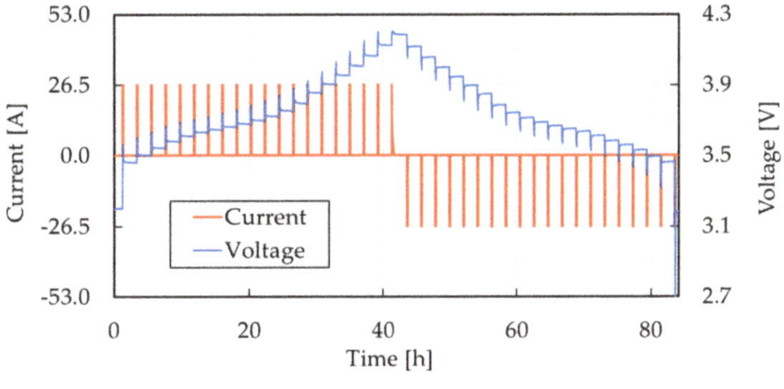

Figure 7. Step-response test data at 0.5 and 25 °C.

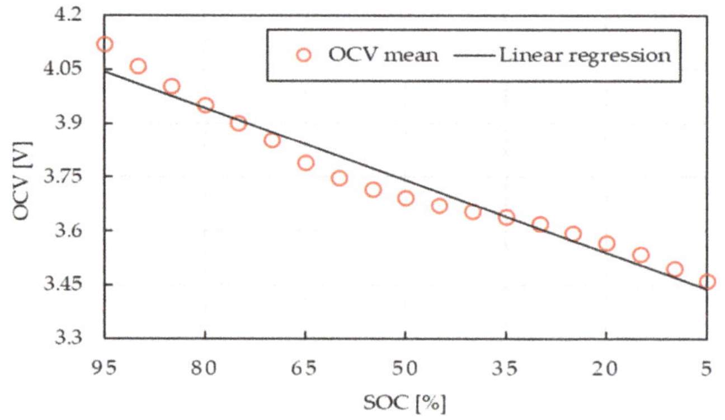

Figure 8. OCV vs. SOC characteristic.

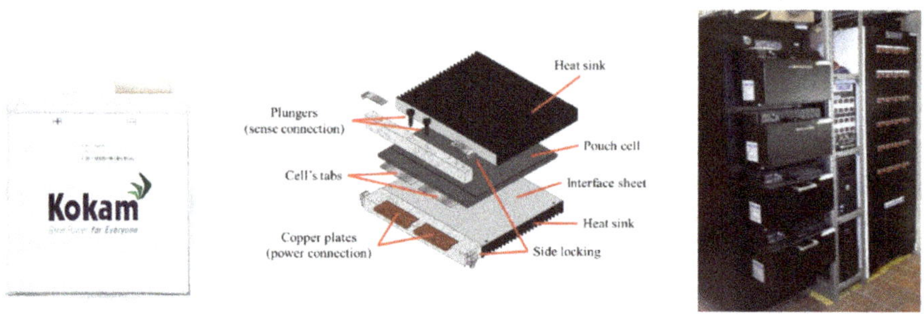

(a) (b) (c)

Figure 9. Test setup: (**a**) Kokam 53 Ah SLPB pouch cell; (**b**) 3-D representation of one of the cells and its holder; (**c**) from right to left, Maccor 4000 Series automated battery tester, FRA, and thermal chambers cabinets.

On the other hand, a statistical model of cell-to-cell variations is built to estimate the value of the series resistance and the cell capacity for each cell in the module at different aging states, as described in next Section 2.1.2.

2.1.2. Statistical Model of Cell-to-Cell Variations

Estimation or prediction of cell-to-cell variations over the battery life is particularly challenging. Manufacturing tolerances and uneven temperature distributions are key factors behind imbalances in electrical parameters such as capacity, internal resistance, or self-discharge rate. While there is a myriad of modeling and aging studies at the single-cell level, only a very few focus on the diversion of aging of large groups of lithium-ion cells. Moreover, just two of them present experimental results for large-capacity cells, as summarized in Table 1.

Table 1. Overview of experimental studies on cell-to-cell variations in the literature.

Capacity [1] [Ah]	Sample Size [–]	Chemistry	Format	Ageing State	Reference	Year
70	96	C/LFP	Prismatic	BOL	Zheng [29]	2011
53	208	C/NMC	Pouch	BOL	Barreras [28]	2017
5.3	198	C/NCA+NMC	-	BOL	An [30]	2016
5	216	C/NMC	Prismatic	BOL	Rothgang [31]	2014
4.4	96	C/LFP	Cylindrical	BOL	Paul [32]	2013
3.35	2	C/NCA	Cylindrical	MOL	An [33]	2015
3	1100	C/LFP	Cylindrical	BOL	Rumpf [34]	2017
3	248	-	-	BOL	Zou [35]	2018
2.9	356	C/NCA	Cylindrical	BOL, EOL	Baumann [36]	2018
2.8	51	C/LCO+NMC	Cylindrical	BOL	Devie [37]	2018
2.8	112	C/NCA	Cylindrical	BOL, MOL	Campestrini [38]	2016
1.95	2392	C/NMC	Cylindrical	BOL, MOL	Schuster [39]	2015
1.9	10	C/LMNC+LMO	Cylindrical	BOL	Dubarry [40]	2011
1.85	48	C/NMC	Cylindrical	BOL	Baumhöfer [41]	2014
0.3	100	C/LCO	Cylindrical	BOL	Dubarry [42]	2010
~0.3	60	C/LCO	Cylindrical	-	Shin [43]	2013

[1] Articles sorted in descending order of cell capacity.

In this study, statistical data obtained by Barreras et al. [28] through screening tests conducted on 208 Kokam SLPB 53 Ah cells tested at Beginning-of-Life (BOL) is used to parametrize the statistical model of cell-to-cell variation at BOL. The cell capacities are derived from the discharge capacities measured during a 1C full discharge at constant current and room temperature. The internal resistances are derived from measurements of the so-called direct current resistance (DCR) around room temperature and 50% SOC. The DCR parameterization is interesting in this context because it accounts for pure ohmic plus charge-transfer polarization effects, i.e., the dominant effects in an EV battery pack during driving [44]. The two histograms of the relative frequency of discharge capacity and the DCR resistance at BOL are shown in Figure 10. These histograms are fitted to normal distributions, which are the de facto statistical models of cell-to-cell variations at BOL [28].

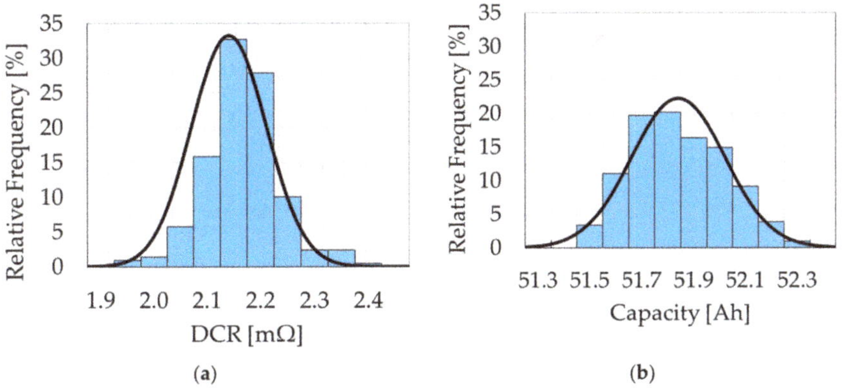

Figure 10. Histograms and normal distributions of the relative frequency of (**a**) the DCR and (**b**) the 1C discharge capacity for 208 Kokam SLPB 53 Ah lithium-ion pouch cells at BOL [5,14].

On the other hand, in the absence of more statistical data from large format cells at other aging states in the literature, the Middle-of-Life (MOL) and End-of-Life (EOL) parameter distributions are calculated through certain manipulation of the BOL distributions based on insights from the literature, following the same approach of Pinto et al. [14]. Then, as summarized in Figure 11, based on these statistical models of parameter distributions, a mixed aging scenario with substantial cell-to-cell variations is generated for our model.

Figure 11. Mixed aging scenario for cell-to-cell variations considered for the 8S module in this study [14].

Table 2 presents the resistance and capacity values derived for each cell of the 8S module under this scenario. The values are presented in vectors of 8 elements and normalized to the mean value of the internal resistance at the BOL (2.09 mΩ), and the nominal cell capacity (53 Ah), respectively. Further information on the methods used to determine these datasets in [14].

Table 2. Normalized internal resistances and capacities for the 8S module.

Normalized R_j	Normalized $\bar{Q}_j$
[1.602 2.955 2.882 1.636 0.999 1.428 0.973 1.487]	[0.934 0.883 0.874 0.925 0.977 0.921 0.976 0.934]

In addition, for all the BEV driving scenarios we assumed the next dataset for the initial SOC of each cell in the 8S module: φ = [0.925 0.935 0.932 0.930 0.931 0.922 0.930 0.938]. The complement of this dataset is assumed for all the fast charging scenarios, i.e., $(1 - \varphi)$.

2.1.3. Thermal Model

With respect to the thermal model, we assumed a lumped-parameter model and considered that: (1) the heat generation is evaluated based on Joule losses; (2) the cell heat capacity is constant; (3) the thermal resistance from cell to ambient is constant and formulated based on heat convection; (4) there is certain conductive heat transfer from one cell to the next/previous.

This is represented by means of a simple lumped TECM for each cell with links to neighboring cells, as illustrated in Figure 12, where T_j [K] is the actual temperature of cell j, T_{amb} [K] is the ambient temperature, C_p [J/K] is the cell heat capacity, R_{cond} [K/W] is the conductive thermal resistance between adjacent cells, and R_{conv} [K/W] is the convective thermal resistance between each cell and the environment. For each cell j, the TECM is coupled with the corresponding EECM through the heat generation term, $\dot{Q}_{gen,j}$ [W], which represents Joule losses and is expressed as:

$$\dot{Q}_{gen,j} = R_j i_j^2(t) \tag{4}$$

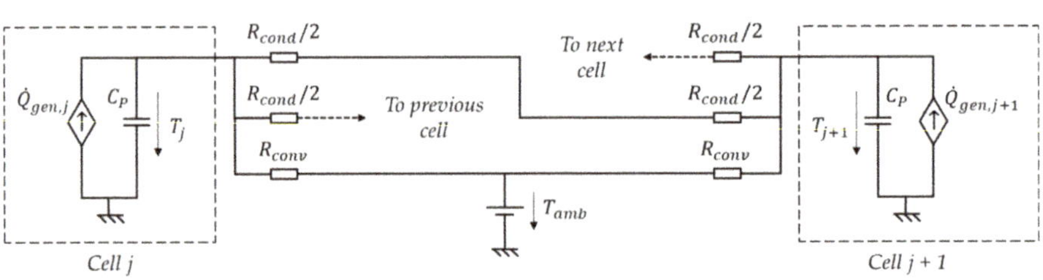

Figure 12. TECM of cell j with links to neighboring cells.

Regarding heat capacity, is assumed that $C_P = 1032$ J/K. This value is derived from the cell mass (1.2 kg) and the specific heat capacity (860 J/kgK)) determined by means of flash lamp thermal characterization tests by Barreras et al. [45,46]. The tests are conducted at room conditions on a similar cell (Kokam SLPB 40 Ah pouch cell), following ASTM E1461 standard procedures [47].

From the same references [45,46], the in-plane thermal conductivity is obtained. This value is used in combination with the cell dimensions, the tab dimensions, and the thermal conductivity of the tabs, to estimate the cell-to-cell conductive thermal resistance from the next equations:

$$R_{cond} = 2R_{cond,cell} + R_{cond,tabs} \tag{5}$$

where $R_{cond,cell}$ [K/W] is the in-plane conductive thermal resistance of a cell, and $R_{cond,tabs}$ [K/W] is the conductive thermal resistance of the cell tabs that physically connect adjacent cells, and are derived as follows:

$$R_{cond,cell} = \frac{d/2}{k_{||}A_{||}} \tag{6}$$

where $d = 0.222$ m is the height of the cell, $k_{||} = 33$ W/mK is the in-plane thermal conductivity [45,46], and $A_{||} = 0.01908$ m^2 is the averaged cross-section of the cell calculated from the datasheet.

$$R_{cond,tabs} = \frac{L/2}{k_{tabs}A_{tabs}} \tag{7}$$

where $L = 0.02$ m is the distance assumed between the tabs of adjacent cells, $k_{tabs} = 237$ W/mK is the thermal conductivity of the tabs, assuming that is equal to the tabulated value of aluminum and neglecting contact resistances between tabs, and $A_{tabs} = 32{\cdot}10^{-5}$ m^2 is the cross-section of two tabs superimposed, calculated from the datasheet.

These calculations result in $R_{cond} = 6.16$ K/W, $R_{cond,cell} = 1.76$ K/W, and $R_{cond,tabs} = 2.64$ K/W, suggesting that both the cell thermal resistances and the tab resistance play an important role in cell-to-cell heat transfer through the tabs of adjacent cells. Unfortunately, there is a lack of results in the literature for other cells, for example, Zhu et al. [48] just estimated a value of 1.711 K/W for large format prismatic cells. In any case, $R_{cond,cell}$ is an intrinsic cell property, which suggests that $R_{cond} > 3.53$ K/W even assuming that $R_{cond,tabs}$ could be minimized by improved thermal design of tab-interconnection.

On the other hand, R_{conv} is derived from:

$$R_{conv} = \frac{1}{hA} \tag{8}$$

where h [W/m^{2}K] is the convective heat transfer coefficient, and A [m^2] is the interface surface area of the solid material. In this study, we estimated $R_{conv} = 0.813$ K/W, assuming $h = 15$ W/m^{2}K and $A = 0.082$ m^2. The former is a typical value for low forced air velocities, and the latter is the sum of the areas of the larger front and rear faces of the cell. We considered a battery module with parallel cooling/heating based on forced air. The arrangement is sketched in Figure 13. In comparison with series cooling/heating, a parallel arrangement provides a more even distribution of the temperatures [14].

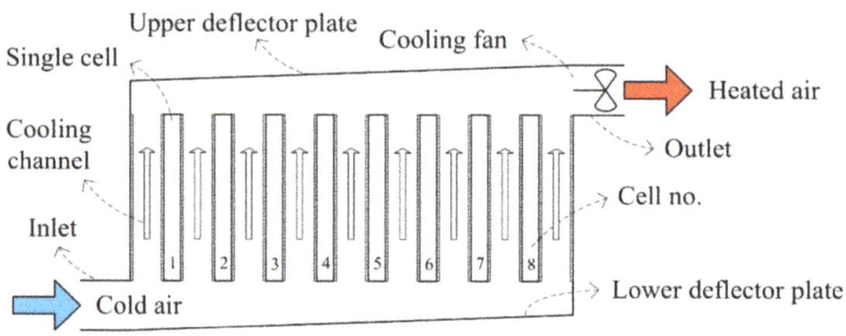

Figure 13. 2-D schematic of the parallel air-cooling arrangement assumed for the 8S module [5].

2.2. Balancing System Model

The overall balancing system topology proposed consists of a set of bi-directional DC/DC converters connected in parallel with each cell, which can move energy from/to each cell to/from the 8S battery module. This is the so-called CPC energy transfer configuration. As shown in Figure 6, the balancing system is represented through a simplified EECM, assuming that all the conduction power losses of each DC/DC converter can be lumped together into an equivalent resistive component, assuming $R_B = 10\ \text{m}\Omega$. On the other hand, for the calculation of the overall balancing system losses, switching, magnetic, and drivers' energy losses are assumed to be constant and equal to $P_{B,loss,other} = 100$ mW for each converter.

2.3. Battery Power Profiles Generation

To test the algorithm under realistic demanding conditions, we have generated both fast charging and aggressive driving power profiles considering an exemplary BEV and the battery specifications.

2.3.1. Aggressive Driving Power Profiles

In a BEV, the battery pack supplies power and energy to the vehicle driveline, which is composed of the electric motor(s), inverter, and mechanical transmission. This load power ($p_{out}\mathrm{x_i}(\mathrm{t}) = -\mathrm{Lx}(\mathrm{t})$) is dependent on numerous factors, including the type of driving cycle the vehicle is operating, the energy efficiency of the driveline's components, as well as the inertial, rolling, and aerodynamic resistance forces that affect the vehicle (see [49] for details). Given that our goals consist in the design and test of the balancing control algorithm, in what follows we will assume that the load power p_{out} is a known disturbance. Its value is computed considering the uCar [49] as a reference vehicle. In this paper, we use the US06 (repeated a certain number of times) as the driving cycle, since we focus on more demanding aggressive driving scenarios, but the model framework also implements other standard and non-standard driving cycles. The driving process is finished when the end-of-discharge condition is detected, i.e., when $\min(\text{SOC}_j) \leq 0.05$, a common limit in real BEVs.

With regard to low voltage violations, they take place when one or several cells show a voltage below the lower voltage threshold set by the application. Here we used the minimum cell voltage stated by the manufacturer to define this limit ($v_j < 2.7$ V), but more restrictive windows are common for safety and reliability reasons. In real-life, the BMS would derate the BEV battery power when a low voltage violation is detected [6,11], either temporarily or permanently by forcing a low-power operating mode, sometimes called "limp home" mode. However, here we decided not to implement such a derating algorithm, to enable a fair comparison between different cases by ensuring the same power levels. Instead, we allowed BEV operation until the aforementioned lower SOC threshold and quantified the operation time with low voltage violations as a key performance indicator.

2.3.2. Fast Charging Power Profiles

The DC fast charging profiles are created based on a constant power constant voltage (CP-CV) strategy, which is common in fast charging protocols. The charging process shifts to the CV stage when $\max(v_j) \geq 4.2$ V. The power level in the CP stage is determined by trial-and-error aiming to charge with the maximum power that does not result in cell over-current considering the limits defined in the manufacturer's datasheet, i.e., $\max(i_j) \leq 106$ A. The power level is determined considering also the maximum balancing currents, which are below 20A for all the cases evaluated. The charging process is finished when the end-of-charge condition is detected, i.e., when $\max(\text{SOC}_j) \geq 0.8$. This is a also common setting in fast charging protocols.

2.4. Multi-Agent Consensus Algorithm

The coordination of distributed system with multiple agents is enabled by a protocol known as consensus algorithms, which allows the agents to exchange information via the communication network. In the case of the battery system, consensus algorithms can achieve information sharing and coordination among different cells.

The coordination of distributed system with multiple agents is enabled by a protocol known as consensus algorithms, which allows the agents to exchange information via the communication network. In the case of a battery pack, each cell in a pack represents one agent. An important feature of the consensus algorithm is that information exchange occurs only between the neighboring cells and that the centralized controller does not exist. For this reason, even if some of the communication links fail, the full information exchange capability can be preserved, where the actual level of resiliency depends on the communication network topology, as discussed in [50].

When every given cell knows the average of these variables, it is possible to design local control protocols that can ensure desired behavior of both the battery pack system and each cell totally independently of the centralized pack-level controller. In this context, a single point of failure is avoided, while the performance can be designed per technical requirements without any limitation.

The consensus algorithm is applied to the controller of every given cell by locally implementing a simple difference of variables of interest (in our case cell SOCs, terminal voltages, and temperatures) between each cell and those of other adjacent cells, which can be expressed as [4]

$$\dot{x}_j(t) = \sum_{k=N_j} \left| x_j(t) - x_k(t) \right| \tag{9}$$

where $x_j(t)$ and $x_k(t)$ are the variables of interest in agent j and k. N_j represents the whole set of neighbors of cell j.

From the equation above, it can be understood that the variables of interest change interactively according to the values of local measurements and the receiving measurements coming from the neighboring cells. Provided that the communication network is fully connected, all the variable values will eventually converge to an average. Therefore, based on the consensus algorithm that information is shared among units, each cell can compute an exact average value of the variables of interest, i.e., cells SOCs, terminal voltages, and temperatures, as shown in [51].

The multi-agent network of the battery pack is modeled as an undirected and connected graph. And the consensus algorithm can be expressed as follows [5,6]:

$$\dot{x}_j(t) = -Lx(t) \tag{10}$$

where $\boldsymbol{L}$ is the graph Laplacian of the network, which formally describes the network's topology. The elements of the matrix are defined as

$$L_{jk} = \begin{cases} -1, & k \in N_j \\ d_j, & k = j \end{cases} \tag{11}$$

where d_j is the number of neighbors of node j, which means the diagonal elements L_{jj} are determined by the number of edges attached to the node.

In this application, $x(t)$ includes the average estimated SOC, or temperature, or terminal voltage, but other variables are also possible. Each cell (or agent) j receives the estimated current average value from neighboring agents and updates its information based on the SOC, temperature, and terminal voltage using three consensus balancing algorithms.

The estimation of SOC is updated as

$$\overline{\mathrm{SOC}}_j(t) = \mathrm{SOC}_j(t) + \sigma_1 \sum\nolimits_{k=N_j} \int \left[\overline{\mathrm{SOC}}_j(t) - \overline{\mathrm{SOC}}_k(t)\right] dt \tag{12}$$

where $\overline{\mathrm{SOC}}_j$ denotes the agent j's average estimated SOC of all agents, SOC_j is the actual SOC of agent j, and σ_1 is the SOC consensus coefficient.

Similarly, the agent j's average temperature and terminal voltage of all agents can be obtained based on the next consensus algorithms:

$$\overline{T}_j(t) = T_j(t) + \sigma_2 \sum\nolimits_{k=N_j} \int \left[\overline{T}_j(t) - \overline{T}_k(t)\right] dt \tag{13}$$

where $\overline{T}_j$ represents the estimation of agent j, T_j is the actual temperature of agent j, and σ_2 is the temperature consensus coefficient, and

$$\overline{v}_j(t) = v_j(t) + \sigma_3 \sum\nolimits_{k=N_j} \int \left[\overline{v}_j(t) - \overline{v}_k(t)\right] dt \tag{14}$$

where $\overline{v}_j$ represents the estimation of agent j, v_j is the actual terminal voltage of agent j, and σ_3 is the voltage consensus coefficient.

Then, the reference balancing current for each cell j, i.e., $i_{B,j}$ [A] in Figure 6, is calculated with a correction term obtained from the cooperation between agents, incorporating the SOC balancing (Equation (12)), temperature balancing (Equation (13)), and/or terminal voltage balancing (Equation (14)), based on multi-agent. As a result, the multi-agent consensus algorithm helps each local unit to converge towards the global average of SOC, terminal voltage, and temperature.

Regarding the value of each consensus coefficient, σ_1, σ_2, and σ_3, they could be formally selected taking into account the topology of the communication network defined by graph Laplacian matrix L (Equation (11)) to ensure fast and stable convergence. Indeed, these parameters affect the second smallest eigenvalue of the L, which should be maximized to ensure the fastest possible convergence, as explained in detail in [52]. However, formal stability analysis falls beyond the scope of this paper, and the interested reader is referred to the literature for more details.

In this paper, we have just followed a sub-optimal trial-and-error selection method to determine the starting values of the consensus parameters. The selection is also supported by sensitivity analysis in the more complex cases of multi-function control.

In addition, we took into account the following considerations: (1) the values of the consensus coefficients determine the convergence rate of the algorithm for each variable of interest, being faster for larger values, (2) the control actions, i.e., the balancing currents, are larger for larger values of the coefficients and larger differences in the variables of interest, and (3) the magnitude of the differences might be quite different for each variable of interest, being particularly lower for the SOCs, as they are defined as dimensionless variables with values between 0 and 1. This means that different values are expected for each consensus parameter, being particularly larger for the SOC case.

It is also worth noting that in ordinary control theory, hardware implementation of the controller is often not considered, and therefore in theory too large consensus coefficients that might lead to extremely large control signals are possible. This might result in hardware or stability issues. In practice, this situation is typically avoided by including saturation in the actuator, as discussed in [53]. Here, we included a saturation block to the controller to limit the balancing currents to ±1C (53 A), but other limits could be explored.

Finally, it is worth noting that, within this framework, static or dynamic values for the consensus coefficients can be applied. For instance, in this paper, beyond the static coefficients, we also propose a dynamic voltage consensus coefficient determined in quadratic proportion to the battery current, i_{out} [A]:

$$\sigma_3 = k_1\left(1 + k_2 i_{out}^2\right) \tag{15}$$

where k_1 and k_2 [–] are constant coefficients, and typically $k_1 \gg k_2$ to avoid controller saturation. This dynamic definition of the voltage consensus coefficient virtually eliminates the widely known problem in regular balancing systems [54] of over-balancing voltages during low current periods or under-balancing voltages during high current peaks.

3. Results

Using our simulation framework, we can easily study a myriad of scenarios with respect to battery states, external conditions, or power demand. In this paper, due to space constraints, we focus only on demonstrating the effectiveness of the algorithm under more demanding scenarios. Hence, we focus on aggressive driving and fast charging conditions (see Section 2.3), in combination with substantial differences in cell-to-cell variations (see Section 2.1.2), and moderate parallel forced air cooling (see Section 2.1.3). Thus, we assumed in all the scenarios analyzed the *mixed aging* cell-to-cell variations scenario. As qualitatively shown in Figure 12, this is the case with a larger variation in cell internal resistances and capacities, since it combines cells with BOL, MOL, and EOL parameterization.

We compare side-by-side the performance of single-controllers for either SOC, voltage, and temperature balancing, and evaluate the results in comparison with the 'no control' case, which is used as a benchmark. We also evaluate multi-function dual controllers, in particular: (1) simultaneous SOC and voltage balancing, and (2) SOC and temperature balancing.

For a detailed comparison, we present plots in time of key variables, but also provide performance metrics in corresponding tables. In the case of multi-function balancing, we also conduct sensitivity analysis, aiming to provide insights on the performance of the controller depending on the tuning of the consensus gains.

The following results are displayed for the aggressive driving scenarios:

- Figure 14 and Table 3: a comparison of single controllers;
- Figure 15 and Table 4: a comparison of single and dual controllers, including voltage balancing using the dynamic voltage consensus gain (Equation (15));
- Figure 16 and Table 5: a sensitivity analysis of dual balancing (SOC and temperature);
- On the other hand, the next results are presented for the fast charging scenarios:
- Figure 17 and Table 6: a comparison of single controllers;
- Figure 18 and Table 7: a sensitivity analysis of dual balancing (SOC and voltage);
- Figure 19: a comparison of single and dual controllers with different fast charging profiles.

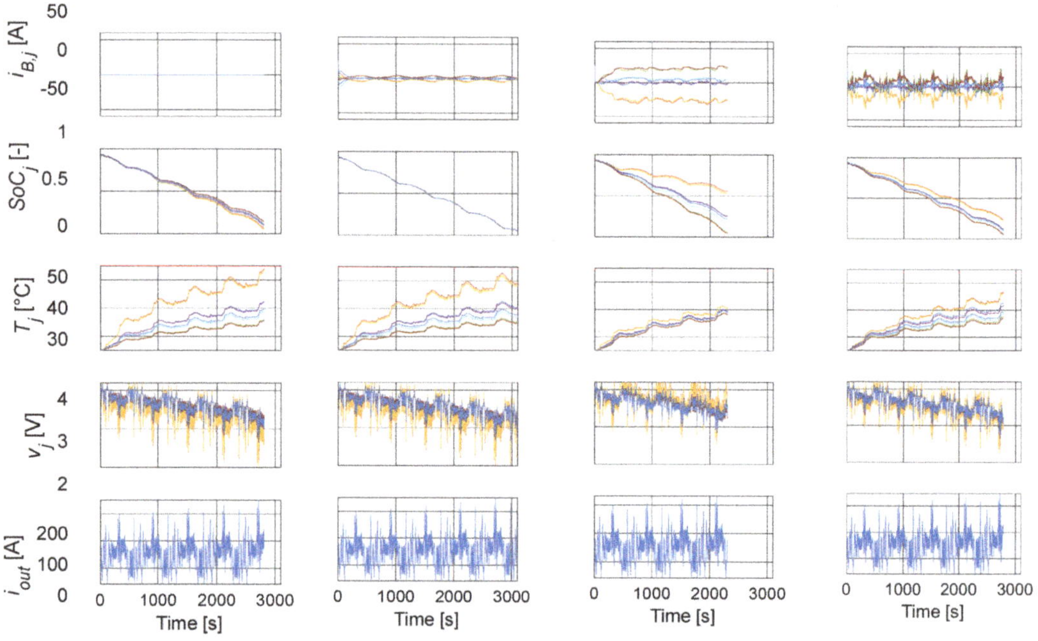

(**a**) No control ($\sigma_1 = \sigma_2 = \sigma_3 = 0$) (**b**) SOC balancing ($\sigma_1 = 2000,\ \sigma_2 = \sigma_3 = 0$) (**c**) Temperature balancing ($\sigma_2 = 20,\ \sigma_1 = \sigma_3 = 0$) (**d**) Voltage balancing ($\sigma_3 = 250,\ \sigma_1 = \sigma_2 = 0$)

Figure 14. Single controllers result in time. End condition: $\min\left(\mathrm{SOC}_j\right) \leq 0.5$. Scenario: aggressive driving (US06), mixed agingg, parallel forced air cooling ($h = 15\ \mathrm{W/m^2K}$).

Table 3. Key metrics of single controllers results shown in Figure 14.

Balancing Scenario	Δv_{rms} [mV]	$\Delta \mathrm{SOC}_{rms}$ [%]	ΔT_{rms} [°C]	$\max(T_j)$ [°C]	$E_{B,loss}$ [Wh]	E_{loss} [Wh]	$v_j < 2.7$ V Time [%]	Simulation Time [s]
No control	94	1.5	3.5	54.2	0	121.1	5.86	2813
SOC ($\sigma_1 = 2000$)	84	0.1	3.3	52.8	0.9	127.0	5.27	3092
Temperature ($\sigma_2 = 20$)	82	7.9	0.7	41.4	13.4	87.9	0.43	2306
Voltage ($\sigma_3 = 250$)	57	3.4	1.8	46.8	5.0	110.5	2.26	2781

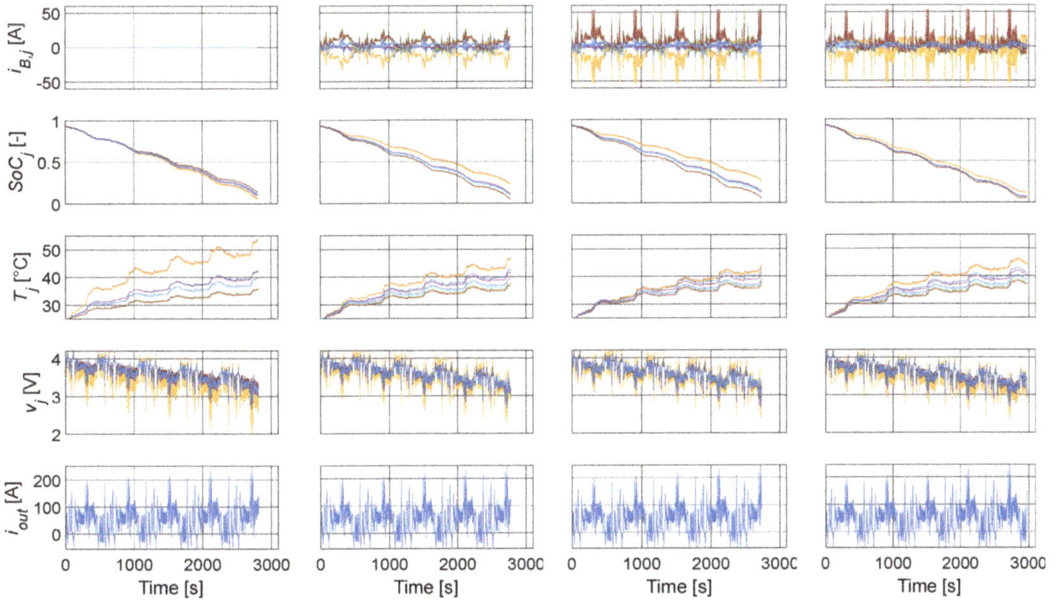

(a) No control ($\sigma_1 = \sigma_2 = \sigma_3 = 0$)

(b) Voltage balancing ($\sigma_3 = 250,\ \sigma_1 = \sigma_2 = 0$)

(c) Voltage balancing ($\sigma_3 = 250 + 0.025\ i_{out}^2$, $\sigma_1 = \sigma_2 = 0$)

(d) Dual balancing (SOC and voltage) ($\sigma_1 = 600,\ \sigma_2 = 0$ $\sigma_3 = 175 + 0.0175\ i_{out}^2$)

Figure 15. Single and dual controllers results in time. End condition: $\min\left(SOC_j\right) \leq 0.5$. Scenario: aggressive driving (US06), mixed agingg, parallel forced air cooling ($h = 15$ W/m^{2}K).

Table 4. Key metrics of single and dual controllers results shown in Figure 15.

Balancing Scenario	Δv_{rms} [mV]	ΔSOC_{rms} [%]	ΔT_{rms} [°C]	$\max(T_j)$ [°C]	$E_{B,loss}$ [Wh]	E_{loss} [Wh]	$v_j < 2.7$ V Time [%]	Simulation Time [s]
No control	94	1.5	3.5	54.2	0	121.1	5.86	2813
SOC ($\sigma_1 = 2000$)	84	0.1	3.3	52.8	0.9	127.0	5.27	3092
Voltage ($\sigma_3 = 250$)	57	3.4	1.8	46.8	5.0	110.5	2.26	2781
Voltage ($\sigma_3 = 250 + 0.025\ i_{out}^2$)	49	4.0	1.2	44.3	8.7	106.8	1.17	2738
Dual (SOC and voltage) ($\sigma_1 = 600, \sigma_3 = 175 + 0.0175\ i_{out}^2$)	59	1.6	1.7	46.2	7.5	116.4	1.38	2973

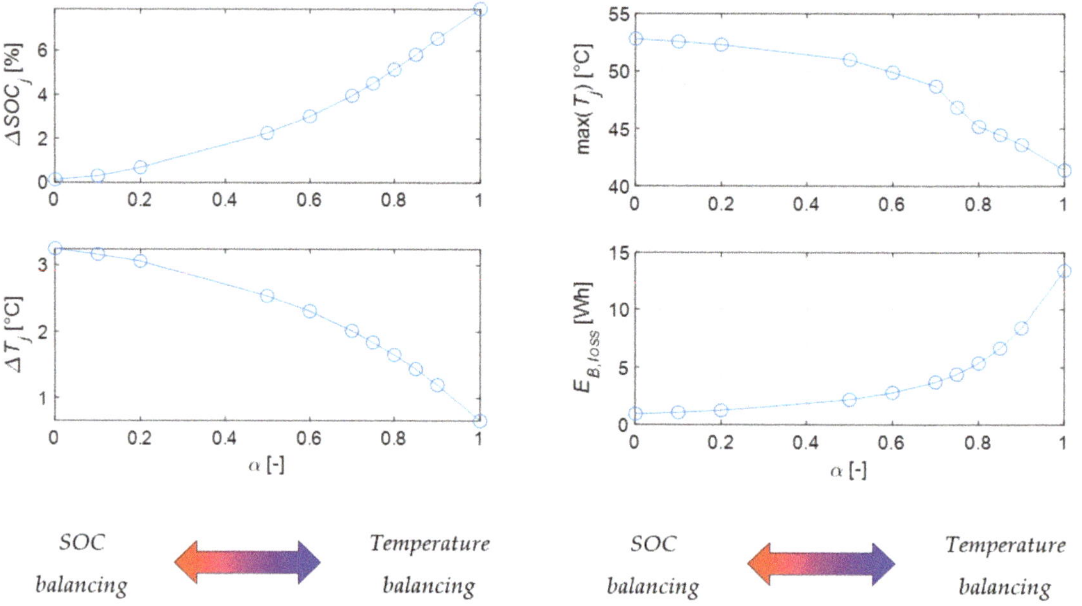

Figure 16. Sensitivity analysis: dual balancing (SOC and temperature): $\sigma_1 = 2000(1-\alpha)$, $\sigma_2 = 20\alpha$, $\sigma_3 = 0$. End condition: $\min\left(\text{SOC}_j\right) \leq 0.5$. Scenario: aggressive driving (US06), mixed aging, parallel forced air cooling ($h = 15\ \text{W/m}^2\text{K}$).

Table 5. Key metrics of dual balancing (SOC and temperature) results shown in Figure 16.

α [–].	Δv_{rms}[mV]	ΔSOC_{rms} [%]	ΔT_{rms} [°C]	$max(T_j)$ [°C]	$E_{B,loss}$ [Wh]	E_{loss} [Wh]	$v_j < 2.7$ V Time [%]	Simulation Time [s]
0	84	0.1	3.3	52.8	0.9	127.0	5.27	3092
0.2	78	0.7	3.1	52.3	1.2	124.3	4.47	3063
0.5	70	2.2	2.5	51.0	2.2	117.4	3.63	2834
0.8	64	5.1	1.7	45.2	5.3	102.0	1.59	2696
1	82	7.9	0.7	41.4	13.4	87.9	0.43	2306

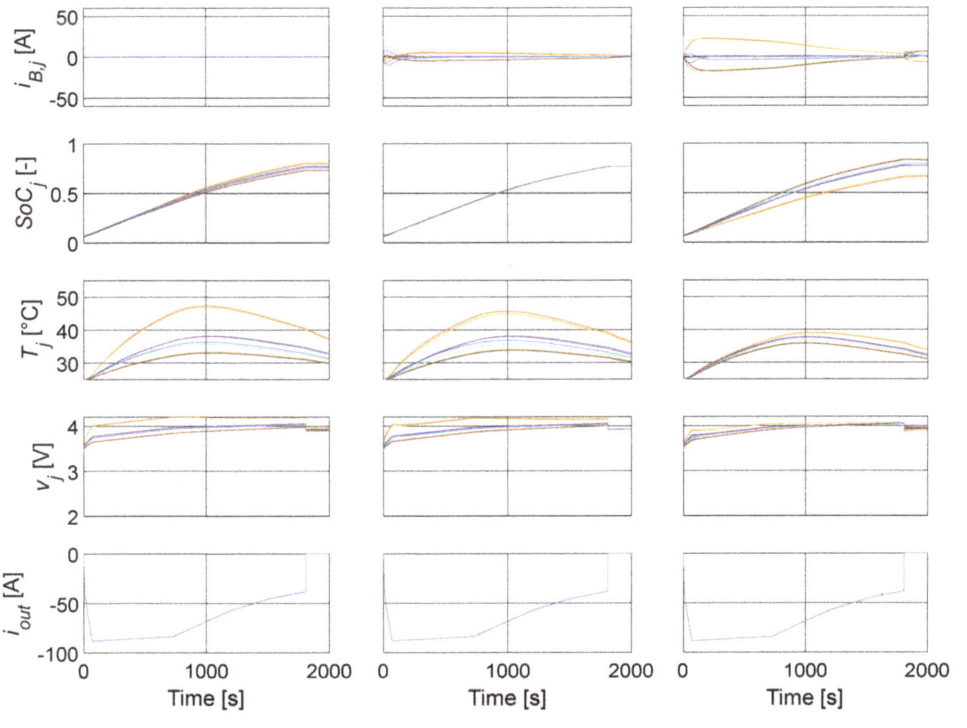

(a) No control ($\sigma_1 = \sigma_2 = \sigma_3 = 0$)

(b) SOC balancing ($\sigma_1 = 2000,\ \sigma_2 = \sigma_3 = 0$)

(c) Voltage balancing ($\sigma_3 = 250,\ \sigma_1 = \sigma_2 = 0$)

Figure 17. Single controllers results in time. Scenario: fast charging, mixed aging, parallel forced air cooling ($h = 15$ W/m^{2}K). All cases using the CP-CV charge profile of the 'no control' case.

Table 6. Key metrics of single controllers results shown in Figure 17.

Balancing Scenario	Δv_{rms} [mV]	ΔSOC_{rms} [%]	ΔT_{rms} [°C]	$max(T_j)$ [°C]	$E_{B,loss}$ [Wh]	E_{loss} [Wh]
No control	86	1.5	3.3	47.3	0	71.5
SOC ($\sigma_1 = 2000$)	74	0.1	2.7	45.5	0.7	70.0
Voltage ($\sigma_3 = 250$)	31	3.9	0.8	38.9	4.0	65.7

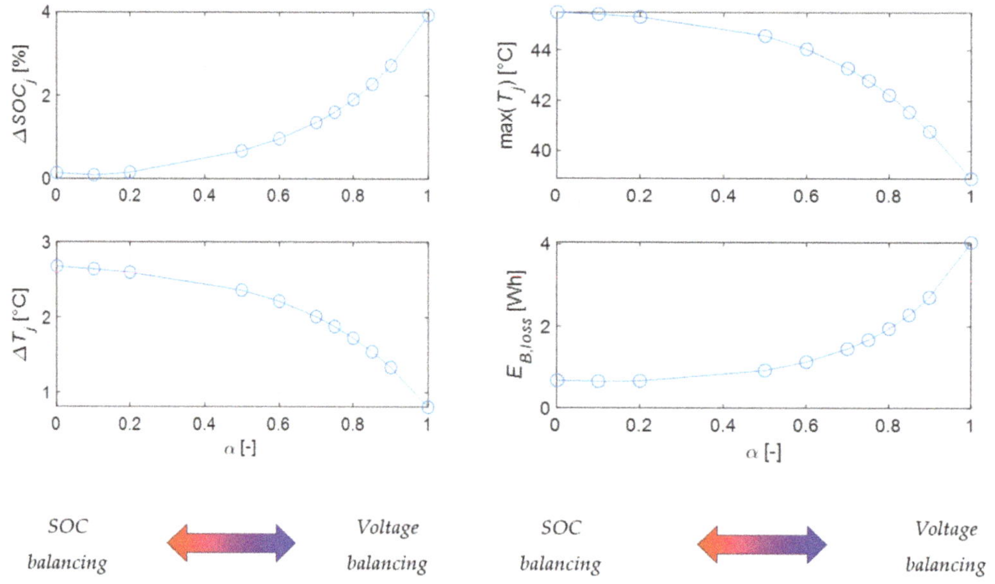

Figure 18. Sensitivity analysis: dual balancing (SOC and voltage): $\sigma_1 = 2000(1-\alpha)$, $\sigma_2 = 0$, $\sigma_3 = 250\alpha$. Scenario: fast charging, mixed aging, parallel forced air cooling ($h = 15\ \mathrm{W/m^2K}$). All cases using the CP-CV charge profile of the 'no control' case.

Table 7. Key metrics of dual balancing (SOC and voltage) results of Figure 18.

α [–]	Δv_{rms} [mV]	ΔSOC_{rms} [%]	ΔT_{rms} [°C]	$\max(T_j)$ [°C]	$E_{B,loss}$ [Wh]	E_{loss} [Wh]
0	74	0.1	2.7	45.5	0.7	70.0
0.2	72	0.2	2.6	45.3	0.7	69.8
0.5	66	0.7	2.4	44.6	0.9	69.4
0.8	52	1.9	1.7	42.2	1.9	68.1
1	31	3.9	0.8	38.9	4.0	65.7

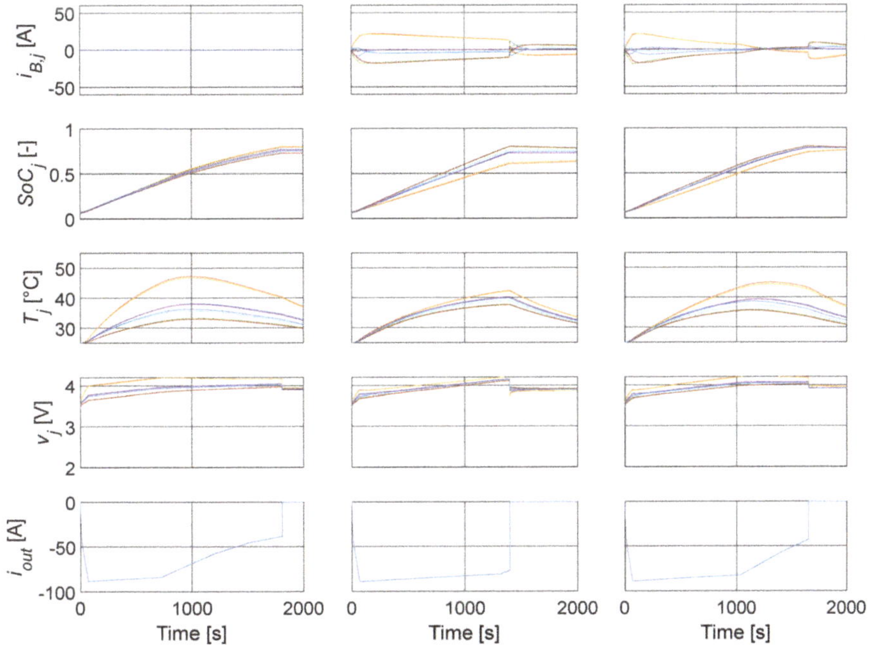

(**a**) No control ($\sigma_1 = \sigma_2 = \sigma_3 = 0$)

(**b**) Voltage balancing ($\sigma_3 = 250,\ \sigma_1 = \sigma_2 = 0$)

(**c**) Dual balancing (SOC and voltage) ($\sigma_1 = 600,\ \sigma_2 = 0$ $\sigma_3 = 175 + 0.0175\ i_{out}^2$)

Figure 19. Single and dual controllers results in time. Scenario: fast charging, mixed aging, parallel forced air cooling ($h = 15\ \text{W/m}^2\text{K}$). Each case using its own CP-CV charge power profile optimized to reduce charging times.

3.1. Aggressive Driving Results

From Figure 14 and Table 3, we learn that near-perfect SOC balancing ($\Delta SOC_{rms} < 0.1\%$) is possible with relatively low currents and can also extend driving range (+10%). In practice, it might be limited by the error in the SOC estimation, which is not considered here. However, conventional SOC balancing cannot solve the problem of low voltage violations, overheating, and thermal unbalance.

On the other hand, temperature and voltage balancing might induce larger SOC to unbalance (7.9 and 3.4%, respectively), which might shorten the driving range (−18% and −1%) but actually minimizes the occurrence of low voltage violations and thus the potential need for derating during driving. In addition, both temperature and voltage balancing can reduce the maximum temperatures (in 12.8 and 7.4 °C) and the cell-to-cell temperature differences. The peak balancing currents are around 20 A, i.e., 50% higher than for SOC balancing, but below 10% of the peak battery current.

The magnitude of these balancing effects is dictated by the values of the consensus gains. To illustrate that, in Figure 15 and Table 4, we analyze the impact of implementing a dynamic voltage consensus coefficient (Equation (15)). In this case, we can see that the balancing action is more pronounced during high power peaks. To keep the balancing power level within reasonable limits, we limited the balancing currents to 53A (or 1C). It is observed that the dynamic gain boosts the performance of the voltage balancing controller, with a minor increase in SOC unbalances and balancing losses. Moreover, when used in combination with SOC balancing in a dual controller fashion, we can get the advantages of both controllers while minimizing their caveats.

As previously stated, the magnitude of the balancing effects is dictated by the values of the consensus gains. To illustrate the importance of appropriate tuning for the consensus gains, in Figure 16 and Table 5 we present a sensitivity analysis for a dual balancing case. We considered SOC and temperature balancing, but similar studies can be conducted for other scenarios. To conduct the parameter sweep we re-defined the values of the SOC and temperature consensus gains as a function of a dimensionless variable α, which takes values from 0 to 1. This is expressed as $\sigma_1 = 2000(1-\alpha)$ and $\sigma_2 = 20\alpha$. In extreme cases, the dual controller behaves as a single controller, as either σ_1 or σ_2 are equal to zero. As illustrated in Figure 16, different non-linear trends are observed in the metrics, including an irregular pattern for the maximum temperatures. These results highlight the importance of developing selection methods to balance the trade-offs of each controller.

3.2. Fast Charging Results

From Figure 17 and Table 6, we learn that the algorithm is also effective during fast charging scenarios. The SOC balancing controller can equalize SOC smoothly, and the voltage balancing controller is able not only to equalize terminal voltages but also to reduce both the maximum temperatures (>8 °C) and the cell-to-cell temperature differences. However, it increases the SOC unbalance.

As pointed out earlier for the aggressive driving scenarios, a balance between the pros and cons of single controllers can be achieved by combining them in a multi-function controller and making an appropriate selection of the consensus gains. This possibility is investigated in the sensitivity analysis presented in Figure 18 and Table 7 for a dual controller that performs SOC and voltage balancing.

For a fair comparison, the results displayed in Figures 17 and 18, and their corresponding Tables 6 and 7, are obtained using in all cases the same fast-charging power profile. The maximum power levels are determined considering the maximum terminal voltages of the 'no control' case (Figure 17a), and the maximum balancing currents of the SOC and voltage balancing cases (Figure 17b,c).

However, as can be observed in Figure 17b,c, and thanks to the balancing actions, it would be possible to charge the battery at higher power levels in those cases without violating the upper voltage limit. That is particularly true for the voltage balancing case (Figure 17c).

This is investigated in the cases shown in Figure 19, where the battery is charged using its own CP-CV charge profiles to optimize the charging time. In comparison with the 'no control' case (Figure 19a), which requires 1811 s to bring the battery up to the end-of-charge condition, the battery with voltage balancing (Figure 19b) and dual balancing (SOC and voltage) (Figure 19c) can be charged up to the same condition in only 1400 s and 1655 s, respectively. This enables a charging time reduction of 22.7% and 8.6%, respectively.

4. Discussion

Here we discuss other qualitative aspects of this work. Firstly, we would like to acknowledge that both the EECM and the TECM proposed are relatively simple, but they are sufficient to validate the novel balancing control algorithm proposed. This is particularly true when coupled with a rational distribution of cell-to-cell parameter variations. Indeed, an equivalent modeling methodology has already been followed and validated experimentally by de Castro et al. [20].

In addition, this relatively simple electro-thermal model can be formulated with convex functions in a convex set. This means that in future work the control for the balancing system can be evaluated offline using convex optimization, following the pioneering methodology developed by Pinto et al. [14]. This enables the comparison of the sub-optimal online results obtained with the novel consensus algorithm, with the global optimal results provided by convex optimization.

Apart from that, we would like to highlight the flexibility of the algorithm, which can be easily modified to equalize other variables of interest. This can include variables related to battery internal states and degradation mechanisms, such as the surface concentration of electrode particles, or the rates of degradation derived from more or less complex degradation models. Thus, it would be possible to equalize battery lifetime at the cell level with this framework if such variables or states can be monitored, diagnosed, and/or predicted up to some extent, which is doable with the current state-of-the-art.

As a final remark, it is worth noting that the model framework is extremely fast—one study case can be run in less than a second. This is very convenient when running sensitivity analysis of key variables of interest, such as the consensus gains, and opens the door to generate large amounts of synthetic data. This large amount of data could be useful, for example, in combination with big data analytics to uncover hidden patterns, correlations, and other valuable insights.

5. Conclusions

This study presents for the first time a multi-objective distributed control for balancing systems, based on a multi-agent consensus algorithm. In our view, the algorithm's outstanding effectiveness, scalability, flexibility, and ease of implementation and tuning make it an ideal candidate for industry adoption. Beyond its distributed architecture, our new consensus algorithm offers key advantages over current centralized and de-centralized algorithms in the literature, such as (1) model-based, since it does not demand prediction of future variables and does not need a model of the system; (2) machine learning, since it does not demand any data for training and validation, and has extremely low memory and computational requirements; and (3) fuzzy logic, since it offers a systematic implementation approach.

Our results show that our novel multi-functional balancing can boost the performance of batteries with substantial cell-to-cell differences under the most demanding operating conditions, i.e., aggressive driving and DC fast charging (2C). Indeed, it is well known that under high power demand battery performance might be limited by overheating and excessive unbalance in cells SOC, terminal voltages, or temperatures. However, we demonstrate that a well-tuned consensus algorithm can generate balancing current setpoints that virtually eliminate these limitations, achieving a good balance between the pros and cons of single controllers.

In comparison with no control, driving times are increased (>10%), charging times reduced (>22%), maximum cell temperatures decreased (>10 °C), temperature differences lowered (~3 °C rms), and the occurrence of low voltage violations during driving reduced (>5×). This minimizes the need for power derating and enhances the driver experience. In a broader sense, it might also have an impact on the thermal management design, by reducing the demand for intensive and uniform cooling.

The benefits of multi-functional balancing are particularly interesting in the case of (1) EVs oriented to motorsport (e.g., Formula E), (2) any EVs that contemplate aggressive driving operating modes, such as the Tesla ludicrous mode, (3) all EVs that offer supercharging, and (4) batteries with significant cell-to-cell differences. In future work, we would like to evaluate the impact of the control in battery lifetime and to investigate the applicability within battery energy storage applications.

Author Contributions: Conceptualization, J.V.B. and R.d.C.; Formal analysis: J.V.B., and R.d.C.; Methodology: J.V.B., R.d.C. and T.D.; Visualization, J.V.B. and R.d.C.; Writing—original draft preparation, J.V.B., R.d.C., Y.W. and T.D.; writing—review and editing, J.V.B. All authors have read and agreed to the published version of the manuscript.

Funding: J.V.B. was funded by the EPSRC Faraday Institution Multi-Scale Modelling Project (EP/S003053/1, grant number FIRG003).

Data Availability Statement: The data presented in this study are available on request from the corresponding author.

Conflicts of Interest: The authors declare no conflict of interest.

References

1. The Paris Agreement, United Nations Framework Convention on Climate Change (UNFCCC). Available online: https://unfccc.int/process-and-meetings/the-paris-agreement/the-paris-agreement (accessed on 26 October 2020).
2. International Energy Agency (IEA). *Global EV Outlook 2020*; International Energy Agency (IEA): New York, NY, USA, 2020.
3. BloombergNEF. *BNEF Electric Vehicle Report (EVO) Report 2020*; BloombergNEF: New York, NY, USA, 2020.
4. Diaz, L.B.; He, X.; Hu, Z.; Restuccia, F.; Marinescu, M.; Barreras, J.V.; Patel, Y.; Offer, G.J.; Rein, G. Meta-Review of Fire Safety of Lithium-Ion Batteries: Industry Challenges and Research Contributions. *J. Electrochem. Soc.* **2020**, *167*, 090559. [CrossRef]
5. Barreras, J.V. Practical Methods in Li-ion Batteries: For Simplified Modeling, Battery Electric Vehicle Design, Battery Management System Testing and Balancing System Control. Ph.D. Thesis, The Faculty of Engineering and Science, Aalborg University, Aalborg, Denmark, 2017. [CrossRef]
6. Sun, Y.; Saxena, S.; Pecht, M. Derating Guidelines for Lithium-Ion Batteries. *Energies* **2018**, *11*, 3295. [CrossRef]
7. Sowe, J.; Few, S.; Barreras, J.V.; Schimpe, M.; Wu, B.; Nelson, J.; Candelise, C. How Can Insights from Degradation Modelling Inform Operational Strategies to Increase the Lifetime of Li-Ion Batteries in Islanded Mini-Grids? *ECS Meet. Abstr.* **2020**, *MA2020-02*, 3780. [CrossRef]
8. Schimpe, M.; Barreras, J.V.; Wu, B.; Offer, G.J. Novel Degradation Model-Based Current Derating Strategy for Lithium-Ion-Batteries. *ECS Meet. Abstr.* **2020**, *MA2020-02*, 3808. [CrossRef]
9. Schimpe, M.; Barreras, J.V.; Wu, B.; Offer, G.J. Battery Degradation-Aware Current Derating: An Effective Method to Prolong Lifetime and Ease Thermal Management. *J. Electrochem. Soc.* **2021**, *168*, 060506. [CrossRef]
10. Pelletier, S.; Jabali, O.; Laporte, G.; Veneroni, M. Battery degradation and behaviour for electric vehicles: Review and numerical analyses of several models. *Transp. Res. Part B Methodol.* **2017**, *103*, 158–187. [CrossRef]
11. Barreras, J.V.; Raj, T.; Howey, D.A. Derating Strategies for Lithium-Ion Batteries in Electric Vehicles. In Proceedings of the IECON 2018—44th Annual Conference of the IEEE Industrial Electronics Society, Washington, DC, USA, 21–23 October 2018; pp. 4956–4961.
12. Fleischer, C.; Sauer, D.U. Simulative comparison of balancing algorithms for active and passive cell balancing systems for lithium-ion batteries. In Proceedings of the Advanced Automotive Battery Conference (AABC), Detroit, MI, USA, 15–19 June 2014.
13. Barreras, J.V.; Frost, D.; Howey, D. Smart Balancing Systems: An Ultimate Solution to the Weakest Cell Problem? *IEEE Veh. Technol. Soc. Newsl.* **2018**. Available online: https://dokumen.tips/documents/capacitors-in-power-electronics-applications-reliability-and-.html (accessed on 10 June 2021).
14. Pinto, C.; Barreras, J.V.; Schaltz, E.; Araújo, R.E. Evaluation of Advanced Control for Li-ion Battery Balancing Systems Using Convex Optimization. *IEEE Trans. Sustain. Energy* **2016**, *7*, 1703–1717. [CrossRef]
15. Daowd, M.; Omar, N.; Bossche, P.V.D.; Van Mierlo, J. Passive and active battery balancing comparison based on MATLAB simulation. In Proceedings of the 2011 IEEE Vehicle Power and Propulsion Conference, Chicago, IL, USA, 6–9 September 2011.
16. Altaf, F.; Egardt, B.; Mardh, L.J. Load Management of Modular Battery Using Model Predictive Control: Thermal and State-of-Charge Balancing. *IEEE Trans. Control. Syst. Technol.* **2016**, *25*, 47–62. [CrossRef]
17. Altaf, F.; Johannesson, L.; Egardt, B. Simultaneous Thermal and State-of-Charge Balancing of Batteries: A Review. In Proceedings of the 2014 IEEE Vehicle Power and Propulsion Conference (VPPC), Coimbra, Portugal, 27–30 October 2014; pp. 1–7.
18. Barreras, J.V.; Pinto, C.; De Castro, R.; Schaltz, E.; Andreasen, S.J.; Araújo, R.E. Multi-Objective Control of Balancing Systems for Li-Ion Battery Packs: A Paradigm Shift? In Proceedings of the 2014 IEEE Vehicle Power and Propulsion Conference (VPPC), Coimbra, Portugal, 27–30 October 2014; pp. 1–7.
19. Pinto, C.; De Castro, R.; Barreras, J.V.; Araujo, R.E.; Howey, D.A. Smart Balancing Control of a Hybrid Energy Storage System Based on a Cell-to-Cell Shared Energy Transfer Configuration. In Proceedings of the 2018 IEEE Vehicle Power and Propulsion Conference (VPPC), Chicago, IL, USA, 27–30 August 2018; pp. 1–6.
20. De Castro, R.P.; Pinto, C.; Barreras, J.V.; Araujo, R.E.; Howey, D.A. Smart and Hybrid Balancing System: Design, Modeling, and Experimental Demonstration. *IEEE Trans. Veh. Technol.* **2019**, *68*, 11449–11461. [CrossRef]
21. De Castro, R.; Pereira, H.; Araujo, R.E.; Barreras, J.V.; Pangborn, H.C. Multi-Layer Control for Hybrid Balancing System. In Proceedings of the 2021 5th IEEE Conference on Control Technology and Applications (CCTA), San Diego, CA, USA, 9–11 August 2021.
22. Docimo, D.J.; Fathy, H.K. Multivariable State Feedback Control as a Foundation for Lithium-Ion Battery Pack Charge and Capacity Balancing. *J. Electrochem. Soc.* **2016**, *164*, A61–A70. [CrossRef]
23. Sui, Y.; Song, S. A Multi-Agent Reinforcement Learning Framework for Lithium-ion Battery Scheduling Problems. *Energies* **2020**, *13*, 1982. [CrossRef]
24. Ma, Y.; Duan, P.; Sun, Y.; Chen, H. Equalization of Lithium-Ion Battery Pack Based on Fuzzy Logic Control in Electric Vehicle. *IEEE Trans. Ind. Electron.* **2018**, *65*, 6762–6771. [CrossRef]
25. Olfati-Saber, R.; Murray, R. Consensus Problems in Networks of Agents with Switching Topology and Time-Delays. *IEEE Trans. Autom. Control.* **2004**, *49*, 1520–1533. [CrossRef]
26. Abhinav, S.; Binetti, G.; Davoudi, A.; Lewis, F.L. Toward consensus-based balancing of smart batteries. In Proceedings of the 2014 IEEE Applied Power Electronics Conference and Exposition—APEC 2014, Fort Worth, TX, USA, 16–20 March 2014; pp. 2867–2873.
27. Ouyang, Q.; Chen, J.; Zheng, J.; Fang, H. Optimal Cell-to-Cell Balancing Topology Design for Serially Connected Lithium-Ion Battery Packs. *IEEE Trans. Sustain. Energy* **2018**, *9*, 350–360. [CrossRef]
28. Barreras, J.V.; Raj, T.; Howey, D.A.; Schaltz, E. Results of Screening over 200 Pristine Lithium-Ion Cells. In Proceedings of the 2017 IEEE Vehicle Power and Propulsion Conference (VPPC), Belfort, France, 11–14 December 2017; pp. 1–6.

29. Zheng, Y.; Han, X.; Lu, L.; Li, J.; Ouyang, M. Lithium ion battery pack power fade fault identification based on Shannon entropy in electric vehicles. *J. Power Sources* **2013**, *223*, 136–146. [CrossRef]
30. An, F.; Chen, L.; Huang, J.; Zhang, J.; Li, P. Rate dependence of cell-to-cell variations of lithiumion cells. *Sci. Rep.* **2016**, *6*, 35051. [CrossRef]
31. Rothgang, S.; Baumhöfer, T.; Sauer, D.U. Diversion of Aging of Battery Cells in Automotive Systems. In Proceedings of the 2014 IEEE Vehicle Power and Propulsion Conference (VPPC), Coimbra, Portugal, 27–30 October 2014; pp. 1–6.
32. Paul, S.; Diegelmann, C.; Kabza, H.; Tillmetz, W. Analysis of ageing inhomogeneities in lithium-ion battery systems. *J. Power Sources* **2013**, *239*, 642–650. [CrossRef]
33. An, F.; Huang, J.; Wang, C.; Li, Z.; Zhang, J.; Wang, S.; Li, P. Cell sorting for parallel lithium-ion battery systems: Evaluation based on an electric circuit model. *J. Energy Storage* **2016**, *6*, 195–203. [CrossRef]
34. Rumpf, K.; Naumann, M.; Jossen, A. Experimental investigation of parametric cell-to-cell variation and correlation based on 1100 commercial lithium-ion cells. *J. Energy Storage* **2017**, *14*, 224–243. [CrossRef]
35. Zou, H.; Zhan, H.; Zheng, Z. A Multi—Factor Weight Analysis Method of Lithiumion Batteries Based on Module Topology. In Proceedings of the 2018 International Conference on Sensing, Diagnostics, Prognostics, and Control (SDPC), Xi'an, China, 15–17 August 2018; pp. 61–66.
36. Baumann, M.; Wildfeuer, L.; Rohr, S.; Lienkamp, M. Parameter variations within Li-Ion battery packs—Theoretical investigations and experimental quantification. *J. Energy Storage* **2018**, *18*, 295–307. [CrossRef]
37. Devie, A.; Baure, G.; Dubarry, M. Intrinsic Variability in the Degradation of a Batch of Commercial 18650 Lithium-Ion Cells. *Energies* **2018**, *11*, 1031. [CrossRef]
38. Campestrini, C.; Keil, P.; Schuster, S.F.; Jossen, A. Ageing of lithium-ion battery modules with dissipative balancing compared with single-cell ageing. *J. Energy Storage* **2016**, *6*, 142–152. [CrossRef]
39. Schuster, S.F.; Brand, M.J.; Berg, P.; Gleissenberger, M.; Jossen, A. Lithium-ion cell-to-cell variation during battery electric vehicle operation. *J. Power Sources* **2015**, *297*, 242–251. [CrossRef]
40. Dubarry, M.; Truchot, C.; Cugnet, M.; Liaw, B.Y.; Gering, K.; Sazhin, S.; Jamison, D.; Michelbacher, C. Evaluation of commercial lithium-ion cells based on composite positive electrode for plug-in hybrid electric vehicle applications. Part I: Initial characterizations. *J. Power Sources* **2011**, *196*, 10328–10335. [CrossRef]
41. Baumhöfer, T.; Brühl, M.; Rothgang, S.; Sauer, D.U. Production caused variation in capacity aging trend and correlation to initial cell performance. *J. Power Sources* **2014**, *247*, 332–338. [CrossRef]
42. Dubarry, M.; Vuillaume, N.; Liaw, B.Y. Origins and accommodation of cell variations in Li-ion battery pack modeling. *Int. J. Energy Res.* **2010**, *34*, 216–231. [CrossRef]
43. Shin, D.; Poncino, M.; Macii, E.; Chang, N. A statistical model of cell-to-cell variation in Li-ion batteries for system-level design. In Proceedings of the International Symposium on Low Power Electronics and Design (ISLPED), Beijing, China, 4–6 September 2013; pp. 94–99.
44. Barreras, J.V.; Pinto, C.; De Castro, R.; Schaltz, E.; Swierczynski, M.J.; Andreasen, S.J.; Araújo, R.E. An improved parametrization method for Li-ion linear static Equivalent Circuit battery Models based on direct current resistance measurement. In Proceedings of the 2015 International Conference on Sustainable Mobility Applications, Renewables and Technology (SMART), Kuwait City, Kuwait, 23–25 November 2015; pp. 1–9.
45. Barreras, J.V.; Fleischer, C.; Christensen, A.E.; Swierczynski, M.J.; Schaltz, E.; Andreasen, S.J.; Sauer, D.U. An Advanced HIL Simulation Battery Model for Battery Management System Testing. *IEEE Trans. Ind. Appl.* **2016**, *52*, 5086–5099. [CrossRef]
46. Barreras, J.V.; Swierczynski, M.J.; Schaltz, E.; Andreasen, S.J.; Fleischer, C.; Sauer, D.U.; Christensen, A.E. Functional analysis of Battery Management Systems using multi-cell HIL simulator. In Proceedings of the 2015 Tenth International Conference on Ecological Vehicles and Renewable Energies (EVER), Monte Carlo, Monaco, 31 March–2 April 2015; pp. 1–10.
47. ASTM E1461-13 Standard Test Method for Thermal Diffusivity by the Flash Method. Available online: https://www.astm.org/Standards/E1461.htm (accessed on 13 October 2020).
48. Zhu, C.; Li, X.; Song, L.; Xiang, L. Development of a theoretically based thermal model for lithium ion battery pack. *J. Power Sources* **2013**, *223*, 155–164. [CrossRef]
49. Araújo, R.E.; De Castro, R.; Pinto, C.; Melo, P.; Freitas, D. Combined Sizing and Energy Management in EVs with Batteries and Supercapacitors. *IEEE Trans. Veh. Technol.* **2014**, *63*, 3062–3076. [CrossRef]
50. Meng, L.; Dragicevic, T.; Roldan-Perez, J.; Vasquez, J.C.; Guerrero, J. Modeling and Sensitivity Study of Consensus Algorithm-Based Distributed Hierarchical Control for DC Microgrids. *IEEE Trans. Smart Grid* **2016**, *7*, 1504–1515. [CrossRef]
51. Olfati-Saber, R.; Fax, J.A.; Murray, R. Consensus and Cooperation in Networked Multi-Agent Systems. *Proc. IEEE* **2007**, *95*, 215–233. [CrossRef]
52. Xiao, L.; Boyd, S. Fast linear iterations for distributed averaging. *Syst. Control. Lett.* **2004**, *53*, 65–78. [CrossRef]
53. Gu, G. Linear Feedback Control—Analysis and Design with MATLAB (by Dingyu Xue et al; 2007) [Bookshelf]. *IEEE Control. Syst.* **2009**, *29*, 128–129. [CrossRef]
54. Wen, B.S. *Cell Balancing Buys Extra Run Time and Battery Life*; Power Management, Analog Applications Journal; Texas Instruments Incorporated: New York, NY, USA, 2009.

energies

Article

Adaptive Square-Root Unscented Kalman Filter-Based State-of-Charge Estimation for Lithium-Ion Batteries with Model Parameter Online Identification

Quan Ouyang, Rui Ma, Zhaoxiang Wu, Guotuan Xu and Zhisheng Wang *

College of Automation Engineering, Nanjing University of Aeronautics and Astronautics, Nanjing 210016, China; ouyangquan@nuaa.edu.cn (Q.O.); maruinuaa@nuaa.edu.cn (R.M.); wuzhaoxiang@nuaa.edu.cn (Z.W.); xuguotuan@126.com (G.X.)

* Correspondence: wangzhisheng@nuaa.edu.cn; Tel.: +86-025-5283-2301

Received: 13 August 2020; Accepted: 17 September 2020; Published: 22 September 2020

Abstract: The state-of-charge (SOC) is a fundamental indicator representing the remaining capacity of lithium-ion batteries, which plays an important role in the battery's optimized operation. In this paper, the model-based SOC estimation strategy is studied for batteries. However, the battery's model parameters need to be extracted through cumbersome prior experiments. To remedy such deficiency, a recursive least squares (RLS) algorithm is utilized for model parameter online identification, and an adaptive square-root unscented Kalman filter (SRUKF) is designed to estimate the battery's SOC. As demonstrated in extensive experimental results, the designed adaptive SRUKF combined with RLS-based model identification is a promising SOC estimation approach. Compared with other commonly used Kalman filter-based methods, the proposed algorithm has higher precision in the SOC estimation.

Keywords: lithium-ion batteries; state-of-charge estimation; adaptive square-root unscented Kalman filter; recursive least squares

1. Introduction

Rechargeable lithium-ion batteries have been widely used in numerous applications due to their superior power performance, long life cycle, and so forth [1,2]. The state-of-charge (SOC) is an important and fundamental parameter that indicates the remaining available capacity of the battery [3]. Inaccurate SOC estimation can easily lead to the battery's overdischarge or overcharge even resulting in an explosion. It hence calls for promising strategies to accurately estimate the SOC of the battery's.

Two commonly utilized non-model-based SOC estimation approaches are the ampere-hour counting algorithm and open circuit voltage (OCV) method, respectively. Ampere-hour counting algorithm uses the time integration of the current as the indicator of the change of the SOC of the battery. Although this method is simple to implement, measurement errors accumulate since the SOC calculation process is open-loop based. Generally, it needs frequent calibrations to eliminate the battery's SOC estimation error. Derived from the identified relationship between the battery's OCV and SOC [4], the SOC can be obtained through measuring the OCV of the battery. However, the OCV can only be measured by cutting the battery off from the external circuit, which could inconvenience the battery user.

In order to get more accurate estimated SOC, model-based SOC estimation strategies attract great research attention. Currently, the commonly used battery models can be classified into the following three categories: electrochemical mechanism models [5,6], equivalent circuit models [7], and neural network models [8]. Among them, the equivalent circuit models, utilizing the circuits to simulate the dynamic characteristics of the battery, have reasonable computational complexity and great flexibility in battery materials and sizes, which are the most suitable to be utilized for the SOC estimation. Extended Kalman filter (EKF) algorithms are the most widely studied model-based SOC estimation approaches [9,10]. Adaptive EKFs are proposed in [11] and [12] to estimate the SOC with noise covariances online update. However, the nonlinear battery model needs to be linearized in these EKF algorithms, which will reduce the SOC estimation accuracy of the battery. The unscented Kalman filter (UKF) [13] and adaptive UKF [14] are proposed as the improvements of the EKFs for the SOC estimation. The first-order linearization is not required in the UKFs. Instead, they utilize a nonlinear transformation, called unscented transform (UT), for the propagation of the battery states' mean and covariance. The square-root UKF (SRUKF) can be treated as an enhanced version of the UKF with the attached advantage that the positive semi-definiteness of the state covariance can be guaranteed [15]. In addition, it needs less computational complexity for state estimation. In [16], based on a radial basis function neural network battery model, an SRUKF is utilized for the SOC estimation. A square-root spherical UKF is designed for SOC estimation for the battery in Nanosatellite [17]. An adaptive SRUKF method is designed in [18] to estimate the battery's SOC. A proportional-integral observer is proposed for the SOC estimation of the battery in electric drive vehicles [19]. A neural network-based SOC observer is proposed in [20] with the convergence proved utilizing Lyapunov stability analysis. A disturbance observer-based strategy is designed in [21] to estimate the SOC, which can reduce the computational burden without reducing the estimation accuracy.

The above SOC estimation strategies [9–14,16–20] all run on a battery model with parameters known. However, the model parameters must be extracted by cumbersome prior experiments. It will be more convenient if the model parameters are not necessary to be identified in advance for the designed SOC estimation algorithm. Therefore, an interesting and important question can be investigated: how to design a battery's SOC estimation strategy with the model parameters treated as unknown values. In [22], a recursive least squares (RLS) algorithm is utilized to determine the model parameters of the battery, and a linear observer is adopted to estimate the battery's SOC. However, the model noise information is not considered in this method, and the SOC estimation accuracy is reduced. The battery's SOC and model parameters are simultaneously estimated by utilizing an iterated EKF in [23]. A UKF algorithm is developed to estimate the SOC with another adaptive UKF utilized for the online identification of the model parameters of the battery in [24]. Its experimental results indicate that this UKF algorithm has higher SOC estimation accuracy compared with the EKF.

In this paper, an equivalent circuit model is utilized to describe the dynamics of the battery, where the model parameters are considered to be unknown. An RLS algorithm is utilized for model parameter online identification. Next, an adaptive SRUKF based on this online identified battery model is developed to estimate the battery's SOC, where a noise statistic estimator is utilized for the noise information online update. The main novelties of this work are summarized in the following two aspects.

(1) The model parameters are treated as unknown variables, and an RLS algorithm is utilized to estimate them online. This can avoid unnecessary experiments prior to SOC estimation for parameter extraction.

(2) Extensive experiments demonstrate the effectiveness of the proposed charging strategy showing that the proposed adaptive SRUKF can provide higher SOC estimation accuracy compared with other commonly used Kalman filter-based methods.

The rest of this manuscript is arranged as follows. In Section 2, the battery's equivalent circuit model is presented. In Section 3, an RLS algorithm is utilized for the online model parameter identification, and an adaptive SRUKF is developed to estimate the SOC of the battery. Related experimental results are included in Section 4, and concluding remarks are provided in Section 5.

2. Battery Model Development

2.1. Battery Equivalent Circuit Model

A commonly used equivalent circuit model [25,26] is utilized to simulate the lithium-ion battery's dynamics, which can strike an excellent balance between accuracy and simplicity. As illustrated in Figure 1, the capacitor C_b denotes the battery's full charged capacity, and the resistor R_0 represents the battery's energy losses during the charging and discharging process. The RC network (R_t, C_t) characterizes the battery's transient voltage-current response. The voltage across C_b represents the battery's SOC quantitatively, with 0–1 V corresponding to 0–100% of the SOC [20]. The mapping from the SOC to the OCV of the battery is denoted as

$$V_{OC} = g(SOC), \tag{1}$$

where V_{OC} and SOC are the battery's OCV and SOC, respectively; $g(\cdot)$ is a nonlinear function. Based on the Kirchhoff's laws of current and voltage, the dynamics of the battery is obtained as

$$\begin{aligned} S\dot{O}C &= -\frac{\eta_0}{C_b} I_B \\ \dot{V}_t &= -\frac{1}{R_t C_t} V_t + \frac{1}{C_t} I_B \\ V_B &= V_{OC} - R_0 I_B - V_t, \end{aligned} \tag{2}$$

where V_t, V_B, and I_B are, respectively, the voltage across the capacitor C_t, the terminal voltage, and the current of the battery; η_0 is the Coulomb coefficient.

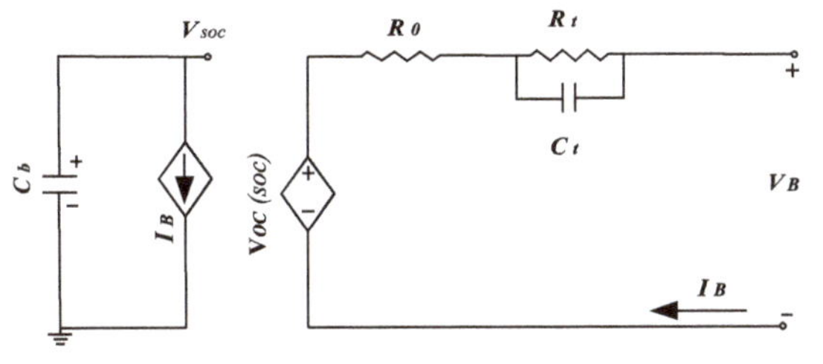

Figure 1. Equivalent circuit model of the battery.

2.2. Model Discretization

With the current I_B assumed to be constant during each sampling period, and the battery model (2) can be discretized as follows:

$$\begin{aligned} SOC(k) &= SOC(k-1) - \frac{T}{C_b} I_B(k-1) \\ V_t(k) &= e^{-\frac{T}{R_t C_t}} V_t(k-1) + (1 - e^{-\frac{T}{R_t C_t}}) R_t I_B(k-1) \\ V_B(k) &= g(SOC(k)) - R_0 I_B(k) - V_t(k), \end{aligned} \tag{3}$$

with T the sampling period. To simplify the notations, with the process and measurement noise considered, the battery model (3) can be expressed in the following form

$$\begin{aligned} x(k) &= A(\theta)x(k-1) + B(\theta)u(k-1) + w(k) \\ y(k) &= h(x(k), \theta, u(k)) + v(k), \end{aligned} \tag{4}$$

with

$$A(\theta) = \begin{bmatrix} 1 & 0 \\ 0 & \theta_2 \end{bmatrix}, \quad B(\theta) = \begin{bmatrix} -\frac{T}{C_b} \\ \theta_1 - \theta_1\theta_2 \end{bmatrix}$$
$$h(x(k), \theta, u(k)) = g(x_1(k)) - x_2(k) - \theta_3 u(k)$$

where the system output is $y(k) \triangleq V_B \in \mathbb{R}$; $u(k) \triangleq I_B \in \mathbb{R}$ is the system input; the states $x(k) \in \mathbb{R}^2$ and parameters $\theta \in \mathbb{R}^3$ are defined as $x(k) = [x_1(k), x_2(k)]^T \triangleq [SOC(k), V_t(k)]^T$ and $\theta = [\theta_1, \theta_2, \theta_3]^T \triangleq [R_t, e^{-\frac{T}{R_t C_t}}, R_0]^T$, respectively; $w(k)$ and $v(k)$ are the process noise and the measurement noise of the battery model, which are usually assumed as uncorrelated zero-mean Gaussian white noises that satisfy

$$\begin{aligned} w(k) &\sim N(0, Q^2) \\ v(k) &\sim N(0, R^2), \end{aligned} \tag{5}$$

with $Q^2 \in \mathbb{R}^{2\times 2}$, and $R^2 \in \mathbb{R}$ representing the covariances of the process and measurement noises. In practice, the noise statistic is correlated with the operating environment of the battery, which is generally unknown.

2.3. Observability Analysis

Referring to [27], the battery model (4) satisfies the observability condition, if the following observability matrix

$$\vartheta = \begin{bmatrix} C(k) \\ C(k+1)A(\theta) \end{bmatrix} \tag{6}$$

with

$$C(k) = \begin{bmatrix} \frac{\mathrm{d}g(x_1(k))}{\mathrm{d}x_1} & -1 \end{bmatrix}$$

is full rank for all x.

As C_b is much larger than the value of $I_B(k)$, the SOC change is small enough in each sampling step. Hence, $\frac{\mathrm{d}g(x_1(k))}{\mathrm{d}x_1}$ can be simplified to a constant $k_0 > 0$ [28,29]. By substituting (4) into (6), it can be derived that

$$\vartheta = \begin{bmatrix} k_0 & -1 \\ k_0 & -\theta_1 \end{bmatrix}. \tag{7}$$

Since k_0 is a positive constant and θ_2 is less than 1 as θ_2 is an exponential function, ϑ is full rank. Hence, it can be concluded that if the model parameter vector θ is known, the battery system is observable, therefore the battery's SOC can be estimated.

The model parameters are assumed to be known in numerous SOC estimation algorithms in the literature such as [9–14]. However, the RC parameters θ in the equivalent circuit model of the battery should be identified through cumbersome and inconvenient prior experiments [30]. It motivates us to propose an SOC estimation strategy by treating the parameters θ in (4) as unknown, and online estimating the model parameters and the SOC of the battery simultaneously.

3. Battery SOC and Model Parameter Estimation

Based on the model developed above, an RLS algorithm is utilized for the online model parameter identification, and an adaptive SRUKF algorithm is designed to estimate the SOC of the battery.

3.1. Model Parameter Identification with Rls Algorithm

Referring to [28,29], it can be assumed that the OCV of the battery remains constant during a short sampling period that satisfies $g(x_1(k)) - g(x_1(k-1)) = 0$. With defining $y_1(k) = y(k) - y(k-1)$ and $u_1(k) = u(k) - u(k-1)$, from (4), it yields

$$y_1(k) = U^T(k)\Phi + v_1(k), \tag{8}$$

with

$$\Phi \triangleq [\theta_2, -\theta_3, \ \theta_2\theta_3 - (1-\theta_2)\theta_1]^T$$
$$U(k) \triangleq [y_1(k-1), u_1(k), u_1(k-1)]^T,$$

where $v_1(k)$ is the noise. Then, an RLS is proposed for the battery model parameters online identification as follows:

$$\hat{\Phi}(k) = \hat{\Phi}(k-1) + \Gamma(k)(y_1(k) - U^T(k)\Phi(k-1)) \tag{9}$$

where $\hat{\Phi}(k) = [\hat{\phi}_1(k), \hat{\phi}_2(k), \hat{\phi}_3(k)]^T \in \mathbb{R}^3$ is the estimation of Φ; $\Gamma(k) \in \mathbb{R}^3$ is the designed gain vector that can be updated as

$$\begin{aligned} \Gamma(k) &= \frac{P(k-1)U(k)}{\lambda + U^T(k)P_\theta(k-1)U(k)} \\ P_\theta(k) &= \frac{P_\theta(k-1) - \Gamma(k)U^T(k)P_\theta(k-1)}{\lambda}, \end{aligned} \tag{10}$$

with λ denoting the forgetting factor that is normally chosen as a value between 0.98 and 0.995. Based on (9) and (10), the estimated battery model parameter vector $\hat{\theta}(k)$ is calculated that

$$\hat{\theta}(k) = [\frac{-\hat{\phi}_1(k)\hat{\phi}_2(k) - \hat{\phi}_3(k)}{1-\hat{\phi}_1(k)}, \hat{\phi}_1(k), -\hat{\phi}_2(k)]^T. \tag{11}$$

3.2. Adaptive Square-Root Ukf Based Soc Estimation

Different from the traditional UKF, the square-root of the state covariance matrix is directly propagated in the SRUKF method, avoiding the need to calculate the square-root of the state covariance at each iteration step. The SRUKF is more advantageous with better numerical properties and guaranteed positive semi-definiteness of the state covariance matrix [31]. Due to these advantages, based on the battery model with the estimated model parameters, an adaptive SRUKF algorithm is adopted here to estimate the battery's SOC, where an adaptive noise statistic estimator is utilized to update the noise statistic information at each sampling instant to get rid of its dependence on priori noise information. The algorithmic process of the adaptive SRUKF algorithm can be divided into four steps: initialization, prediction, measurement update, and noise statistic estimation, which are provided as follows:

Step 1: Initialization. The initial estimated state vector is set as $\hat{x}(0) = E[x(0)]$ and the initial square-root of the state covariance matrix is $S(0) = cholupdate\{E[(x(0) - \hat{x}(0))(x(0) - \hat{x}(0))^T]\}$, where $E[\cdot]$ denotes the mean and $cholupdate\{\cdot\}$ represents the Cholesky factorization [32]. The initial estimated covariances of the process and measurement noises are set as $\hat{Q}^2(0) = Q_0^2$ and $\hat{R}^2(k) = R_0^2$, respectively.

Step 2: Prediction. The UT technique is utilized for the propagation of the states' mean and covariance, where sigma points are chosen to approximate the mean and the covariance of the actual distribution. A set of $2n+1$ sigma points are selected in the following way

$$\begin{cases} \chi_0(k-1) = \hat{x}(k-1), \\ \chi_i(k-1) = \hat{x}(k-1) + (\sqrt{n+\lambda}S_i(k-1)), \quad i = 1, \cdots, n, \\ \chi_i(k-1) = \hat{x}(k-1) - (\sqrt{n+\lambda}S_{i-n}(k-1)), \ i = n+1, \cdots, 2n \end{cases} \tag{12}$$

where $\hat{x}(k-1) \in \mathbb{R}^n$ denotes the estimated state vector at time step $k-1$, n represents the state vector's dimension , $S_i(k-1)$ $(1 \le i \le n)$ is the i-th column of the matrix $S(k-1) \in \mathbb{R}^{n \times n}$ with $S(k-1)$ the square-root matrix of the state covariance, λ is a scalar that can be calculated as

$$\lambda = \alpha^2(n+\sigma) - n, \tag{13}$$

where α and σ are the tuning parameters determining the spread of the sigma points. With utilizing $\hat{\theta}(k-1)$ to replace θ in (4), each point in (12) is updated through the battery model as

$$\chi_i(k|k-1) = A(\hat{\theta}(k-1))\chi_i(k-1) + B(\hat{\theta}(k-1))u(k-1), \tag{14}$$

for $0 \leq i \leq 2n$. Then, the the a priori estimation of the state vector $\hat{x}(k|k-1)$ is computed as follows:

$$\hat{x}(k|k-1) = \sum_{i=0}^{2n} W_i^{(m)} \chi_i(k|k-1) \tag{15}$$

with the weights

$$\begin{cases} W_0^{(m)} &= \frac{\lambda}{\lambda+n}, \\ W_i^{(m)} &= \frac{1}{2(\lambda+n)} \quad i = 1, \cdots, 2n. \end{cases}$$

The a priori square-root of the state covariance $S(k|k-1)$ is updated as:

$$\begin{aligned} S_1(k|k-1) &= qr\{[\sqrt{W_1^{(c)}}(\chi_1(k|k-1) - \hat{x}(k|k-1)) \cdots \\ &\qquad \sqrt{W_{2n}^{(c)}}(\chi_{2n}(k|k-1) - \hat{x}(k|k-1))\ \hat{Q}(k)]^T\} \\ S(k|k-1) &= cholupdate\{S_1^T(k|k-1),\ \chi_0(k|k-1) - \hat{x}(k|k-1),\ W_0^{(c)}\} \end{aligned} \tag{16}$$

with the weights

$$\begin{cases} W_0^{(c)} &= \frac{\lambda}{\lambda+n} + (1-\alpha^2+\beta), \\ W_i^{(c)} &= \frac{1}{2(\lambda+n)} \qquad i = 1, \cdots, 2n \end{cases},$$

where $qr\{\cdot\}$ denotes the QR decomposition [32]; $\hat{Q}(k)$ denotes the estimation of the square-root of the process noise covariance matrix; β is a constant, which is usually chosen as 2 for the Gaussian distribution of the states [33]. Note that the QR decomposition of the matrix $D = [\sqrt{W_1^{(c)}}(\chi_1(k|k-1) - \hat{x}(k|k-1)) \cdots \hat{Q}(k-1)]^T \in \mathbb{R}^{3n\times n}$ is given by $D = Q_1R_1$, where $Q_1 \in \mathbb{R}^{3n\times 3n}$ is an orthogonal matrix and $R_1 \in \mathbb{R}^{3n\times n}$ is an upper triangular matrix [32]. Here, $qr\{D\}$ denotes the QR decomposition of the matrix D with $\bar{R}_1 \in \mathbb{R}^{n\times n}$ returned, where $\bar{R}_1$ is the upper triangular part of R_1, i.e., $S_1(k|k-1) = \bar{R}_1$ in (16). The Cholesky factor of the rank 1 update $P \pm \nu\xi\xi^T$ is represented as $S_2 = cholupdate\{S_1, \xi, \pm\nu\}$, where S_1 is the original Cholesky factor of P with $P = S_1S_1^T$. In (16), $S(k|k-1) = cholupdate\{\cdot\}$ denotes the Cholesky factor of $S_1^T(k|k-1)S_1(k|k-1) + W_0^{(c)}\xi\xi^T$ with $\xi = \chi_0(k|k-1) - \hat{x}(k|k-1)$. Note that $S_1^T(k|k-1)$ is used in (16) since $\bar{R}_1$ denotes the transpose of the Cholesky factor of DD^T with $\bar{R}_1^T\bar{R}_1 = \bar{R}_1^TQ^TQ\bar{R}_1 = DD^T$. From (16), it yields that the corresponding priori sate covariance matrix $P(k|k-1)$ is

$$\begin{aligned} P(k|k-1) &= S(k|k-1)S^T(k|k-1) = D^TD + W_0^{(c)}\xi\xi^T \\ &= \sum_{i=0}^{2n} W_i^{(c)}(\chi_i(k|k-1) - \hat{x}(k|k-1))(\chi_i(k|k-1) - \hat{x}(k|k-1))^T + \hat{Q}^2(k), \end{aligned} \tag{17}$$

which is consistent with that of the UKF [24]. Note that $S(k|k-1)$ is utilized rather than the square-root of $P(k|k-1)$ in the designed SRUKF since it is more efficient to calculate the QR decomposition and Cholesky factorization in (16) than performing the square-root of $P(k|k-1)$ directly in the UKF [15]. The estimated output is computed by propagating the sigma points based on the model (4) as follows:

$$\hat{y}(k|k-1) = \sum_{i=0}^{2n} W_i^{(m)} \Upsilon_i(k|k-1), \tag{18}$$

with

$$\Upsilon_i(k|k-1) = h(\chi_i(k|k-1), \hat{\theta}(k), u(k)), \tag{19}$$

where $\hat{y}(k|k-1)$ is the estimated output through the SRUKF, and $\hat{\theta}(k)$ are updated based on (9)–(11).

Step 3: Measurement update. Similarly, the square-root of the measurement covariance $S_y(k)$ can be calculated by

$$S_y(k) = cholupdate\{[S_{y1}^T(k), \Upsilon_0(k|k-1) - \hat{y}(k|k-1), W_0^{(c)}]\}, \tag{20}$$

with

$$S_{y1}(k) = qr\{[\sqrt{W_1^{(c)}}(\Upsilon_1(k|k-1) - \hat{y}(k|k-1)) \cdots \sqrt{W_{2n}^{(c)}}(\Upsilon_{2n}(k|k-1) - \hat{y}(k|k-1))\ \hat{R}(k)]^T\}, \tag{21}$$

with $\hat{R}(k)$ denoting the estimated square-root of the measurement noise covariance. The cross-correlation covariance $P_{xy}(k)$ are obtained as

$$P_{xy}(k) = \sum_{i=0}^{2n} W_i^{(c)}[\chi_i(k|k-1) - \hat{x}(k|k-1)][\Upsilon_i(k|k-1) - \hat{y}(k|k-1)]. \tag{22}$$

Then, the estimated state vector $\hat{x}(k)$ is updated as follows:

$$\hat{x}(k) = \hat{x}(k|k-1) + L(k)(y(k) - \hat{y}(k|k-1)), \tag{23}$$

with the filter gain vector $L(k)$ updated as

$$L(k) = P_{xy}(k)(S_y^{-1}(k))^T S_y^{-1}(k), \tag{24}$$

where $S_y^{-1}(k)$ denotes the inverse of $S_y(k)$. Based on (20) and (24), the square-root matrix of the state covariance is updated that

$$S(k) = cholupdate\{S(k|k-1), L(k)S_y(k), -1\}. \tag{25}$$

Step 4: Noise statistic estimation. A moving window of the output estimation error sequence is utilized for the noise statistic estimation. Referring to [14,24], the covariances of the process and measurement noises are updated as

$$\begin{aligned} \hat{Q}^2(k) &= L(k)\Big(\sum_{j=k-L+1}^{k}(y(j) - \hat{y}(j|j-1))(y(j) - \hat{y}(j|j-1))^T\Big)L^T(k) \\ \hat{R}^2(k) &= \sum_{j=k-L+1}^{k}(y(j) - \hat{y}(j|j-1))(y(j) - \hat{y}(j|j-1))^T \\ &\quad + \sum_{i=1}^{2n} W_i^{(c)}(\Upsilon_i(k|k-1) - y(k))(\Upsilon_i(k|k-1) - y(k))^T, \end{aligned} \tag{26}$$

where L is the size of the selected moving window used for covariance matching.

4. Experimental Results

To validate the performance of the designed adaptive SRUKF with RLS-based model parameters online identification on battery SOC estimation, an IFP36130155-36Ah lithium-ion battery with a nominal capacity of 36 Ah [26] was chosen for the experiment. The experiment was conducted at room temperature. The mapping from the battery's SOC to its OCV was as shown in Figure 2. The ITECH dc electronics Load IT8512 was programmed to simulate the actual working condition of the battery with the recorded signals of the current and the corresponding terminal voltage shown in Figures 3 and 4, where the sampling rate is 1 Hz. In the RLS algorithm, the initial estimated model parameter vector was randomly selected as $\hat{\theta}_0 = [0.001, 0.95, 0.02]^T$, λ and $P_\theta(0)$ are set as 0.99 and $diag\{0.1, 0.1, 0.1\}$, respectively. For the proposed adaptive SRUKF algorithm, the initial estimated state vector was randomly selected as $\hat{x}(0) = [70\%,\ 0]^T$. The initial square-root state estimation covariance matrix and the initial estimated square-root covariances of the process and measurement noises were randomly set as $S_0 = diag\{0.01,\ 0.01\}$, $Q_0 = diag\{0.01,\ 0.01\}$, and $R_0 = 0.05$, respectively. The parameters L, α, and β were selected as 100, 0.85, and 2, respectively.

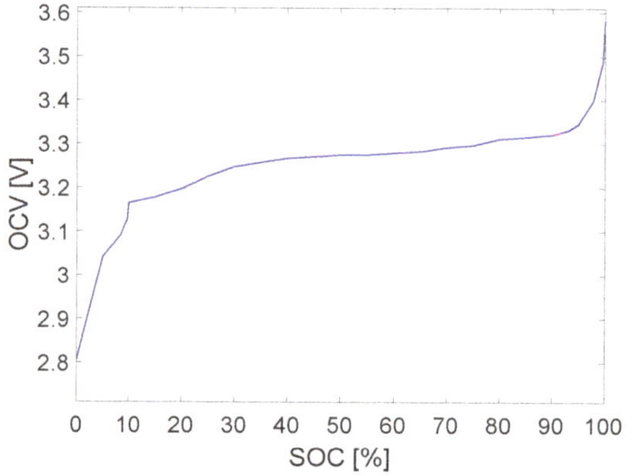

Figure 2. Open circuit voltage (OCV) versus state-of-charge (SOC) of the battery [26].

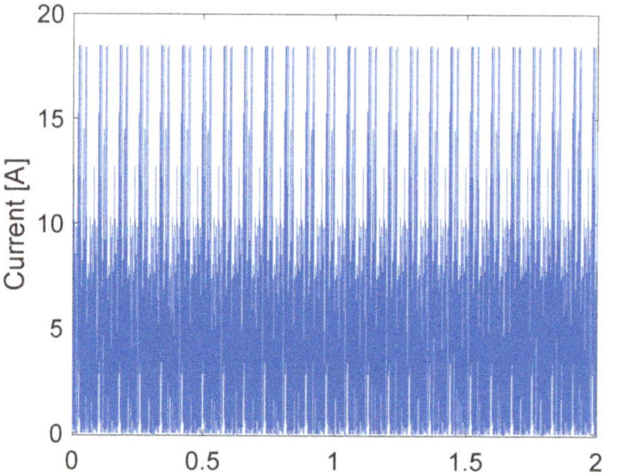

Figure 3. Applied current signal.

The experimental results in terms of the estimated model parameters $\theta_i (1 \leq i \leq 3)$ are demonstrated in Figures 5–7. The corresponding estimated C_t is shown in Figure 8. As a comparison, referring to [26], the parameter vector θ in the battery equivalent circuit model is calculated offline through utilizing the least-squares method with the results of $\hat{\theta} = [0.00631, 0.9838, 0.0224]^T$.

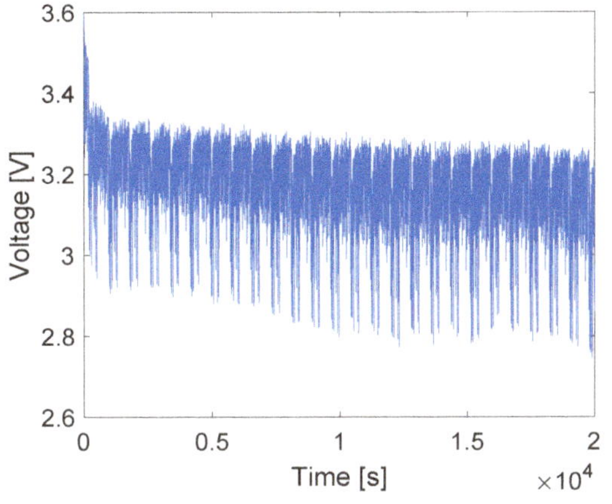

Figure 4. Terminal voltage of the battery.

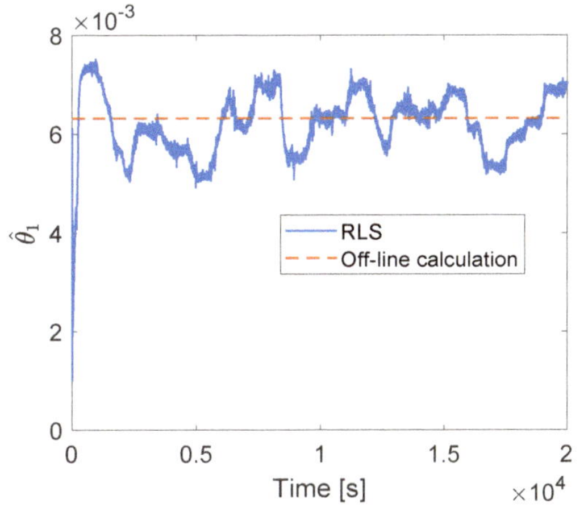

Figure 5. Comparison of the estimated results of parameter θ_1.

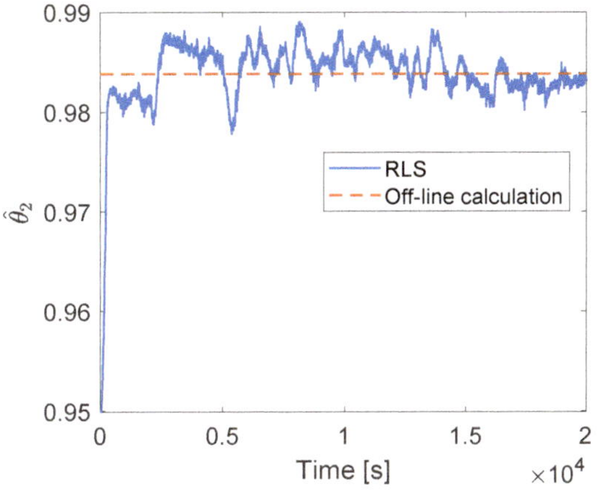

Figure 6. Comparison of the estimated results of parameter θ_2.

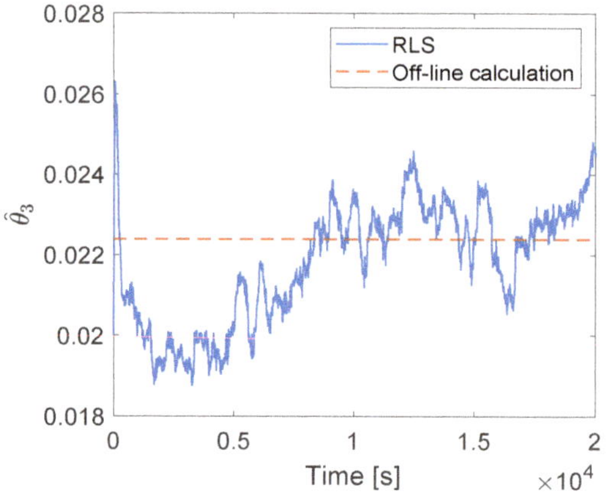

Figure 7. Comparison of the estimated results of parameter θ_3.

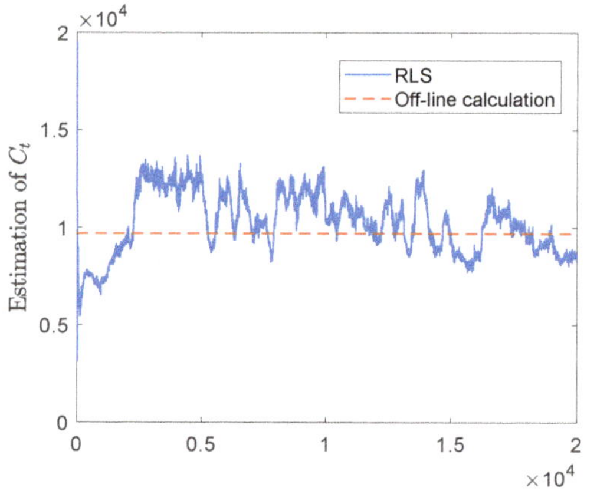

Figure 8. Comparison of the estimated results of C_t.

The actual and estimated SOCs by the adaptive SRUKF with RLS-based model parameters online identification are illustrated in Figure 9. The battery's actual SOC is gotten with using the ampere-hour counting method, where its initial SOC is obtained through measuring the initial OCV of the battery [20]. The SOC estimation error of the designed approach, calculated by subtracting the actual SOC from the estimated one, is illustrated in Figure 10. The statistic results in terms of the mean and root mean square (RMS) of the SOC estimation error are 0.13% and 0.98%, respectively, in which the mean and RMS errors are calculated as follows:

$$
\begin{aligned}
mean &= \frac{1}{N}\sum_{k=1}^{N}|x_1(k) - \hat{x}_1(k)| \\
RMS &= \sqrt{\frac{1}{N}\sum_{k=1}^{N}(x_1(k) - \hat{x}_1(k))^2}
\end{aligned}
\tag{27}
$$

with N denoting the data number. To demonstrate the superior performance of the designed adaptive SRUKF with RLS-based model parameters online identification method, the estimation results of the SRUKF and the adaptive EKF with model parameters identified by RLS, and the adaptive SRUKF with off-line calculated model parameters are given as comparisons. The comparison of the measured and estimated terminal voltages is shown in Figure 11 and Table 1. It demonstrates that the methods with model parameters online identification can provide a smaller voltage estimation error compared with those with model parameter offline calculation. Their comparison results of the SOC estimation error are listed in Figure 10 and Table 2. It shows that higher estimation accuracy can be achieved by the proposed estimation method than other commonly used Kalman filter-based methods, which demonstrates its promising performance in terms of the SOC estimation.

Table 1. Statistics of the terminal voltage simulation errors.

	Mean	RMS
Adaptive EKF + RLS	0.0108 V	0.0194 V
SRUKF + RLS	0.0082 V	0.0111 V
Adaptive SRUKF + offline calculation	0.0117 V	0.0635 V
Adaptive SRUKF + RLS	0.0081 V	0.0115 V

Table 2. Statistical results of the SOC estimation errors.

	Mean	RMS
Adaptive EKF + RLS	1.14%	2.22%
SRUKF + RLS	1.91%	2.38%
Adaptive SRUKF + offline calculation	1.08%	1.99%
Adaptive SRUKF + RLS	0.13%	0.98%

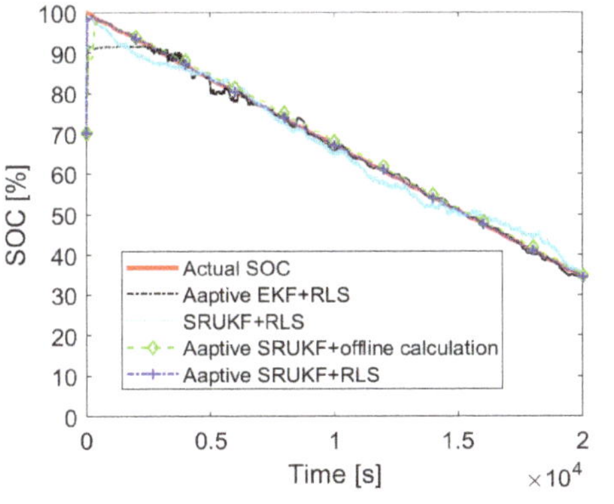

Figure 9. Actual and estimated SOC curves.

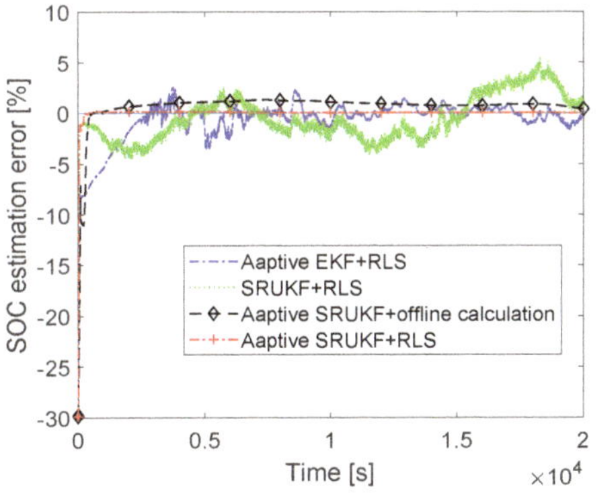

Figure 10. SOC estimation error of the battery.

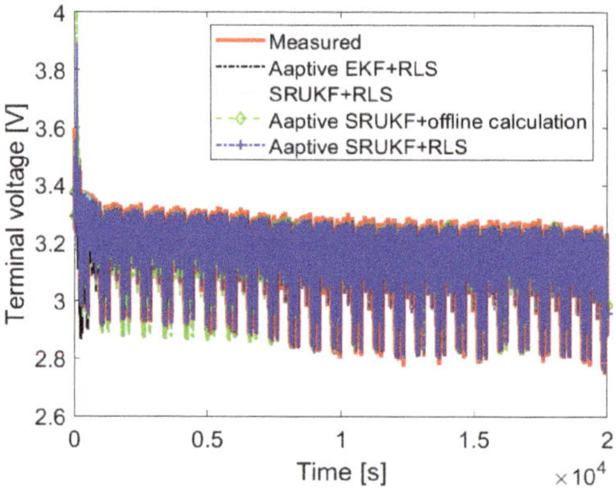

Figure 11. Comparison of the measured and simulated terminal voltages.

5. Conclusions

Accurate SOC estimation is important for the operation optimization of the battery. Based on a battery equivalent circuit model with unknown parameters, an RLS algorithm is utilized for model parameters online identification, and an adaptive SRUKF is proposed to estimate the battery's SOC, where a noise statistic estimator is utilized for the noise information online update. Extensive experimental results are provided to validate the performance of the proposed method showing that the mean and RMS of the SOC estimation error can be 0.13% and 0.98%, respectively. Higher accuracy in the SOC estimation is demonstrated compared with other commonly-used Kalman filter-based methods.

Author Contributions: Conceptualization, Q.O. and Z.W. (Zhisheng Wang); Methodology, Q.O.; Software, R.M.; Validation, Q.O., G.X. and R.M.; Formal Analysis, Z.W. (Zhaoxiang Wu); Investigation, R.M.; Resources, Z.W. (Zhaoxiang Wu); Data Curation, G.X.; Writing—Original Draft Preparation, Q.O.; Writing—Review & Editing, Q.O.; Visualization, R.M.; Supervision, Z.W. (Zhisheng Wang); Project Administration, Z.W. (Zhisheng Wang); Funding Acquisition, Q.O. All authors have read and agreed to the published version of the manuscript.

Funding: This work was supported in part by the National Natural Science Foundation of China (Grant No. 61903189), and in part by the Open Research Project of the State Key Laboratory of Industrial Control Technology, Zhejiang University, China (No. ICT20053).

Conflicts of Interest: The authors declare no conflict of interest.

References

1. Tsujikawa, T.; Yabuta, K.; Arakawa, M.; Hayashi, K. Safety of Large-Capacity Lithium-Ion Battery and Evaluation of Battery System for Telecommunications. *J. Power Sources* **2013**, *244*, 11–16. [CrossRef]
2. Ouyang, Q.; Han, W.; Zou, C.; Xu, G.; Wang, Z. Cell Balancing Control For Lithium-Ion Battery Packs: A Hierarchical Optimal Approach. *IEEE Trans. Ind. Informatics* **2020**, *16*, 5065–5075. [CrossRef]
3. Cheng, K.W.E.; Divakar, B.P.; Wu, H.; Ding, K. Battery-Management System(BMS) and SOC Development for Electrical Vehicles. *IEEE Trans. Veh. Technol.* **2011**, *60*, 76–88. [CrossRef]
4. Lee, S.; Kim, J.; Lee, J.; Cho, B.H. State-of-Charge and Capacity Estimation of Lithium-Ion Battery Using A New Open-Circuit Voltage versus State-of-Charge. *J. Power Sources* **2008**, *185*, 1367–1373. [CrossRef]
5. Moura, S.J.; Argomedo, F.B.; Klein, R.A. Mirtabatabaei and M. Krstic, Battery State Estimation for a Single Particle Model With Electrolyte Dynamics. *IEEE Trans. Control. Syst. Technol.* **2017**, *25*, 453–468. [CrossRef]
6. Corno, M.; Bhatt, N.; Savaresi, S.M.; Verhaegen, M. Electrochemical Model-Based State of Charge Estimation for Li-Ion Cells. *IEEE Trans. Control. Syst. Technol.* **2015**, *23*, 117–127. [CrossRef]
7. Chen, M.; Rincon-Mora, G.A. Accurate Electrical Battery Model Capable of Predicting Runtime and I-V Performance. *IEEE Trans. Energy Convers.* **2006**, *21*, 504–511. [CrossRef]

8. Kang, L.; Zhao X.; Ma, J. A New Neural Network Model for the State-of-Charge Estimation in the Battery Degradation Process. *Appl. Energy* **2014**, *121*, 20–27. [CrossRef]
9. Chen, Z.; Fu, Y.; Mi, C.C. State of Charge Estimation of Lithium-Ion Batteries in Electric Drive Vehicles Using Extended Kalman Filtering. *IEEE Trans. Veh. Technol.* **2013**, *62*, 1020–1030. [CrossRef]
10. Domenico, D.D.; Prada, E.; Creff, Y. An Adaptive Strategy for Li-Ion Battery Internal State Estimation. *Control. Eng. Pract.* **2013**, *21*, 1851–1859. [CrossRef]
11. He, H.; Xiong, R.; Zhang, X.; Su, F. State-of-Charge Estimation of the Lithium-Ion Battery Using An Adaptive Extended Kalman Filter based on An Improved Thevenin Model. *IEEE Trans. Veh. Technol.* **2011**, *60*, 1461–1469.
12. Xiong, R.; Gong, X.; Mi, C.C.; Sun, F. A Robust State-of-Charge Estimator for Multiple Types of Lithium-Ion Batteries Using Adaptive Extended Kalman Filter. *J. Power Sources* **2013**, *243*, 805–816. [CrossRef]
13. Tian, Y.; Xia, B.; Sun, W.; Xu, Z. A Modified Model based State of Charge Estimation of Power Lithium-Ion Batteries Using Unscented Kalman Filter. *J. Power Source* **2014**, *270*, 619–626. [CrossRef]
14. Sun, F.; Hu, X.; Zou, Y.; Li, S. Adaptive Unscented Kalman Filter for State of Charge Estimation of A Lithium-Ion Battery for Electric Vehicles. *Energy* **2011**, *36*, 3531–3540. [CrossRef]
15. Huang, M.; Li, W.; Yan, W. Estimating parameters of synchronous generators using square-root unscented Kalman filter. *Electr. Power Syst. Res.* **2010**, *80*, 1137–1144. [CrossRef]
16. Gholizade-Narm, H.; Charkhgard, M. Lithium-ion battery state of charge estimation based on square-root unscented Kalman filter. *IET Power Electron.* **2013**, *6*, 1833–1841. [CrossRef]
17. Aung, H.; Low, K.S.; Goh, S.T. State-of-Charge Estimation of Lithium-Ion Battery Using Square Root Spherical Unscented Kalman Filter (Sqrt-UKFST) in Nanosatellite. *IEEE Trans. Power Electron.* **2015**, *30*, 4774–4783. [CrossRef]
18. Liu, S.; Cui, N.; Zhang, C. An Adaptive Square Root Unscented Kalman Filter Approach for State of Charge Estimation of Lithium-Ion Batteries. *Energies* **2017**, *10*, 1345.
19. Xu, J.; Mi, C.C.; Cao, B.; Deng, J.; Chen, Z.; Li, S. The State of Charge Estimation of Lithium-Ion Batteries Based on a Proportional-Integral Observer. *IEEE Trans. Veh. Technol.* **2014**, *63*, 1614–1621.
20. Chen, J.; Ouyang, Q.; Xu, C.; Su, H. Neural Network-Based State of Charge Observer Design for Lithium-Ion Batteries. *IEEE Trans. Control. Syst. Technol.* **2018**, *26*, 313–320. [CrossRef]
21. Messier, P.; Nguyen, B.; LeBel, F.; Trovao, J.P.F. Disturbance observer-based state-of-charge estimation for Li-ion battery used in light electric vehicles. *J. Energy Storage* **2020**, *27*, 101144. [CrossRef]
22. Roscher, M.A.; Bohlen, O.S.; Sauer, D.U. Reliable State Estimation of Multicell Lithium-ion Battery Systems. *IEEE Trans. Energy Convers.* **2011**, *26*, 737–743. [CrossRef]
23. Fang, H.; Wang, Y.; Sahinoglu, Z.; Wada, T.; Hara, S. State of Charge Estimation for Lithium-Ion Batteries: An Adaptive Approach. *Control. Eng. Pract.* **2014**, *25*, 45–54. [CrossRef]
24. Partovibakhsh, M.; Liu, G. An Adaptive Unscented Kalman Filtering Approach for Online Estimation of Model Parameters and State-of-Charge of Lithium-Ion Batteries for Autonomous Mobile Robots. *IEEE Trans. Control. Syst. Technol.* **2015**, *23*, 357–363. [CrossRef]
25. Chen, C.; Xiong, R.; Shen, W. A Lithium-ion Battery-in-the Loop Approach to Test and Validate Multiscale Dual H Infinity Filters for State-of-Charge and Capacity Estimation. *IEEE Trans. Power Electron.* **2018**, *33*, 332–342. [CrossRef]
26. Ouyang, Q.; Chen, J.; Zheng, J. State-of-Charging Observer Design for Batteries With On-line Model Parameter Identification: A Robust Approach. *IEEE Trans. Power Electron.* **2020**, *35*, 5820–5831. [CrossRef]
27. Reif, K.; Gunther, S.; Yaz, E.; Unbehauen, R. Stochastic Stability of the Discrete-Time Extended Kalman Filter. *IEEE Trans. Autom. Control.* **1999**, *44*, 714–728. [CrossRef]
28. Rahimi-Eichi, H.; Baronti, F.; Chow, M.Y. Online Adaptive Parameter Identification and State-of-Charge Coestimation for Lithium-Polymer Battery Cells. *IEEE Trans. Ind. Electron.* **2014**, *61*, 2053–2061. [CrossRef]
29. Chen, X.; Shen, W.; Cao, Z.; Kapoor, A. A novel approach for state of charge estimation based on adaptive switching gain sliding mode observer in electric vehicles. *J. Power Sources* **2014**, *246*, 667–678. [CrossRef]
30. Han, J.; Kim, D.; Sunwoo, M. State-of-Charge Estimation of Lead-Acid Batteries Using An Adaptive Extended Kalman Filter. *J. Power Sources* **2009**, *188*, 606–612. [CrossRef]
31. Haykin, S.S. *Kalman Filtering and Neural Networks*; Wiley: New York, NY, USA, 2001.

32. Van der Merwe, R.; Wan, E.A. The square-root unscented Kalman filter for state and parameter-estimation. In Proceedings of the IEEE International Conference on Acoustics, Speech, and Signal Processing, Salt Lake City, UT, USA, 7–11 May 2001.
33. Julier, S.; Uhlmann, J.; Durrant-Whyte, H.F. A new method for the nonlinear transformation of means and covariances in filters and estimators. *IEEE Trans. Autom. Control.* **2000**, *45*, 477–482. [CrossRef]

energies

Article

Switching Frequency Determination of SiC-Inverter for High Efficiency Propulsion System of Railway Vehicle

Joon-Hyoung Ryu [1], June-Hee Lee [1] and June-Seok Lee [2,*]

1 Propulsion System Research Team, Korea Railroad Research Institute, 176, Cheoldo Bangmulgwan-ro, Uiwang-si 16105, Gyeonggi-do, Korea; jhryu@krri.re.kr (J.-H.R.); juneh99@krri.re.kr (J.-H.L.)
2 School of Electronics and Electrical Engineering, Dankook University, Yongin-si 16890, Korea
* Correspondence: ljs@dankook.ac.kr; Tel.: +82-31-8005-3628

Received: 17 August 2020; Accepted: 22 September 2020; Published: 24 September 2020

Abstract: This paper suggests the reasonable switching frequency determination method for achieving highest efficiency of the railway propulsion system consisting the silicon carbide (SiC) inverter and permanent magnet synchronous motor (PMSM). The SiC power device allows increasing the switching frequency of the inverter because it has the small switching power loss. The total efficiency is taken into account for determining the switching frequency of SiC inverter in this paper. In the efficiency analysis of SiC inverter and PMSM, the PMSM drive control is considered with the hybrid switching method combined the synchronous PWM and asynchronous PWM. The result of the analysis shows the efficiency curve of propulsion system depending on the switching frequency. The switching frequency having the minimum power loss of propulsion system is selected based on the extracted power loss curve.

Keywords: propulsion inverter; PMSM drive; efficiency analysis; SiC inverter

1. Introduction

Much research on the efficiency improvement of railway system has been performed for last decade [1–3]. In the railway vehicle, it has been trying to apply new technologies such as the permanent magnet synchronous motor (PMSM) and the silicon carbide (SiC) power device to the propulsion system. The PMSMs have 5–8% efficiency higher than induction motors [1,4], and SiC power devices reduce the switching loss to less than half loss of Silicon (Si) power device [5–7]. It is obvious that the propulsion system consisting the SiC inverter and PMSM, which is shown in Figure 1, guarantees the high efficiency.

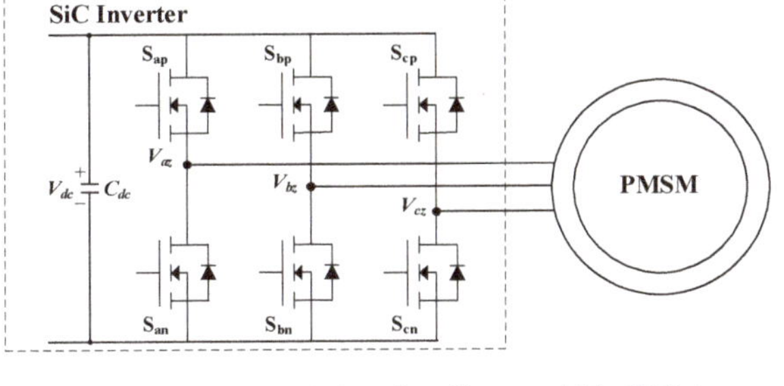

Figure 1. Railway propulsion system consisting the silicon carbide (SiC) inverter and permanent magnet synchronous motor (PMSM).

In the railway industry, the induction motor has been selected as the traction motor for several decades [1,2,8]. The main reason is that the induction motor has the high reliability and superior maintainability beside it is inexpensive compared to PMSMs. However, since the totally enclosed structure improves the reliability and maintainability of PMSM, the number of the cases where the totally-enclosed PMSM is applied to the railway vehicle increases gradually [3,4,9–13]. In addition, the PMSMs show the higher efficiency than that of induction motor. Nowadays, the PMSMs have become attractive for railway operation company to save the operating cost. Along with the interest in PMSMs, the research on the PMSM drive control has been introduced in many paper [11–18]. The research theme in the railway vehicle can be divided into several areas: torque control, sensorless control, restarting control, and fault tolerant control. Nevertheless, these papers are based on the high efficiency operation of inverter. The existing PWM method for induction motor is applied to drive PMSM [13,14]. In the existing PWM method for induction motor, the six-step operation appears at propulsion inverter. Although the six-step operation has the large torque ripple, it is attractive PWM method to reduce the switching loss of power device.

Nowadays, the voltage class of SiC power module is announced to 3300 V [5,6,19]. Therefore, the studies applying the high voltage SiC power module into various applications are introducing in lots of papers [19–29]. The SiC power module with 1700 V class has been applied for the auxiliary converter of railway vehicle [20,25]. The switching frequency of 50 kHz is used in the resonant converter consisting 1700 V class SiC power module in [20]. In the catenary-free tram, the SiC power module with 1700 V class and hybrid SiC power with 1700 V class for the DC–DC converter used in the energy storage system are compared to Si module with 1700 V class [21]. Here, the switching frequency of 6 kHz in hybrid SiC power module has the same power loss with the Si module with the switching frequency of 2 kHz. The SiC power modules with various voltage classes have been considered in the traction inverter of railway vehicle [19,22–24,26–28]. Initially, the hybrid SiC power modules where the SiC freewheeling diode combines with silicon (Si)-IGBT were tested [19,22–24,26,27]. The mass and volume comparison results between the inverters consisting Si-IGBT modules and SiC power modules in the same switching frequency, which is respectively low switching frequency, 800 Hz, are shown in [22,24,26]. The propulsion inverter consisting of the hybrid SiC power modules can decrease the mass and volume until 60% of the conventional Si-IGBT inverter. Nonetheless, it can lead to the power loss reduction of 35% compared to the conventional Si-IGBT inverter [24]. The full SiC power modules enhance the efficiency improvement of propulsion inverter [28,29]. Since the SiC power module with 1200 V class is applied in [28], the switching frequency could be increased to 10 kHz. Although the SiC power module with 3300 V class is used in [29], the number of switching pulse increases in only the synchronous mode of PWM method. Generally, in the railway application, the SiC power module has been used to reduce the mass and volume of inverter in keeping the switching frequency as low value.

In Figure 1, the efficiency of propulsion system is determined by each efficiency of the PMSM and the SiC inverter. This paper focuses on the switching frequency because the switching frequency can influence both efficiencies of PMSM and SiC inverter. The effect of switching frequency on both the induction motor and Si-IGBT inverter is introduced in [24,29]. Ref. [29] used the triple pulse PWM method instead of the six-step operation to increase the efficiency of induction motor. Ref. [24] has proved that the six-step operation aggravates the efficiency of induction motor.

This paper proposes the reasonable switching frequency determination method for achieving highest efficiency of the railway propulsion system consisting the PMSM and SiC inverter. The efficiency analysis is implemented by considering the existing PWM strategies which are introduced in Section 2 briefly. In the efficiency analysis, the efficiencies of SiC inverter and PMSM are calculated through the co-simulation of SiC inverter (PSIM software, Powersim Inc., Rockville, MD, USA) and PMSM (J-MAG software, JSOL Corporation, Tokyo, Japan) to consider the interaction of them. Simulation results are based on the determination of the switching frequency of SiC inverter [30–32]. Based on the power loss curve of the propulsion system, the switching frequency with highest efficiency can be determined for propulsion inverter.

2. PWM Strategy for SiC Inverter

This paper focusses on how to maximize the efficiency of propulsion system. Therefore, the PWM strategy for the SiC inverter is also aimed to achieve the high efficiency of the SiC inverter. The existing PWM strategy used in the propulsion inverter guarantees maximizing the efficiency of the SiC inverter [14]. This strategy uses not only linear voltage modulation but also over-voltage modulation and six-step modulation. These modulations can be classified by magnitude of V_o ($|V_o|$) and dc-link voltage (V_{dc}) [15]. Figure 2 shows the output voltage (V_o) of existing Si-IGBT inverter depending on speed of railway vehicle. By considering that the maximum speed is 80 km/h, the six-step modulation relatively appears at low speed. Even though the six-step modulation leads to large torque ripple with vibration, it is enough to be ignored in railway vehicle. The six-step modulation appears over half of whole speed range in the existing PWM strategy. It means that the switching loss of the propulsion inverter can be minimized. The conventional PWM methods [15,16] guarantee the ideal change between modulations as $|V_o|$ increases or $|V_o|$ decreases. Generally, the reference voltage without any change is used to generate the desired $|V_o|$ in the linear voltage modulation where the V_o is lower than $V_{dc}/\sqrt{3}(=0.5774\ V_{dc})$. However, in the over-voltage and six step modulations, the changed reference voltage is used to generate the desired $|V_o|$. This paper applies the existing PWM strategy for the SiC inverter.

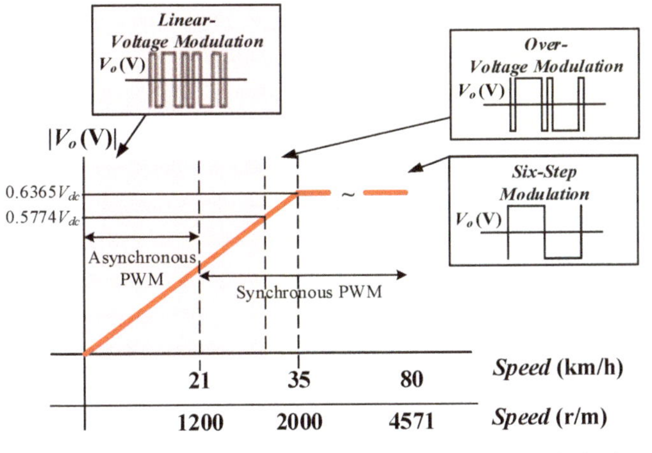

Figure 2. The output voltage (V_o) of SiC inverter depending on the speed of railway vehicle.

The six-step modulation is not affected by the swishing frequency (f_{sw}). The number of the switching is determined by only the fundamental frequency of V_o in the six-step modulation. Therefore, the inverter can show the same efficiency in six-step modulation regardless of f_{sw}. In addition, the f_{sw} is defined as K(integer) x the fundamental frequency of V_o in the synchronous PWM. It means that f_{sw} is not changeable parameter and the efficiency is fixed as the same as that of six-step modulation if K is fixed. On the other hand, the variation of f_{sw} influences the efficiency of inverter during the linear-voltage modulation in the asynchronous PWM. As f_{sw} increases, the efficiency of inverter decreases. Consequently, it is needed to analyze the efficiency of propulsion system during the speed where the asynchronous PWM is applied.

3. Permanent Magnet Synchronous Motor Control

3.1. Permanent Magnet Synchronout Motor

This paper considers the 332-kW interior permanent magnet synchronous motor (IPMSM) as the traction motor for railway vehicle. Its specifications and cross section are shown in Table 1 and Figure 3.

The B-H curves of materials (35PN230 and NdFeB) can be identified through the material datasheets. The lamination thickness of stator core and rotor core is 0.35 mm; the magnet is not laminated.

Table 1. PMSM specifications.

Specification	Value	Specification	Value
Type	IPMSM	Pole/Slot	6/27
Rated Current (I_{phase})	236 A_{rms}	Stator Core Material	35PN230
Rated Torque	1522 Nm	Rotor Core Material	35PN230
Rated Speed	2080 r/m	Magnet Material	NdFeB
Stator an external diameter	500 mm	Rotor an external diameter	280 mm
Stator an internal diameter	283 mm	Rotor an internal diameter	80 mm
Air gap length	1.5 mm	Laminated length (height)	235 mm

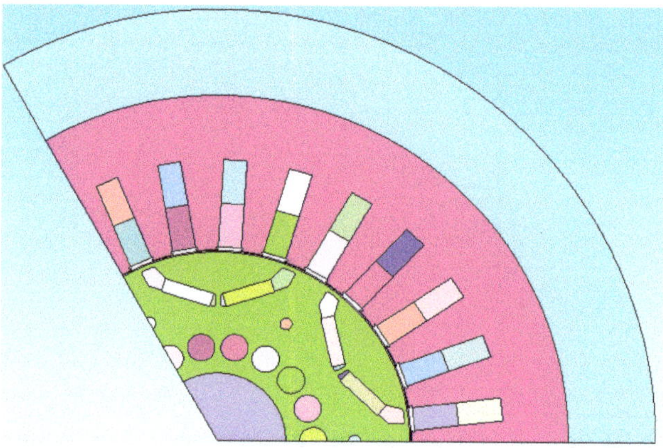

Figure 3. The cross section of interior permanent magnet synchronous motor (IPMSM) used in this paper.

3.2. Control Method

Maximum torque per ampere (MTPA) operation is required to control the IPMSM. Lots of papers have suggested control methods for the MTPA operation [33–35]. This paper uses the current table-based control method [35]. The control block diagram is shown in Figure 4.

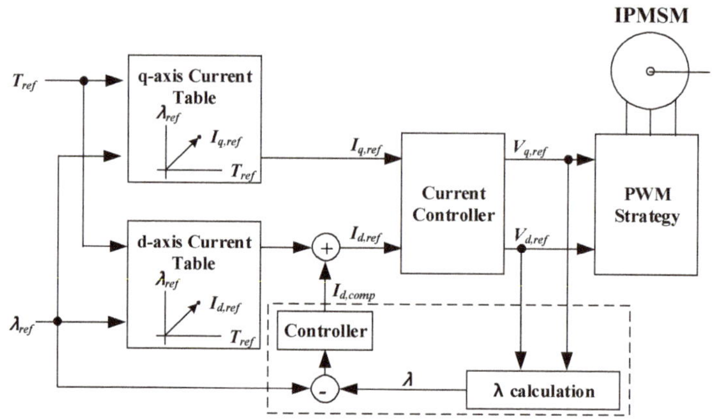

Figure 4. Control block diagram for driving the IPMSM.

The d-axis current ($I_{d,ref}$) and q-axis current ($I_{q,ref}$) are determined based on the MTPA tables according to the reference flux (λ_{ref}) and reference torque (T_{ref}), respectively. To eliminate the flux error, the compensation of the d-axis current ($I_{d,comp}$) is used. To deal with the whole speed range with six-step modulation, the additional considerations (flux-weakening method and maximum torque per flux) should be reflected into the control method. However, this paper applies the simple control method of Figure 4 for the MTPA operation because the effect on the variation of f_{sw} exists in the low speed range where the MTPA operation is needed.

4. Loss Analysis of Propulsion System

The loss analysis of the propulsion system can be divided into two parts: SiC inverter and IPMSM. Two parts affect each other as shown in Figure 5. The non-sinusoidal electromotive force of IPMSM affects the current control of IPMSM; therefore, it leads the low harmonics frequency components in the inverter currents. Finally, the low harmonics frequency components can increase the loss of IPMSM. In addition, the current ripple generated by f_{sw} affects the power loss of IPMSM. Therefore, this paper conducts the loss analysis through the co-simulation between the SiC inverter (Power SIM software) and the IPMSM (J-MAG software) to consider the interaction of them. It is obvious that the simulation results are reliable in the comparison and analysis [30–32].

The loss analysis of propulsion system is implemented in the valid speed range (below 1200 r/m) where the effect of f_{sw} variation exists.

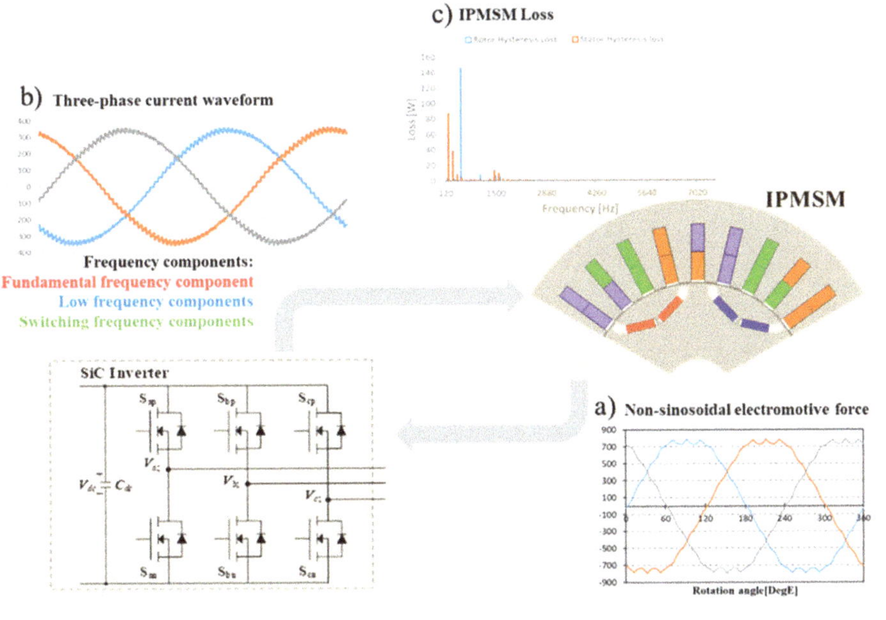

Figure 5. Interaction between SiC inverter and IPMSM.

4.1. SiC Inverter

The SiC inverter consists of six SiC MOSFETs. The 3300 V/800 A 2in1 SiC module where two SiC MOSFETs are connected to configure one leg, which is MSM800FS33ALT manufactured by HITACHI, is selected in this paper. The gate on and off resistances which decide E_{on} (Turn-on Loss per Pulse), E_{off} (Turn-off Loss per Pulse), and E_{rr} (Reverse Recovery Loss per Pulse) are 1 Ω and 1.5 Ω, respectively. The additional data for power loss analysis of SiC MOSFET is shown in datasheet of MSM800FS33ALT.

It is assumed that the SiC MOSFETs are closely connected to the DC-link capacitor to mitigate the effect of stray inductance enough [36,37].

The operation conditions of the propulsion inverter are as follows: V_{dc} is 1500 V, the control period is 1/(2xf_{sw}), and T_{ref} is 1522 Nm.

Figure 6 shows the three-phase current waveforms depending on f_{sw} at 400 r/m. The $|V_o|$ at 400 r/m is 188 V. The currents do not seem the completed sinusoidal waveform. There are several frequency components in currents: fundamental frequency component, low harmonics frequency components, and f_{sw} components. Nevertheless, it is confident that as f_{sw} increases, the current ripple caused by f_{sw} is reduced.

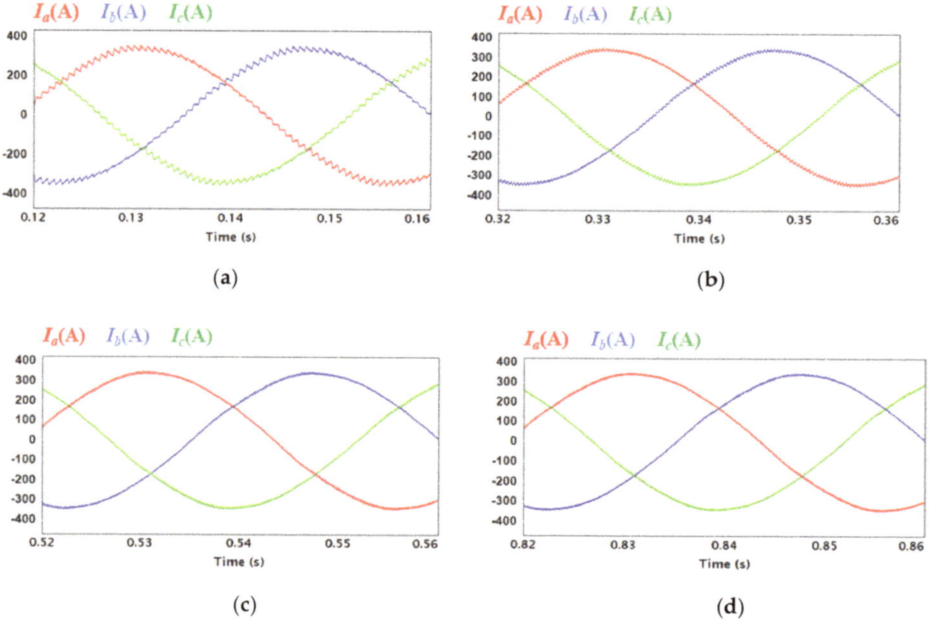

Figure 6. Three-phase current waveforms depending on f_{sw} at 400 r/m: (**a**) 750 Hz; (**b**) 1500 Hz; (**c**) 2250 Hz; (**d**) 3000 Hz.

Figure 7 shows the efficiency of one SiC MOSFET depending on the f_{sw} and speed (r/m). Since the SiC MOSFET consists of MOSFET and Diode in one device, the loss analysis is implemented in two parts separately. As f_{sw} increases, only the switching losses of MOSFET and Diode increase as shown in Figure 7a. One the other hand, the speed variation does not give enough impact on the total loss as shown in Figure 7b. The speed variation of IPMSM leads to the change of $|V_o|$. As the speed increase, $|V_o|$ increases. The much current flows through MOSFET in the high $|V_o|$ compared to Diode. It is identified that conduction loss of Diode is reduced as the speed increases.

Figure 8 shows the total loss of SiC inverter depending on f_{sw}. The inverter loss at 3000 Hz is more than twice inverter loss at 750 Hz. In terms of SiC inverter, the low f_{sw} is required to reduce the power loss.

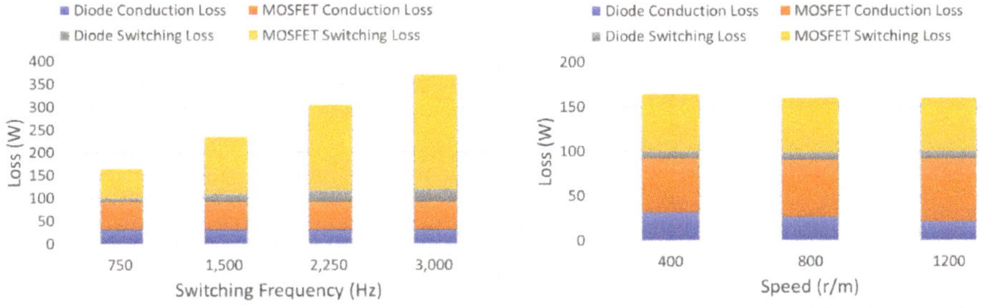

(**a**) (**b**)

Figure 7. Loss analysis of SiC MOSFET depending on f_{sw}: (**a**) 400 r/m; (**b**) 750 Hz.

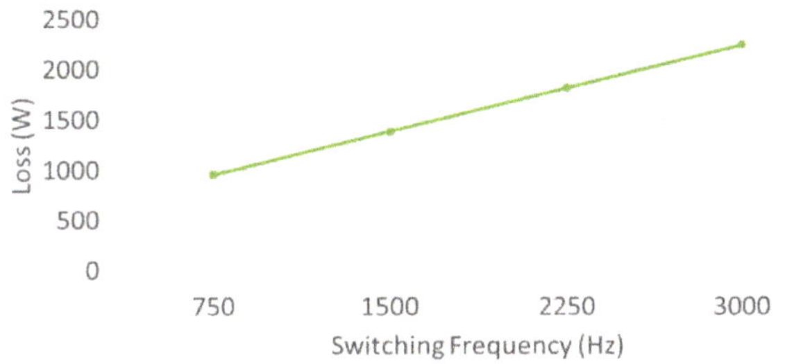

Figure 8. Total loss of SiC inverter depending on f_{sw}.

4.2. IPMSM

The finite elements method (FEM) model of IPMSM, which is Figure 9, is developed to analyze the power loss of IPMSM. The currents with ripple components of Figure 6 are injected to IPMSM through the co-simulation to reveal the effect of current ripple components on the power loss. The conditions for loss analysis are the same as those of loss analysis of SiC inverter. In the loss analysis, the electrical resistivity of NdFeB magnet is 1.80×10^{-5} Ω·m, which is used to analyze the eddy current loss of magnet. The phase resistance (R_{coil}) of stator coil for the copper loss analysis is 0.021882 Ω. The iron loss of stator and rotor cores is estimated by applying the loss density curve of Figure 10.

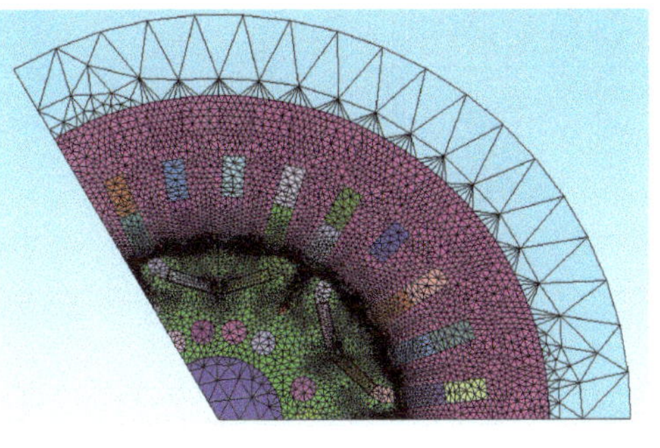

Figure 9. The finite elements method (FEM) model of 322 kW IPMSM considered in this paper.

To verify the estimated power loss of IPMSM from co-simulation, it is compared to the result of the power loss calculation equation. The loss components are divided into the coil copper loss (P_{copper}), the core iron loss (P_{iron}), and the magnet eddy current loss ($P_{m,eddy}$). In this paper, P_{copper} and P_{iron} are calculated and compared to estimated power loss from J-MAG software. Although $P_{m,eddy}$ is not considered in the verification, the several researches have proved its accuracy [38–40].

To calculate P_{copper}, the skin effect is ignored because the stator coil consists of many thin coils with the small dimension [41]. Therefore, P_{copper} can be calculated as

$$P_{copper} = 3I_{phase}^2 R_{coil} \quad (1)$$

The loss curve for P_{iron} is often supplied by the core material manufactures [42]. The P_{iron} is represented as

$$P_{Iron} = P_{Loss}(f_B) \quad (2)$$

where f_B is a frequency component of the magnet flux density and P_{Loss} is loss density curve of Figure 10.

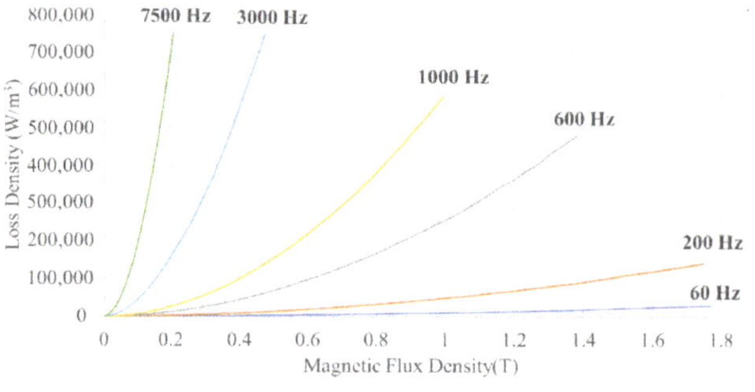

Figure 10. The loss density of stator and rotor core material (0.35 mm) depending on the magnetic flux density.

Figure 11 shows accuracy of the estimated power losses of IPMSM from J-MAG software and calculated power losses from (1) and (2). These values are expressed as per unit. Two copper losses are similar, and its error ratio is under 2%. In addition, the error ratio is about 5% in the iron loss comparison results.

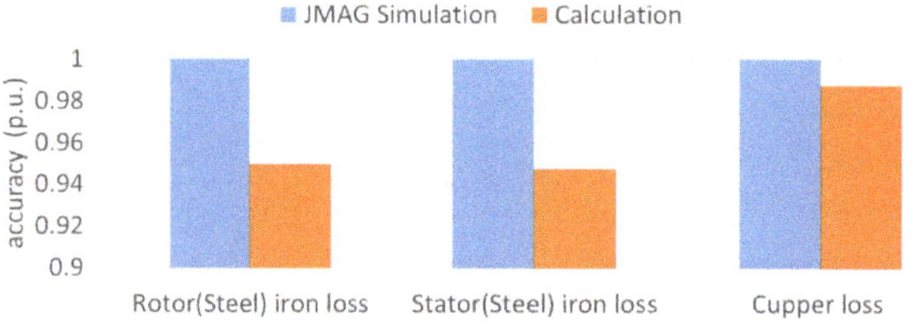

Figure 11. Comparison of estimated power losses and calculated power losses.

Figures 12 and 13 show injected currents and the power loss analysis result of IPMSM at 1200 r/m. The magnitude of current ripple is largest at f_{sw} of 750 Hz and is smallest at f_{sw} of 3000 Hz. The difference between current ripples at f_{sw} of 2250 Hz and f_{sw} of 3000 Hz is relatively small. The f_{sw} over 2000 Hz has less impact on the current ripple as shown in Figure 12. The IPMSM power loss can be divided into four losses as shown in Figure 13. The stator winding has the copper loss. The steel, which is the material of the stator and rotor, results in the stator and rotor iron losses containing the hysteresis and eddy current losses. Only the eddy current loss is considered in the magnet. High f_{sw} reduces the magnitude of current ripple as shown in Figure 12, and it mainly influences on the three losses such as stator iron loss, rotor iron loss, and magnet eddy current loss. Three losses decrease as the magnitude of the ripple current decreases. However, the effect of loss reduction is not large enough since f_{sw} is about 2000 Hz. It can be estimated from the similar ripple currents of 2250 Hz and 3000 Hz. On the other hand, the copper loss does not have the enough impact by the variation of f_{sw}.

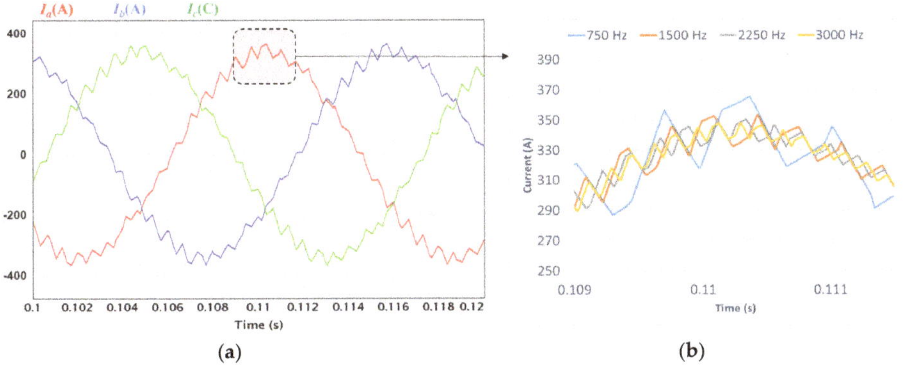

Figure 12. The currents injected to IPMSM at 1200 r/m: (**a**) injected three-phase current at 750 Hz; (**b**) a-phase current depending on switching frequency.

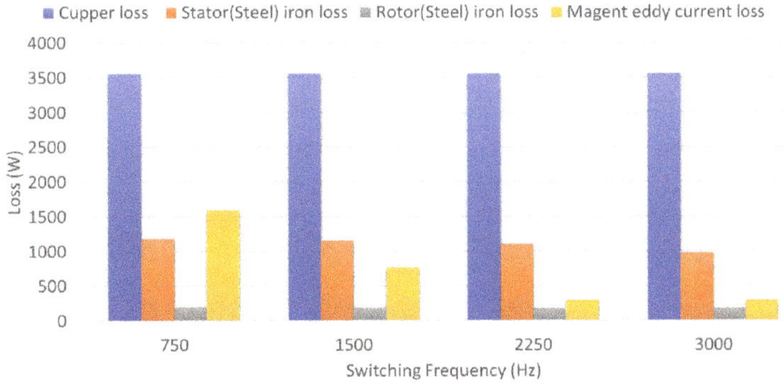

Figure 13. The power loss analysis result of IPMSM at 1200 r/m depending on f_{sw}.

Figure 14 shows the hysteresis losses of stator and rotor in the frequency domain depending on f_{sw} at 1200 r/m. It is identified that the low harmonics frequency components are almost same regardless of f_{sw} and f_{sw} components have the different value each other. f_{sw} of 750 Hz leads to the current ripple of 1500 Hz and its value is larger than that of 3000 Hz. Consequently, the high f_{sw} reduces the power loss of IPMSM and it is opposite to the condition for the low power loss of SiC inverter.

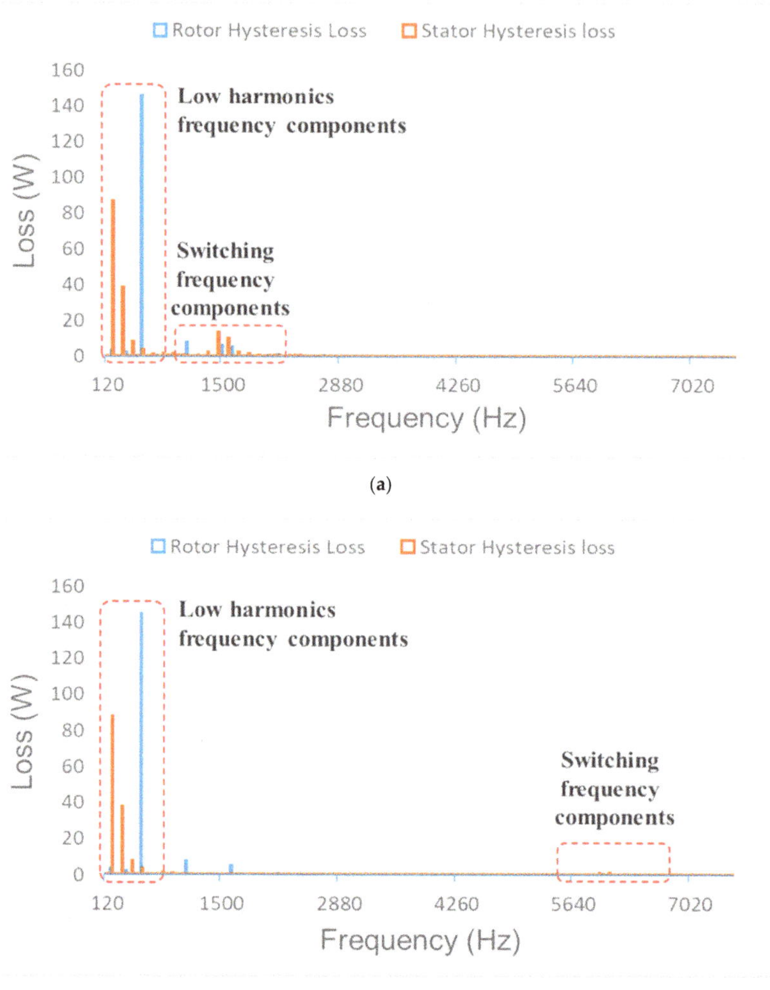

Figure 14. The hysteresis losses of IPMSM in frequency domain at 1200 r/m: (**a**) f_{sw} is 750 Hz; (**b**) f_{sw} is 3000 Hz.

4.3. f_{sw} Determination for Propulstion System

The power loss of propulsion system depending on the speed (r/m) of IPMSM can be represented by considering the results of Figures 8 and 13. Figure 14 show the total power loss of the propulsion system. As mentioned in Sections 4.1 and 4.2, the power loss of SiC inverter increases and IPMSM decreases as f_{sw} increases. Figure 15 shows the individual power losses and the total loss. The total power loss curve of propulsion system has the minimum loss point between 750 Hz and 3000 Hz. This point is 2250 Hz and guarantees maximizing the efficiency of the propulsion system.

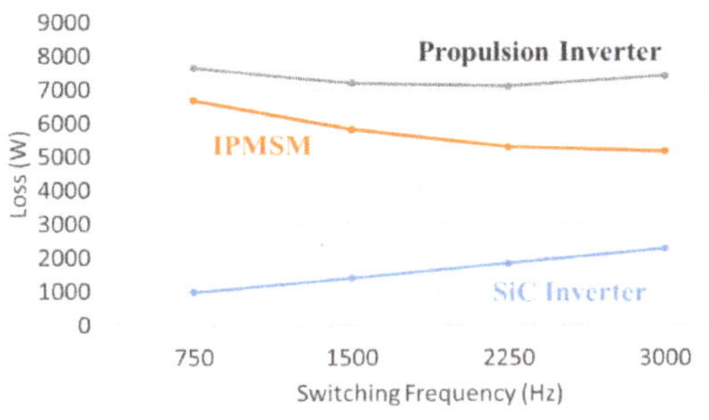

Figure 15. The power loss analysis result of the propulsion system as 1200 r/m.

Figure 16 shows the total power loss of propulsion system depending on speed (r/m). The f_{sw} for the minimum loss point at 400 r/m, 800 r/m, and 1200 r/m is 750 Hz, 1500 Hz, and 2250 Hz, respectively. The power loss of IPMSM at the low speed is small relatively compared to the power loss of SiC inverter. It means the power loss of SiC inverter is dominant at low speed. Therefore, as the speed increases, f_{sw} for the minimum loss point increases. In addition, it can be estimated that the switching frequency upper than 3000 Hz aggravates the power efficiency of propulsions system under 1200 r/m.

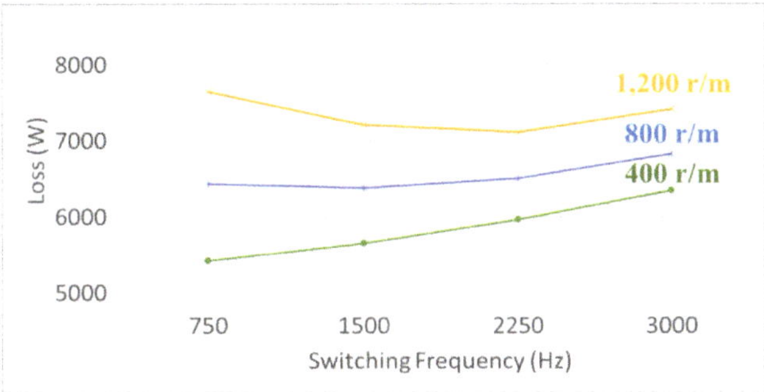

Figure 16. The total power loss of the propulsion system depending on speed (r/m).

5. Conclusions

This paper determines the switching frequency (f_{sw}) guaranteeing the minimum power loss of the propulsion system consisting SiC inverter and 332 kW IPMSM. The co-simulation was implemented and the effect of the low harmonic frequency and f_{sw} components was reflected in the power loss analysis. As f_{sw} increases, the power loss of SiC inverter increases because the number of switching increases, but the power loss of IPMSM decreases because the magnitude of current ripple caused by f_{sw} is reduced. In addition, it was identified that as the motor speed increases, f_{sw} for maximum efficiency operation of propulsion inverter, increases.

The implemented process suggests that both inverter and IPMSM should be considered in determining f_{sw} of inverter to maximize the efficiency of the propulsion system. If the switching device or IPMSM specification is changed, the results of Figure 16 are not valid. However, it is important and meaningful that f_{sw} can be determined through the process implemented in this paper.

Author Contributions: Conceptualization, project administration, writing—original draft preparation, J.-H.R.; data curation and software, J.-H.L.; supervision and writing—review and editing, J.-S.L. All authors have read and agreed to the published version of the manuscript.

Funding: This work was supported by the Railroad Technology Research Program through the Ministry of Land, Infrastructure and Transport of Korean Government under Grant 20RTRP-B146008-03.

Conflicts of Interest: The authors declare no conflict of interest.

References

1. Ronanki, D.; Singh, S.A.; Williamson, S. Comprehensive Topological Overview of Rolling Stock Architectures and Recent Trends in Electric Railway Traction Systems. *IEEE Trans. Transp. Electrif.* **2017**, *3*, 724–738. [CrossRef]
2. Yang, X.; Li, X.; Ning, B.; Tang, T. A Survey on Energy-Efficient Train Operation for Urban Rail Transit. *IEEE Trans. Intell. Transp. Syst.* **2016**, *17*, 2–13. [CrossRef]
3. Diao, L.-J.; Tang, J.; Loh, P.C.; Yin, S.; Wang, L.; Liu, Z. An Efficient DSP–FPGA-Based Implementation of Hybrid PWM for Electric Rail Traction Induction Motor Control. *IEEE Trans. Power Electron.* **2018**, *33*, 3276–3288. [CrossRef]
4. Shikata, K.; Kawai, H.; Nomura, H.; Aoki, H.; Fukasawa, S.; Tasaka, Y. PMSM propulsion system for Tokyo Metro. In Proceedings of the 2012 Electrical Systems for Aircraft, Railway and Ship Propulsion, Bologna, Italy, 16–18 October 2012; pp. 1–6. [CrossRef]
5. Negishi, T.; Tsuda, R.; Ota, K.; Iura, S.; Yamaguchi, H. 3.3-kV All-SiC Power Module for Traction System Use. In Proceedings of the International Exhibition and Conference for Power Electronics, Intelligent Motion, Renewable Energy and Energy Management, Nuremberg, Germany, 16–18 May 2017.
6. Mouawad, B.; Hussein, A.; Castellazzi, A. A 3.3 kV SiC MOSFET Half-Bridge Power Module. In Proceedings of the 10th International Conference on Integrated Power Electronics Systems, Stuttgart, Germany, 20–22 March 2018.
7. Ke, H.; Chang, G.; Zhou, W.; Li, C.; Peng, Y.; Dai, X. 3.3kV/500A SiC Power Module for Railway Traction Application. In Proceedings of the International Exhibition and Conference for Power Electronics, Intelligent Motion, Renewable Energy and Energy Management, Shanghai, China, 26–28 June 2018.
8. Skudenly, H.-C.; Weinhardt, M. An Investigation of the Dynamic Response of Two Induction Motors in a Locomotive Truck Fed by a Common Inverter. *IEEE Trans. Ind. Appl.* **1984**, *20*, 173–179. [CrossRef]
9. Broche, C.; Lobry, J.; Colignon, P.; Labart, A. Harmonic reduction in DC link current of a PWM induction motor drive by active filtering. *IEEE Trans. Power Electron.* **1992**, *7*, 633–643. [CrossRef]
10. Oh, S.Y.; Cho, S.-Y.; Han, J.-H.; Lee, J.; Ryu, G.-H.; Kang, D.; Lee, J. Design of IPMSM Rotor Shape for Magnet Eddy-Current Loss Reduction. *IEEE Trans. Magn.* **2014**, *50*, 841–844. [CrossRef]
11. Zhang, H.; Liu, W.; Chen, Z.; Mao, S.; Meng, T.; Peng, J.; Jiao, N. A Time-Delay Compensation Method for IPMSM Hybrid Sensorless Drives in Rail Transit Applications. *IEEE Trans. Ind. Electron.* **2018**, *66*, 6715–6726. [CrossRef]
12. Zhang, H.; Liu, W.; Chen, Z.; Jiao, N.; Zhao, D. Comparison analysis of low-switching-frequency-based IPMSM sensorless drives considering regulators, observer and inverter non-linearity. *IET Electr. Power Appl.* **2019**, *13*, 1022–1031. [CrossRef]
13. Zhao, S.; Huang, X.; Fang, Y.; Zhang, H. DC-link-Fluctuation-Resistant Predictive Torque Control for Railway Traction Permanent Magnet Synchronous Motor in Six-Step Operation. *IEEE Trans. Power Electron.* **2020**, *35*, 10982–10993. [CrossRef]
14. Liu, J.; Zhang, W.; Xiao, F.; Lian, C.; Gao, S. Six-Step Mode Control of IPMSM for Railway Vehicle Traction Eliminating the DC Offset in Input Current. *IEEE Trans. Power Electron.* **2019**, *34*, 8981–8993. [CrossRef]
15. Kwon, Y.-C.; Kim, S.; Sul, S.-K. Six-Step Operation of PMSM with Instantaneous Current Control. *IEEE Trans. Ind. Appl.* **2014**, *50*, 2614–2625. [CrossRef]
16. Bolognani, S.; Zigliotto, M. Novel digital continuous control of SVM inverters in the overmodulation range. *IEEE Ind. Appl.* **1997**, *33*, 525–530. [CrossRef]
17. Yuan, G.; Hou, X.; Zheng, C.; Li, Z.; Yin, Y.; Liang, R. Restarting with speed for IPMSM based on hybrid synchronized PWM control schemes for rail train traction system. In Proceedings of the 2017 20th International Conference on Electrical Machines and Systems (ICEMS), Sydney, Australia, 11–14 August 2017; Institute of Electrical and Electronics Engineers (IEEE): Piscataway, NJ, USA, 2017; pp. 1–5.
18. Wang, W.; Cheng, M.; Zhang, B.; Zhu, Y.; Ding, S. A Fault-Tolerant Permanent-Magnet Traction Module for Subway Applications. *IEEE Trans. Power Electron.* **2014**, *29*, 1646–1658. [CrossRef]

19. Casarin, J.; Ladoux, P.; Chauchat, B.; Dedecius, D.; Laugt, E. Evaluation of high voltage SiC diodes in a medium frequency AC/DC converter for railway traction. In Proceedings of the International Symposium on Power Electronics Power Electronics, Electrical Drives, Automation and Motion, Sorrento, Italy, 20–22 June 2012.
20. Wu, D.; Xiao, C.; Zhang, H.; Liang, W. Development of auxiliary converter based on 1700V/325A full SiC MOSFET for urban rail transit vehicles. In Proceedings of the IEEE Transportation Electrification Conference and Expo, Asia-Pacific (ITEC Asia-Pacific), Chicago, IL, USA, 22–24 June 2017; Institute of Electrical and Electronics Engineers (IEEE): Piscataway, NJ, USA, 2017.
21. Helsper, M.; Ocklenburg, M.A. SiC MOSFET based Auxiliary Power Supply for rail vehicles. In Proceedings of the 20th European Conference on Power Electronics and Applications, Riga, Latvia, 17–21 September 2018.
22. Rujas, A.; López, V.M.; Garcia-Bediaga, A.; Berasategi, A.; Nieva, T.; López-Martín, V. Railway traction DC–DC converter: Comparison of Si, SiC-hybrid, and full SiC versions with 1700 V power modules. *IET Power Electron.* **2019**, *12*, 3265–3271. [CrossRef]
23. Rujas, A.; Lopez, V.M.; Villar, I.; Nieva, T.; Larzabal, I. SiC-hybrid based railway inverter for metro application with 3.3kV low inductance power modules. In Proceedings of the 2019 IEEE Energy Conversion Congress and Exposition (ECCE), Baltimore, MD, USA, 29 September–3 October 2019; Institute of Electrical and Electronics Engineers (IEEE): Piscataway, NJ, USA, 2019; pp. 1992–1997.
24. Ishikawa, K.; Ogawa, K.; Yukutake, S.; Kameshiro, N.; Kono, Y. Traction Inverter that Applies Compact 3.3 kV/1200 A SiC Hybrid Module. In Proceedings of the International Power Electronics Conference, Hiroshima, Japan, 18–21 May 2014.
25. Sugiyama, A.; Okitsu, H.; Ishimaru, T.; Kishimoto, Y.; Kadooka, S.; Sawakami, T. Verification Tests of 3.3kV SiC (Silicon Carbide) Hybrid IGBT Inverter for Nishi-Nippon Railroad Co., Ltd 3000 Series. In Proceedings of the International Conference on Electrical Systems for Aircraft, Railway, Ship Propulsion and Road Vehicles, Aachen, Germany, 3–5 March 2015.
26. Makishima, S.; Fujimoto, K.; Kondo, K. The Direct Benefit of SiC Power Semiconductor Devices for Railway Vehicle Traction Inverters. In Proceedings of the International Power Electronics Conference, Niigata, Japan, 20–24 May 2018.
27. Sato, K.; Kato, H.; Fukushima, T. Development of SiC Applied Traction System for Shinkansen High-speed Train. In Proceedings of the International Power Electronics Conference, Niigata, Japan, 20–24 May 2018.
28. Yıldırım, D.; Akşit, M.H.; Yolaçan, C.; Pul, T.; Ermiş, C.; Aghdam, B.H.; Çadırcı, I.; Ermiş, M. Full-Scale Physical Simulator of All SiC Traction Motor Drive with Onboard Supercapacitor ESS for Light-Rail Public Transportation. *IEEE Ind. Electron.* **2020**, *67*, 6290–6301. [CrossRef]
29. Kogure, H.; Ishikawa, K.; Kohno, Y.; Sakai, T.; Ishigaki, T. Development of Low Loss Inverter System Adopted Lower Harmonic Losses Technology and Ultra Compact Inverters Adopted High Power Density SiC Module. In Proceedings of the 20th European Conference on Power Electronics and Applications, Riga, Latvia, 17–21 September 2018.
30. Bober, P.; Ferkova, Z. Comparison of an Off-Line Optimized Firing Angle Modulation and Torque Sharing Functions for Switched Reluctance Motor Control. *Energies* **2020**, *13*, 2435. [CrossRef]
31. Nakata, T.; Sanada, M.; Morimoto, S.; Inoue, Y. Automatic Design of IPMSMs Using a Genetic Algorithm Combined with the Coarse-Mesh FEM for Enlarging the High-Efficiency Operation Area. *IEEE Trans. Ind. Electron.* **2017**, *64*, 9721–9728. [CrossRef]
32. Di, C.; Petrov, I.; Pyrhonen, J.J. Modeling and Mitigation of Rotor Eddy-Current Losses in High-Speed Solid-Rotor Induction Machines by a Virtual Permanent Magnet Harmonic Machine. *IEEE Trans. Magn.* **2018**, *54*, 1–12. [CrossRef]
33. Kwon, T.-S.; Sul, S.-K. Novel Antiwindup of a Current Regulator of a Surface-Mounted Permanent-Magnet Motor for Flux-Weakening Control. *IEEE Trans. Ind. Appl.* **2006**, *42*, 1293–1300. [CrossRef]
34. Lai, C.; Feng, G.; Tjong, J.; Kar, N.C. Direct Calculation of Maximum-Torque-Per-Ampere Angle for Interior PMSM Control Using Measured Speed Harmonic. *IEEE Trans. Power Electron.* **2018**, *33*, 9744–9752. [CrossRef]
35. Wu, J.; Wang, J.; Gan, C.; Sun, Q.; Kong, W. Efficiency Optimization of PMSM Drives Using Field-Circuit Coupled FEM for EV/HEV Applications. *IEEE Access* **2018**, *6*, 15192–15201. [CrossRef]
36. Letellier, A.; Dubois, M.R.; Trovão, J.P.F.; Maher, H. Calculation of Printed Circuit Board Power-Loop Stray Inductance in GaN or High di/dt Applications. *IEEE Trans. Power Electron.* **2018**, *34*, 612–623. [CrossRef]

37. Letellier, A.; Dubois, M.R.; Trovao, J.P.F.; Maher, H. Gallium Nitride Semiconductors in Power Electronics for Electric Vehicles: Advantages and Challenges. In Proceedings of the IEEE Vehicle Power and Propulsion Conference (VPPC), Montreal, QC, Canada, 19–22 October 2015.
38. Liew, G.S.; Tsang, E.C.Y.; Ertugrul, N.; Soong, W.L. Analysis of a Segmented Brushless PM Machine Utilising Soft Magnetic Composites. In Proceedings of the 33rd Annual Conference of the IEEE Industrial Electronics Society, Taipei, Taiwan, 5–8 November 2007.
39. Han, S.-H.; Jahns, T.; Zhu, Z. Analysis of Rotor Core Eddy-Current Losses in Interior Permanent-Magnet Synchronous Machines. *IEEE Trans. Ind. Appl.* **2009**, *46*, 196–205. [CrossRef]
40. Chai, W.; Lipo, T.; Kwon, B.-I. Design and Optimization of a Novel Wound Field Synchronous Machine for Torque Performance Enhancement. *Energies* **2018**, *11*, 2111. [CrossRef]
41. Gonzalez, D.A.; Saban, D.M. Study of the Copper Losses in a High-Speed Permanent-Magnet Machine with Form-Wound Windings. *IEEE Trans. Ind. Electron.* **2013**, *61*, 3038–3045. [CrossRef]
42. Hargreaves, P.A.; Mecrow, B.C.; Hall, R. Calculation of Iron Loss in Electrical Generators Using Finite-Element Analysis. *IEEE Trans. Ind. Appl.* **2012**, *48*, 1460–1466. [CrossRef]

energies

Article

Experimental Platform for Evaluation of On-Board Real-Time Motion Controllers for Electric Vehicles

Thanh Vo-Duy [1], Minh C. Ta [1,2,*], Bảo-Huy Nguyễn [1,2] and João Pedro F. Trovão [2,3,4]

1 CTI Laboratory for Electric Vehicles, Department of Industrial Automation, Hanoi University of Science and Technology, Hanoi 10000, Vietnam; thanh.voduy@hust.edu.vn (T.V.-D.); Bao.Huy.Nguyen@USherbrooke.ca (B.-H.N.)
2 e-TESC Lab., University of Sherbrooke, Sherbrooke, QC J1K 2R1, Canada; Joao.Trovao@USherbrooke.ca
3 INESC Coimbra, DEEC, University of Coimbra, Polo II, 3030-290 Coimbra, Portugal
4 Polytechnic Institute of Coimbra, IPC-ISEC, DEE, 3030-199 Coimbra, Portugal
* Correspondence: minh.tacao@hust.edu.vn

Received: 31 October 2020; Accepted: 3 December 2020; Published: 6 December 2020

Abstract: Electric vehicles are considered to be a greener and safer means of transport thanks to the distinguished advantages of electric motors. Studies on this object require experimental platforms for control validation purpose. Under the pressure of research, the development of these platforms must be reliable, safe, fast, and cost effective. To practically validate the control system, the controllers should be implemented in an on-board micro-controller platform; whereas, the vehicle model should be realized in a real-time emulator that behaves like the real vehicle. In this paper, we propose a signal hardware-in-the-loop simulation system for electric vehicles that are driven by four independent electric motors installed in wheels (in-wheel motor). The system is elaborately built on the basis of longitudinal, lateral, and yaw dynamics, as well as kinematic and position models, of which the characteristics are complete and comprehensive. The performance of the signal hardware-in-the-loop system is evaluated by various open-loop testing scenarios and by validation of a representative closed-loop optimal force distribution control. The proposed system can be applied for researches on active safety system of electric vehicles, including traction, braking control, force/torque distribution strategy, and electronic stability program.

Keywords: electric vehicle; real-time; simulator; hardware-in-the-loop

1. Introduction

In the time of energy and environment crisis, electric vehicles (EVs) have emerged as a remarkable solution for replacing gasoline transportation system, which mainly causes environmental issues. Moreover, EVs also bring to users safer and more enjoyable transport means. In the energy usage aspect, EVs have much higher efficiency than that of Internal Combustion Engine (ICE) vehicles. As in recent statistical researches, the plug-to-wheel efficiency of EVs is 73–90% as compared to 16–37% tank-to-wheel efficiency of ICE cars and, more interestingly, if all diesel vehicles are converted to electric ones, the CO_2 emissions will be reduced by 80% [1,2]. For those reasons, EVs have become attractive to researchers worldwide.

Thanks to outstanding features of electric motors, EVs have opened various new research topics. Besides traditional but prospective studies on Power Electronics [3,4] and Electric Drives [5], there are many other subjects that are particularly oriented to EVs, for example, Motion Control [6], Energy Management System [7], and Autonomous Driving [8]. Motion control, also called an active safety system, has exploited the major advantages of electric motors installed on EVs very well. With very fast response (just a few milliseconds), electric cars can achieve better driving performance in comparison with ICE vehicles [9]. Furthermore, the compact and simple structure allow for the separate integration of traction motors into wheels that not only increase the flexible movement, but also increases many interesting studies on driving force distribution for four-wheel-driven vehicles [10–13].

The common point of all research resides on the fact that the control performance must be validated in some ways before prototyping. The best approach is to conduct experimental tests on a real vehicle. Such experiments can deliver the most convincing results and totally evaluate the whole system, i.e., the response of the control strategy to the internal state variations inside the vehicle as well as external influences from real operating environment. On the contrary, its drawbacks are also significant, namely high cost, time consuming, and even very dangerous to experimenters. Another approach for overcoming experimental difficulties is to simulate the entire system on a simulation software, e.g., Matlab/Simulink. Following this solution, a mathematical/numerical model of EVs must be developed [14–16]. Testing the control system on a software seems to be reasonable, as the model can be modified in order to adapt to various requirements and the validation can be freely performed as many times as needed without either harm or expenditure. However, yet, numerical simulation is still off-line test, i.e., the results are not validated in real-time, which restricts the evaluated control system to be applicable in real vehicles.

In order to fill the gap between validating the system by real-road experiments and that by off-line software, Hardware-In-the-Loop (HIL) Simulation System is the proper solution. HIL simulation is the technique that uses equivalent models to replace some real parts of the system, connect to the system-under-test, and then implement the experiment for control validation. Contrary to software simulation, the equivalent models are executed in real-time on a hardware platform that must be designed to be synchronously compatible with the entire studied system. For many years, HIL simulation has been used widely because of remarkable advantages. It offers a cost-effective solution for researches and shortens the time of control system development. Experiments on HIL system are safe and they can be repeated with exactly the same conditions, which is impossible or extremely difficult in real-road test.

Developing a HIL simulation system, in fact, is a challenging work. On one hand, it should precisely represent behavior and/or distinguishing features of the studied system, in which all physical principles must be obeyed. Moreover, the simulation system is necessarily designed in order to adapt to research purpose and be open for future works. On the other hand, the HIL system needs to be simple enough to feasibly implement in real-time due to limitations in computational capability of hardware platforms. Development expenditure is also an important factor that must be compromised with mentioned difficulties.

Currently, there are commercial simulation software that can simulate EVs and vehicle control systems, for example, CarSim [17] and veDYNA [18]. Research on EVs can be validated in real-time by these products, but with specific hardware that are supplied by the same company. Such solutions are plausible, because of the compatibility of the system, but, as a result, the investment for both hardware and software is also considerable.

In the literature, HIL systems are well worked out. There are several types of HIL simulation systems that depend on design purpose as well as which subsystem is replaced [19].

- *Signal HIL Simulation*. The purpose of this type is to validate the performance of the developed controller in actual operation. Whereas, the controlled system is simulated based on its mathematical model and implemented on a real-time platform [20]. In order to do that, the equivalent model must act with similar behaviors to what it simulates and, furthermore, the model should imitate the real system to make the controller "reckon" that it is working with a real process. On the controller side, the algorithm should be programmed and compiled in order to run on a chosen processing hardware with all interfaces designed as a real Electronic Control Unit (ECU).
- *Power HIL Simulation* consists of two types, including Full-scale and Reduced-scale power HIL simulation.

Full-scale power HIL system is used in order to validate a part (with control included) of the system before bringing into the real process. In the implementation of this type, the total system is broken down into subsystems. The subsystems-under-test are retained, while the ones are modeled by an equivalent power emulated system [21]. Additionally, the simulated power system is controlled in such a way that it has the same characteristics with the real part that it emulates. The Power HIL simulation is likewise suitable for testing a new subsystem and its control that will be embedded into the available system.

A reduced-scale power HIL system has the same principle with that of full-scale power HIL ones, but with the lower level of simulation power [22–25]. In other words, both the subsystem to be validated and the emulated model will be tested with reduced power in order to avoid the risk of damage. However, the controller for the system is still designed with full-power level to quickly transfer to real process with minimum programming effort. Accordignly, there are some gains that must be used on the signal into/out of the controller, which is called power adaptation as well.

Among this comprehensive classification, signal HIL system exposes some advantages that can be applied not only for vehicle simulation, but also for many other research fields, e.g process control [26], robotics [27], and power system [28]. This kind of HIL shows its dominant features in studies that operate at high power level and/or work in hazardous environments, with dangerous experimental conditions and high investment cost, because it uses numerical models for real-time simulation instead of real/power elements. For EV researches, signal HIL simulation is a proper way for representing the vehicle effectively. In [29], an electronic stability controller for vehicle was developed and tested on a signal HIL system, which achieved good results. Although the developed model of this work only focuses on lateral behavior, the benefit of the real-time simulation system was fully exploited. Some other studies aimed to build HIL system in a signal level in order to validate vehicle control units in several applications, such as for motion control purpose [30,31], for motor control and power supply evaluation [32], or for ECU development [33]. These works have taken into account many hardware platforms, including DS2210 of dSPACE, NI-PXI and CompactRIO of National Instruments. However, models in these researches are still simple, which could not highlight the complicated relationship in motion of EVs and, more importantly, they targeted the single/dual motor drivetrain configuration. Another study in [34] attempted to develop signal HIL system for multi-motor vehicle based on various hardware platforms and the data fusion approach. Nevertheless, the EV in this research is a multi-axle one and, for military purpose, not for passenger cars.

In this paper, we propose a novel experimental platform that is based on the concept of signal HIL simulation system for validation of motion control strategies of EVs in real-time. The proposed system has the following merits.

- The proposed system describes in details the dynamics and kinematics of four in-wheel-motors (4-IWM) EVs in both longitudinal and lateral motions, which are the typical moving directions used in motion control.
- The experimental platform of the paper is well fitted to the development of motion control systems, including traction/braking control, stability control, and driving force distribution strategy particularly for 4-IWM EVs.
- The signal HIL simulation system is built from available, common, and reasonable hardware platforms that encourages research on EVs as safer and greener means of transportation.

The performance of the proposed signal HIL system is evaluated by various testing scenarios. Additionally, an application example of a driving force distribution algorithm is deployed and validated on the proposed platform. All of the evaluation results demonstrate the advantages of the proposed approach.

The remain of the paper is organized, as follows. Section 2 describes the configuration and hardware deployment of the proposed platform. Section 3 develops the mathematical model of the studied EV. Section 4 presents the testing scenarios and evaluation results. Section 5 illustrates an application example on motion control of EVs that is implemented on the proposed signal HIL simulation system. Finally, Section 6 provides conclusions.

2. Experimental Platform Realization

2.1. System Configuration

Generally, a non-autonomous moving vehicle consists of three main constituent entities, i.e., the driver, the control system, and the vehicle itself. The driver intervenes in the motion of the vehicle by giving his/her commands through the steering wheel, accelerator and brake pedals. The duty of the control system is to control the vehicle following the driver's commands and simultaneously keep the vehicle and passengers in safe and comfort states. In order to carry out its job, besides receiving driver's commands, the controller also observes states of the vehicle in order to generate appropriate control signals that are then sent to the traction system to make the movement.

Figure 1 illustrates the comprehensive system configuration, in which all of the entities will be modeled and simulated, including command signals from the driver, the control system, and the EV. Driver's commands are represented by $\delta_{dr}, T_{dr}, T_{br}$ for the steering angle, the driving torque, and the braking torque, respectively, as shown in this figure. As the paper targets four-wheel independently driven vehicles, the control system will be in charge of two tasks, i.e., (*i*) adjust the commands that are received from the driver to anticipate appropriate total set-points for safety reasons of the vehicle and (*ii*) distribute these set-points to the traction motor control systems and the steering system based on a given allocation algorithm. Consequently, the outputs of the control system after the distribution algorithm are denoted by δ_i^* for distributed steering angle, T_{di}^* for distributed driving torque, and T_{bi}^* for distributed braking torque, in which the subscript $i = 1..4$ with respect to wheels of the vehicle.

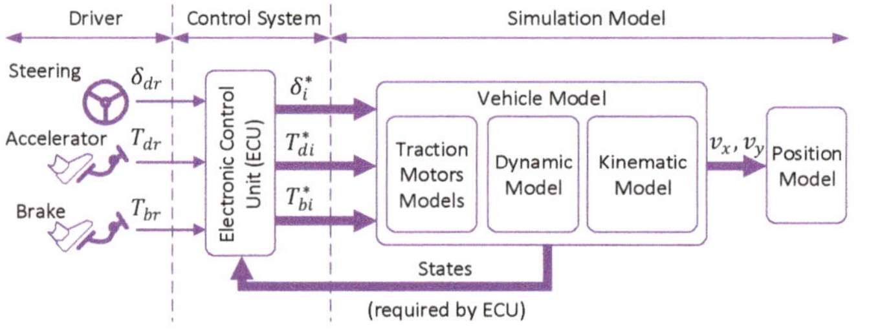

Figure 1. System Configuration for Simulation.

Eventually, the vehicle model will be divided into three parts, which are the powertrain of traction motors, the dynamic and the kinematic models. The position model that is based on longitudinal v_x and lateral v_y velocities are also described. It is noted that some other parts of the vehicle may not be mentioned in the figure, such as steering system, tire model because of the complexity in presentation, but the sophisticated relationship among all of the parts inside the vehicle will be described in detail later.

2.2. Hardware Realization

Based on the proposed configuration, Figure 2 illustrates the entire signal HIL simulation system in terms of hardware implementation. The HIL system adopts two control boards as the main hardware platforms, where each of them covers the specific tasks.

Firstly, the complete EV simulation model is adapted to run on DS1103 PPC Controller Board. This well-known multi-purpose research platform was designed by dSPACE with powerful computing capabilities and various I/O types. Some impressive specifications are 1GHz Central Processing Unit (CPU), 96 MB global SDRAM, 16 ADC channels with 16-bit resolution, 16 DAC channels also with 16-bit resolution, 32 parallel digital Inputs/Outputs (I/Os), and many supported communication protocols. Differently from the simulation on Matlab/Simulink, it should follow some steps for hardware implementation.

- All of the inputs and outputs of the model must be defined with proper I/Os of the DS1103. Unlike in the simulation, where all of the control signals as well as feedback information can be directly linked between the controller and the EV model, as illustrated in Figure 2 and referred to Figure 1, the control signals are transferred through CAN bus communication protocol, other states of the EV are converted into analog signals and exchanged among the platforms. Accordingly, such information must be assigned to appropriate hardware I/Os.
- Typically, the calculated values in the model are represented in floating-point data type, whereas the CAN bus protocol uses unsigned integer in its data field, therefore, data conversion should be taken into account not only for DS1103, but also for all hardware platforms.
- The sample time is the fundamental factor that needs to be carefully chosen. It is the constant iteration of time, in which all computations of the model are repeated. The smaller sample time is used, the more precise model will be but as a trade-off, the higher computational hardware may be required. In this paper, a sample time of 0.5 ms is used.
- After such preparation, the EV model can be complied and downloaded to deploy on the DS1103.

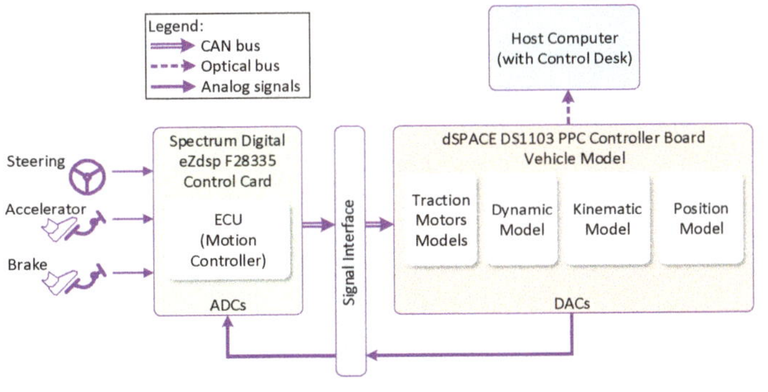

Figure 2. Configuration of the proposed Hardware-in-the-loop simulation system.

Secondly, the eZdsp F28335 Control Card from Spectrum Digital is used for the realization of the control algorithm, i.e., motion controller in this case. The core of the control card is the TMS320F28335 digital signal controller of Texas Instruments that offers the speed of 150 MHz with 32-bit floating point unit on chip. In the proposed HIL system, the eZdsp F28335 imitates the real ECU of the vehicle, which is responsible for the strategy of allocating command torque from driver to the separated motor controllers. In order to do the task, the ECU is connected to the driving system, including the steering wheel, the accelerator and brake pedals, and also collects measured physical data from the vehicle model. The distributed torque values and other control information can be transfer to the EV model on DS1103 through CAN bus. For that reasons, the implementation of control system onto this hardware platform has similar steps as that of DS1103. Based on the requirements of the control algorithm, the number of measured states may varies, so the quantity of analog-to-digital (ADC) and digital-to-analog (DAC) channels on both hardware platforms needs to be defined and assigned appropriately.

Because of different working voltage of the two hardware platforms, the signal interface is designed in order to harmonize the signals exchanged in the system and filter electromagnetic interference (EMI noise) as well. A host computer with installed Control Desk is used in order to collect and store the experimental data. This is the software that is provided together with the DS1103 for human-machine-interface (HMI) design and measurement data acquisition.

Eventually, the complete signal HIL simulation system is connected and described in Figure 3.

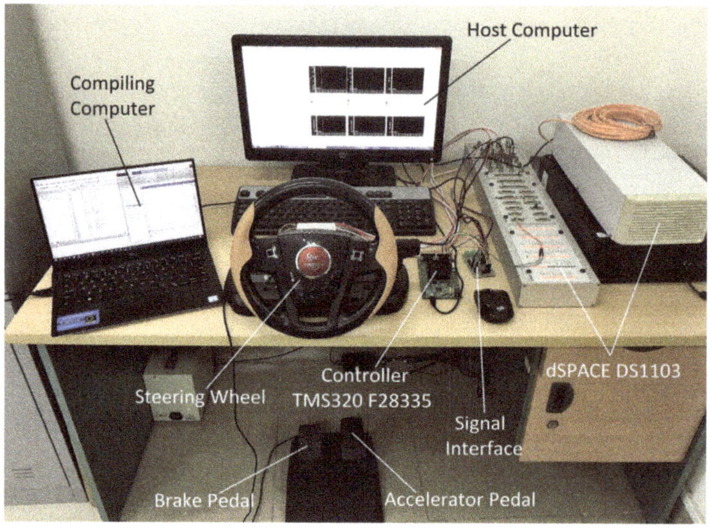

Figure 3. The Complete Hardware-in-the-loop Simulation System.

3. Mathematical Model of 4 In-Wheel-Motor Electric Vehicles

3.1. Vehicle Dynamic Model

A vehicle can be seen as a six degrees of freedom moving object. However, for reasons of simplicity, computational-effectiveness, as well as for representing common vehicles, some assumptions can be made, as follows.

- The vehicle adopts the two-front-wheel steering system. Therefore, both rear wheels are fixed at rear axle as non-steering ones.
- The vehicle is considered to be moving in longitudinal, lateral and yaw directions only, i.e., reduced to three degrees of freedom.

Figure 4 shows how forces apply to the vehicle body as well as wheels and other information related to motion, coordinate systems, and geometric parameters of the vehicle.

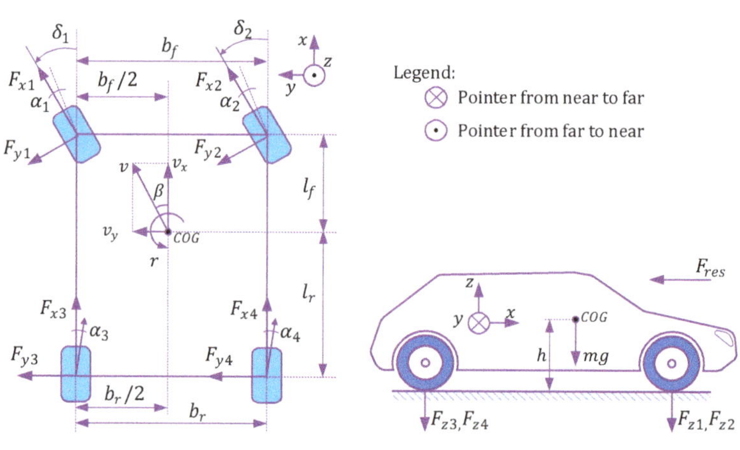

Figure 4. Forces Applied on the Vehicle.

3.1.1. Force and Torque Balance Dynamic Model

A vehicle requires appropriate forces to be applied on wheels for its movement. The combination in amplitude of these forces decides the longitudinal behavior of the vehicle. Moreover, the coupling of different direction forces also creates a torque, which is incorporated with steering system to make the cornering and lateral behavior of the vehicle. Based on Figure 4 and Newton's second law, the force balance equations in longitudinal and lateral directions and torque balance around vertical axis are written as in (1)

$$\begin{aligned} ma_x &= (F_{x1}\cos\delta_1 + F_{x2}\cos\delta_2) - (F_{y1}\sin\delta_1 + F_{y2}\sin\delta_2) + F_{x3} + F_{x4} - F_{res} \\ ma_y &= (F_{x1}\sin\delta_1 + F_{x2}\sin\delta_2) + (F_{y1}\cos\delta_1 + F_{y2}\cos\delta_2) + F_{y3} + F_{y4} \\ J_z\dot{r} &= (l_fF_{x1}\sin\delta_1 + l_fF_{x2}\sin\delta_2) + (l_fF_{y1}\cos\delta_1 + l_fF_{y2}\cos\delta_2) - l_r(F_{y3} + F_{y4}) \\ &\quad - 0.5b_f(F_{x1}\cos\delta_1 - F_{x2}\cos\delta_2) + 0.5b_f(F_{y1}\sin\delta_1 - F_{y2}\sin\delta_2) - 0.5b_r(F_{x3} - F_{x4}) \end{aligned} \tag{1}$$

where, m is the mass of the vehicle, J_z indicates moment of inertia around z-axis, r denotes yaw rate, F_{xi}, F_{yi} are longitudinal and lateral forces applied at wheels, the subscript $i = 1\ldots4$ denotes front-left, front-right, rear-left, and rear-right wheels respectively, a_x, a_y are the acceleration of the vehicle in longitudinal and lateral directions, and parameters l_f, l_r, b_f, b_r describe the geometric characteristic of the vehicle.

Equations in (1) are the dynamic model of the vehicle that contain various unknown variables. Therefore, the rest of this section will deliver all elements to complete the whole dynamic model, which fundamentally contributes to the proposed HIL system.

Neglecting the effect of wind, the total resistance force F_{res} consists of air resistance force F_{air} and rolling resistance force F_{roll}, and can be written, as follows [35].

$$F_{res} = F_{air} + F_{roll} = C_dA_F\frac{\rho}{2}v_x^2 + c_{rr}mg \tag{2}$$

in which, ρ is the air density, A_F is the frontal area of the vehicle, C_d is the aerodynamic drag coefficient, c_{rr} denotes rolling friction coefficient, and g indicates the gravitational acceleration.

3.1.2. Steering Angle Distribution

As aforementioned assumptions, there are only two front wheels steered while cornering. It is common to convert the steering wheel position rotated by the driver to the effective steering angle of the front axle δ_{dr} generated by the rack and pinion system. In the other words, the model takes this effective steering angle as the driver's command instead of steering wheel position into account.

When cornering with a given steering angle of δ_{dr}, the center of gravity (COG) of the car will draw a circle whose radius R can be calculated by (3).

$$R = \sqrt{l_r^2 + (l_f + l_r)^2\cot^2\delta_{dr}} \tag{3}$$

However, because of the width of the axle, the two front wheels will move along two circles that have different radii from that of the CoG. If the steering angles of both wheels are assigned equally, then the vehicle might be unstable. For that reason, a distribution rule [36] can be used, as follows.

$$\begin{cases} \text{if } \delta_{dr} > 0 \rightarrow \delta_1 = \delta_{dr}, \delta_2 = \arctan\dfrac{(R-0.5b_f)\tan\delta_{dr}}{R+0.5b_f} \\ \text{if } \delta_{dr} < 0 \rightarrow \delta_2 = \delta_{dr}, \delta_1 = \arctan\dfrac{(R-0.5b_f)\tan\delta_{dr}}{R+0.5b_f} \\ \text{if } \delta_{dr} = 0 \rightarrow \delta_1 = \delta_2 = \delta_{dr} \end{cases} \tag{4}$$

where, δ_1, δ_2 are the steering angles of the front-left and front-right wheels, respectively.

3.1.3. Tire-Road Behavior Model

Tire-road model describes the tire-road contact characteristic, which is represented by adhesion coefficient μ and its relationship to slip-ratio λ of the vehicle. In the literature, this relationship is non-linear and it can be achieved from the famous Magic Formula [37] or other popular tire models, e.g., the Dugoff and LuGre friction model [38].

In this paper, the Burckhardt tire-road contact friction model is taken into account, since it has sufficient data set for various road types as compared to other mentioned ones. The $\mu - \lambda$ relationship at each wheel is given by Burckhardt, as follows.

$$\mu_{res} = (c_1(1 - e^{-c_2\lambda_{res}}) - c_3\lambda_{res})e^{-c_4\lambda_{res}v}(1 - c_5F_z^2) \tag{5}$$

where, μ_{res} is the adhesion coefficient at the tire-road contact, λ_{res} is the slip-ratio of the considered wheel, F_z denotes the normal force at a wheel of the vehicle, and $c_1 - c_5$ are the road parameter sets. Each type of road has a specific characteristic and can be described by groups of parameters, as listed in Table 1, for typical road conditions. With respect to these data sets, the $\mu - \lambda$ relationship is shown in detail, as illustrated in Figure 5.

Table 1. Burckhard tire-road contact friction model parameter sets for typical road conditions.

Type of Road	c_1	c_2	c_3	c_4	c_5
Dry asphalt	1.2801	23.99	0.52	0.002–0.004	0.00015
Wet asphalt	0.857	33.822	0.347		
Dry concrete	1.1973	25.168	0.5373		
Dry cobblestone	1.3713	6.4565	0.6691		
Wet cobblestone	0.4004	33.7080	0.12.4		
Snow	0.1946	94.129	0.0646		
Ice	0.05	306.39	0		

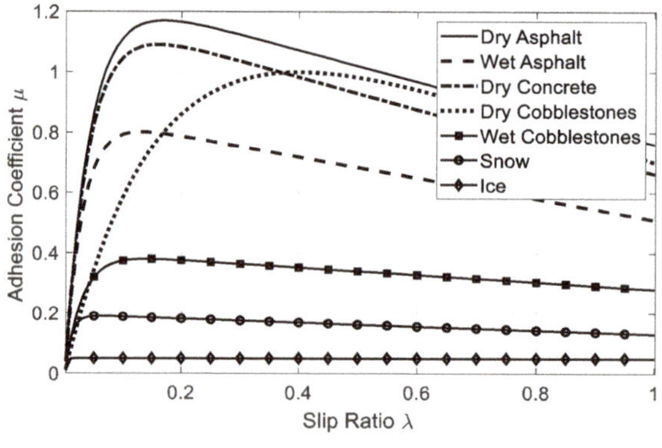

Figure 5. Different $\mu - \lambda$ characteristics by Burckhardt tire-road model.

When cornering, the side force appears that causes the lateral motion of the vehicle. Therefore, the adhesion coefficient at each wheel is also divided into two components, as written in (6).

$$\begin{aligned} \mu_l &= \mu_{res}\frac{\lambda_l}{\lambda_{res}} \\ \mu_s &= k_s\mu_{res}\frac{\lambda_s}{\lambda_{res}} \end{aligned} \tag{6}$$

in which μ_l and μ_s are the adhesion coefficients in longitudinal and lateral directions, respectively, λ_l and λ_s are the slip ratio of the wheel in both of those orientations. In (6), k_s is called attenuation factor, which has the value between 0.9 and 0.95 for low profile tires. Otherwise, k_s is chosen as 1 [36].

3.1.4. Slip-Ratio Determination

Slip-ratio is a fundamental factor in motion control in order to guarantee the safety of a vehicle. In literature, the slip-ratio is the difference between velocity of the car and that of the wheel, which can be written in a simplified way, as follows.

$$\lambda = \frac{\omega R_{eff} - v_x}{\max(\omega R_{eff}, v_x)} \tag{7}$$

where, R_{eff} is the effective radius of the wheel.

As mentioned in (6), the slip-ratio should be in various forms, which are listed in Table 2 with respect to the orientations and accelerating/braking states of the vehicle.

Table 2. Longitudinal and lateral slip-ratio with respect to states of the vehicle.

Slip-Ratio	Braking ($\omega R_{eff} \leq v_W$)	Accelerating ($\omega R_{eff} > v_W$)
λ_l	$\frac{\omega R_{eff} \cos\alpha - v_W}{v}$	$\frac{\omega R_{eff} \cos\alpha - v_W}{\omega R_{eff} \cos\alpha}$
λ_s	$\frac{\omega R_{eff} \sin\alpha}{v_W}$	$\tan\alpha$

where, α is the side-slip angle and v_W is the longitudinal velocity of the specified wheel.

Subsequently, the overall slip ratio is calculated as in (8).

$$\lambda_{res} = \sqrt{\lambda_l^2 + \lambda_s^2} \tag{8}$$

3.1.5. Wheel Longitudinal Velocity

A wheel longitudinal velocity v_W is different from other velocities, e.g., vehicle velocity v, longitudinal velocity v_x, in some specific cases. When moving straight, all of the longitudinal velocities are identically the same in terms of value and orientation. However, when moving in a curve, depending on the tire-road contact adhesion feature, the direction of v_W may not coincide with that of longitudinal axis of the wheel.

The longitudinal velocities of the wheels depend on geometric parameters and some other states of the car, and they can be approximated, as follows [36].

$$\begin{aligned} v_{W1} &= v - r(0.5b_f - l_f\beta) \\ v_{W2} &= v + r(0.5b_f + l_f\beta) \\ v_{W3} &= v - r(0.5b_r + l_r\beta) \\ v_{W3} &= v + r(0.5b_r - l_r\beta) \end{aligned} \tag{9}$$

where, β is the body slip angle, as shown in Figure 4, and it can be determined as in (10).

$$\beta = \arctan\frac{v_y}{v_x} \tag{10}$$

Additionally, the actual velocity at the COG of the vehicle can be obtained, as in (11).

$$v = \sqrt{v_x^2 + v_y^2} \tag{11}$$

3.1.6. Side-Slip Angle Determination

The side-slip angle at a wheel is the one that describes the difference between longitudinal orientations of the wheel and its moving direction, as described in Figure 6. For front-wheel steering cars, the orientations of the front wheels are identically their steering angles. (12) shows the calculation of these slip angles based on the velocities of the vehicle.

$$\begin{aligned} \alpha_1 &= \delta_1 - \arctan\frac{v_y + l_f r}{v_x - 0.5 b_f r} \\ \alpha_2 &= \delta_2 - \arctan\frac{v_y + l_f r}{v_x + 0.5 b_f r} \\ \alpha_3 &= -\arctan\frac{v_y - l_r r}{v_x - 0.5 b_r r} \\ \alpha_4 &= -\arctan\frac{v_y - l_r r}{v_x + 0.5 b_r r} \end{aligned} \quad (12)$$

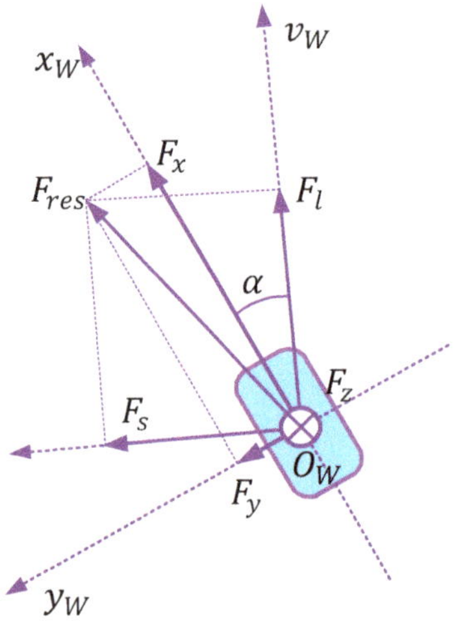

Figure 6. Friction forces acting on a wheel. F_l and F_s are aligned with the actual movement direction of the wheel which is indicated by the vector v_W. Vector F_{res} is the sum of F_l and F_s. The longitudinal and side forces F_x and F_y can be achieved by the projection of F_{res} on the $(Oxy)_W$, which is in association with the wheel coordinate.

3.1.7. Forces Acting on a Wheel

It is the difference between the actual moving direction of the wheel v_W and longitudinal axis of the wheel $(Ox)_W$, which causes the forces F_l and F_s along the v_W orientation and at the right angle to it, respectively. Figure 6 illustrates this circumstance. Based on the adhesion coefficients in the respective directions, these forces can be deduced for each wheel.

$$\begin{aligned} F_l &= \mu_l F_z \\ F_s &= \mu_s F_z \end{aligned} \quad (13)$$

Transforming from direction of movement into wheel coordinate $(Oxy)_W$, we have

$$\begin{aligned} F_x &= F_l \cos\alpha + F_s \sin\alpha \\ F_y &= -F_l \sin\alpha + F_s \cos\alpha \end{aligned} \tag{14}$$

From (5), (6), (8), (14), and the slip-ratios calculated in Table 2, the friction forces acting on each wheel in association with wheel coordinate can be written, as follows.

$$\begin{aligned} F_x &= \left(\mu_{res}\frac{\lambda_l}{\lambda_{res}}\cos\alpha + \mu_{res}k_s\frac{\lambda_s}{\lambda_{res}}\sin\alpha\right)F_z \\ F_y &= \left(-\mu_{res}\frac{\lambda_l}{\lambda_{res}}\sin\alpha + \mu_{res}k_s\frac{\lambda_s}{\lambda_{res}}\cos\alpha\right)F_z \end{aligned} \tag{15}$$

3.1.8. Calculation of Normal Forces

The normal force is necessary for finalizing (5) and (15). In fact, F_z is depend not only on the mass and geometric parameters of the vehicle, but also on the acceleration/deceleration of the car because of the moment of inertia. In accelerating duration, the weight is put more on the rear wheels; thus, with the same adhesion coefficient, the normal forces on these wheels increase with the increase of acceleration. The same manner happens for the front wheels during deceleration. The normal forces are rewritten from [39] for four wheels with consideration of lateral acceleration.

$$\begin{aligned} F_{z1} &= \left(k_{rx} - k_x\frac{a_x}{g}\right)\left(1 - k_{fy}\frac{a_y}{g}\right) \\ F_{z2} &= \left(k_{rx} - k_x\frac{a_x}{g}\right)\left(1 + k_{fy}\frac{a_y}{g}\right) \\ F_{z3} &= \left(k_{fx} + k_x\frac{a_x}{g}\right)\left(1 - k_{ry}\frac{a_y}{g}\right) \\ F_{z4} &= \left(k_{fx} + k_x\frac{a_x}{g}\right)\left(1 + k_{ry}\frac{a_y}{g}\right) \end{aligned} \tag{16}$$

where,

$$k_{rx} = \frac{1}{2}mg\frac{l_r}{l};\quad k_{fx} = \frac{1}{2}mg\frac{l_f}{l};\quad k_x = \frac{1}{2}mg\frac{h}{l}\quad k_{ry} = 2\frac{h}{b_r};\quad k_{fy} = \frac{2h}{b_f};\quad l = l_f + l_r$$

3.2. *Vehicle Kinematic Model*

With the second assumption, which does not deal with the suspension system, the vehicle can be considered as a rigid body object that has its dynamics, as indicated in (17)

$$M\dot{v} + C(v) = \tau \tag{17}$$

where,

$$M = \begin{pmatrix} m & 0 & 0 \\ 0 & m & 0 \\ 0 & 0 & J_z \end{pmatrix};\quad C(v) = \begin{pmatrix} 0 & -mr & 0 \\ mr & 0 & 0 \\ 0 & 0 & 0 \end{pmatrix};\quad \tau = \begin{bmatrix} F_x \\ F_y \\ \tau_z \end{bmatrix};\quad v = \begin{bmatrix} v_x \\ v_y \\ r \end{bmatrix}$$

in which M is the inertia matrix, $C(v)$ is the Colioris matrix and τ_z in the vector, and τ denotes the torque around the vertical axis going through the COG of the vehicle.

Disregarding the torque-related term and taking the first two equations of (17) into account with the consideration of the Newton's second law of motion, we have

$$\begin{aligned} \dot{v}_x &= a_x + r v_y \\ \dot{v}_y &= a_y - r v_x \end{aligned} \tag{18}$$

Equation (18) is the kinematic model that shows the relationship of accelerations, velocities, and yaw rate of a common vehicle.

3.3. Traction Motors Modeling

EVs have taken advantage of electric motors, which feature much better performance than ICE vehicles. Some superior benefits of using electric motors for traction system of a car can be listed here.

- Very fast response. The average response time of an electric motor is around 1–5 ms, which is hundreds of times faster than that of the ICE. Therefore, the vehicle can react quickly to unexpected circumstances in the safety aspect.
- Direct torsion torque control. By measuring and controlling the current of the motor, one can apply torque precisely and very fast to the wheels. This unique characteristic would enhance the safety of the vehicle as well as the excitement of the driver.
- Regenerative braking ability and high energy efficiency. It is only the electric motor that can convert braking/deceleration energy into electricity and charge back to the storage system. Moreover, the global efficiencies of EVs are much higher than that of ICE vehicle, because of the very high efficiency of the motor itself and the absence of some mechanical parts, e.g., gearbox and clutch.
- Distributed traction motors. At the same power range, the electric motor is smaller in volume and simpler in structure than the ICE. This allows for every wheel of the vehicle to be driven by independent and separated motor or even in-wheel/wheel-hub motor, which integrates all of the power converter, brake, cooling system, and motor in one wheel. Such EVs are very flexible, powerful, and promisingly safe.

Thanks to those advantages, the traction electric motors in this paper are simply represented by first-order transfer function. For each wheel, this model can be written, as follows.

$$T_d = \frac{K_m}{\tau_m s + 1} T_d^* \tag{19}$$

in which, K_m is the motor gain from set point to real value, τ_m is the time constant of the motor, T_d indicates the driving torque at the motor shaft.

When running on the road, all of the friction forces as well as resistance forces applied on the vehicle/wheel can be converted into the load of the motor; thus, the torque balance equation around the motor shaft can be figured out by (20).

$$T_d - R_{eff} F_x - k_b T_b = J_x \frac{d\omega}{dt} \tag{20}$$

where, J_x is the inertial torque of the wheel and the rotor and $k_b T_b$ is the braking torque applied on the wheel shaft. In case of the wheel-hub is used, the motor is connected directly to the wheel; thus, $k_b = 1$. Otherwise, k_b is the gain to convert braking force to the motor shaft.

3.4. Vehicle Position Model

Let us define θ the deviation angle between the earth-fixed coordinate and the vehicle body-fixed one, as described in Figure 7. While taking a turn, this angle varies and its value is the integral of the yaw rate r around the vertical axis of the vehicle.

$$\theta = \int_0^t r dt \tag{21}$$

Based on this deviation angle, the velocities of the vehicle that are associated with earth-fixed coordinate v_x^{trans} and v_y^{trans} can be obtained by projecting v_x and v_y onto the respective axes.

$$\begin{aligned} v_x^{trans} &= v_x \cos\theta - v_y \sin\theta \\ v_y^{trans} &= v_x \sin\theta + v_y \cos\theta \end{aligned} \tag{22}$$

Note that the positive rotation direction in (22) is defined by counter-clockwise.

Finally, the position of the vehicle in the two-dimension earth-fixed coordinate is written, as follows.

$$\begin{aligned} x_{pos} &= \int_0^t v_x^{trans} dt \\ y_{pos} &= \int_0^t v_y^{trans} dt \end{aligned} \tag{23}$$

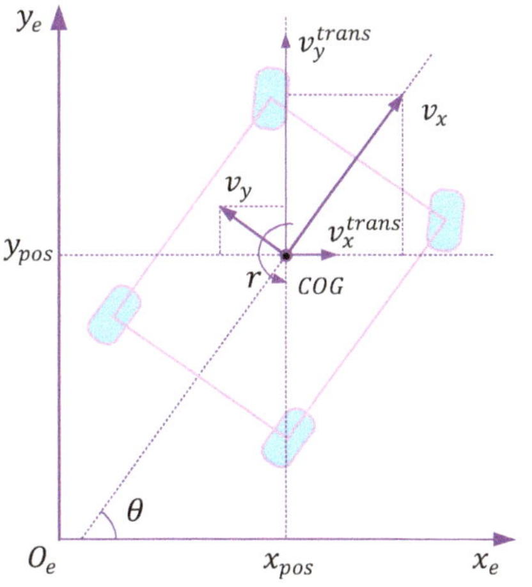

Figure 7. Position of the vehicle in two-dimension earth-fixed coordinate. The velocities v_x and v_y are associated with vehicle body-fixed frame. Sum of the projections of these vectors on the earth-fixed coordinate $(Oxy)_e$ are v_x^{trans} and v_y^{trans}, which are then used for the position model of the vehicle.

3.5. Mathematical Model Implementation Procedure

Figure 8 depicts the flowchart of the program that simulates the vehicle. Because of the limitation in computing power of the simulation hardware, all of the constants should be initialized at the beginning of the program. Additionally, in order to reduce the calculation time, lookup tables can be used for some blocks, for example, adhesion coefficients will be calculated in advance and put in a lookup table instead of repeating (5) in every sampling time. Following that, all of the models and equations are implemented with the procedure that is illustrated in the flowchart. Thereafter, this program is complied and downloaded to run on the chosen hardware platform.

Although all of the equations that contribute to the simulation model of the EV are listed in a given order and the program flowchart is also supplied, the computation procedure of the complete model should be treated carefully, as there exists several recursive calculation loops. Figure 9 presents, in detail, the running process through the equations of the EV model. The following remarks should be noted when realizing this model.

- For each equation in Figure 9, the input vectors indicate the required information for achieving the output vector, which is the result of that equation. Additionally, some equations that are the derivation of the final one may be omitted, i.e., only the most fundamental ones are shown in the flowchart.
- The calculation process is illustrated for one representative wheel, i.e., the front-left wheel, but it is identically applicable for other wheels.
- The flowchart consists of two main blocks, including the wheel model block and vehicle chassis one. The wheel block outputs the longitudinal and lateral forces, which are then incorporated with those of other wheel blocks at (1) to make the accelerations of the vehicle.
- As the front-wheel steering vehicle is mentioned as the studied object, only two steering angles at the front wheels are obtained and delivered to the vehicle chassis block. Therefore, both rear wheel blocks only generate friction forces.
- In order to avoid algebraic loop issues that are caused by the existence of recursive computation, the feedback signals, e.g., states of the vehicle, must be initialized.

In this paper, the simulation model is realized and validated on Matlab/Simulink, while using common transfer functions and S-functions to implement the system on the real-time platform without any further conversion.

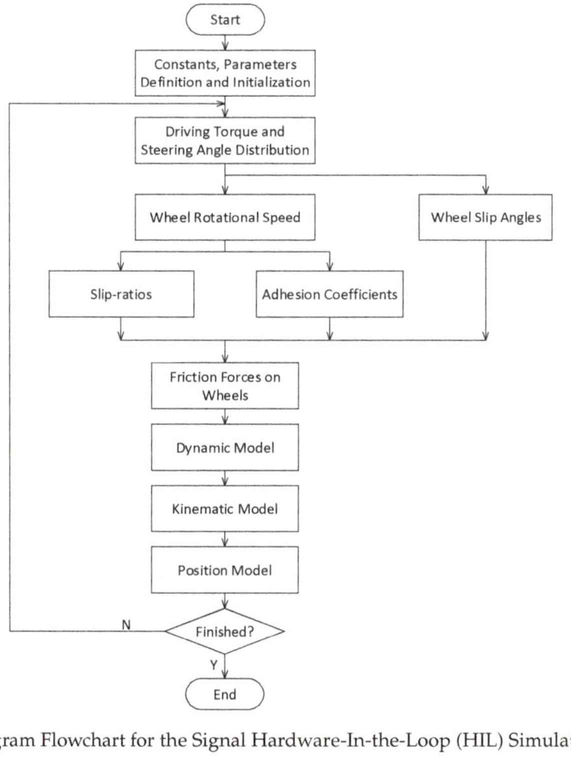

Figure 8. Program Flowchart for the Signal Hardware-In-the-Loop (HIL) Simulation System.

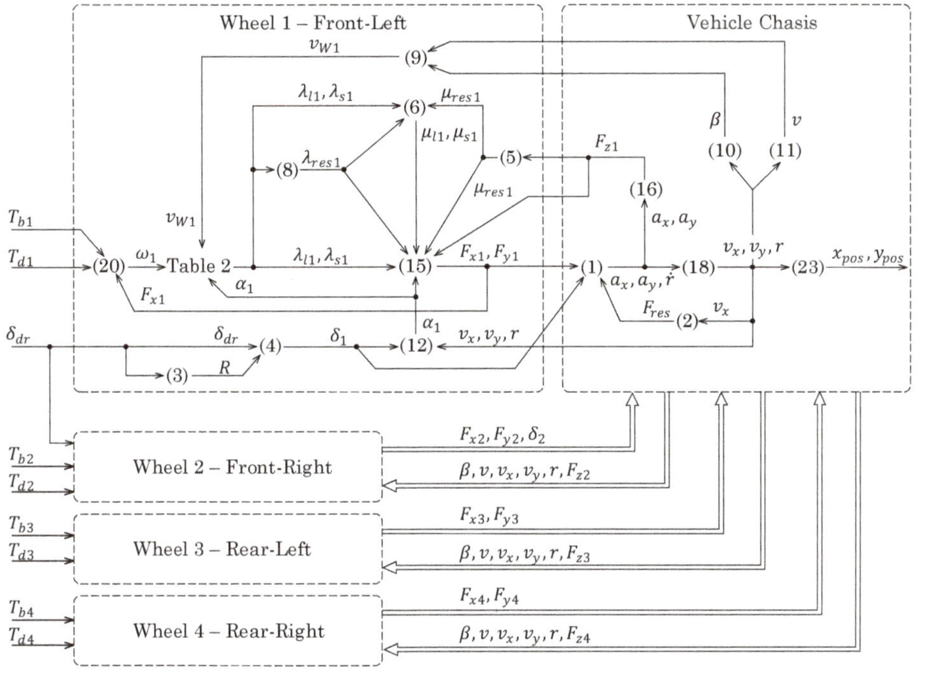

Figure 9. Calculation flow for 4-wheel-driven EV simulation model. For simplification, the representative wheel 1 is shown. The calculation procedure for other wheels is identically the same.

4. Validation Results

4.1. Testing Scenarios

Several representative scenarios are proposed in order to evaluate the performance of the HIL simulation system. For these tests being performed successfully, the system must be properly connected and powered. Additionally, several states or measured signals should be displayed on the screen of the host computer, which can be done by setting up in the Control Desk software. So that the driver (tester) will be able to manage the evaluation as this is the drive-able system.

In the first test case, which is also called Scenario 1, the vehicle moves on the normal road with a good adhesion coefficient μ. The vehicle is accelerated by pushing the accelerator pedal. The driver can adjust the position of the pedal in order to see the change of velocity that aims to describe the longitudinal behavior of the studied vehicle. When the vehicle reaches an arbitrary speed, the driver will turn the steering wheel clockwise and then counter-clockwise continuously to show the lateral behavior of the vehicle.

Scenario 2 is carried out by letting the vehicle moving on the slippery surface with low adhesion coefficient at the right side wheels, whereas the others on the left still run on the adhesive road.

Lastly, in Scenario 3, the vehicle will be run on a poor adhesion surface that covers the whole vehicle's track width.

The scenarios 2 and 3 both represent the typical circumstances in daily transport operation. When driving in such situations, if the control performance is poor, then the vehicle can easily loose the control and consequently the accident may happen. Figure 10 illustrates all of the proposed evaluation scenarios.

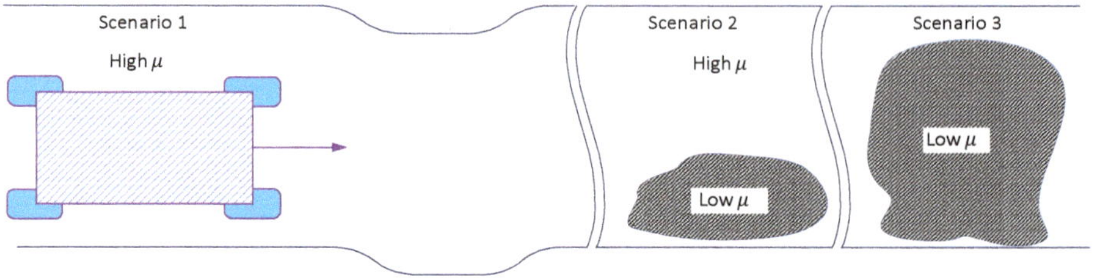

Figure 10. Scenarios for HIL Simulation System Evaluation.

Note that there are no control algorithm applied in these test cases. Hence, in order to ensure the legal model implementation that requires reference torques for four motors at wheels, the command torque of the driver will be distributed equally. In other words, in all three scenarios, the reference torque that is supplied to each motor is one-fourth of the command torque.

$$T_{di}^{*} = \frac{1}{4}T_{dr} \quad (24)$$

Additionally, the high adhesive road that is used in all the scenarios is the dry asphalt and the snow condition is adopted for low adhesive road.

The studied EV is the i-MiEV of Mitsubishi, which has the parameters listed in Table 3.

Table 3. i-MiEV Parameters.

Parameters	Value	Unit
l	2550	mm
l_f	1199	mm
l_r	1351	mm
b_f	1475	mm
b_r	1475	mm
h	559	mm
R_{eff}	300	mm
J_z	900	kgm^2
J_x	2	kgm^2
c_W	0.29	-
A	2.49	m^2
ρ	1.2041	kg/m^3

4.2. Experimental Results

4.2.1. Scenario 1 Results

Figures 11 and 12 show the driving profile that is generated by the driver for the first evaluation scenario, including accelerator, brake pedal position, and steering wheel angle, respectively. The accelerator pedal is initially pushed to 68% for about 20 s and then released to 19%. This position is kept for nearly 10 s, and then the driver increases the pressure again and holds it at 68% to t = 50 s. At this moment, the steering wheel is rotated clockwise, thus the steering angle is changed from zero to negative value. This is to imply that the car will be right cornered. After returning the wheel to zero position, the driver continues to turn left by anti-clockwise rotating the steering wheel. Until t = 70 s, the driver sets the angle to zero in order to control the car going straight. Finally, at t = 85 s, the brake pedal is pressed to slow down the vehicle.

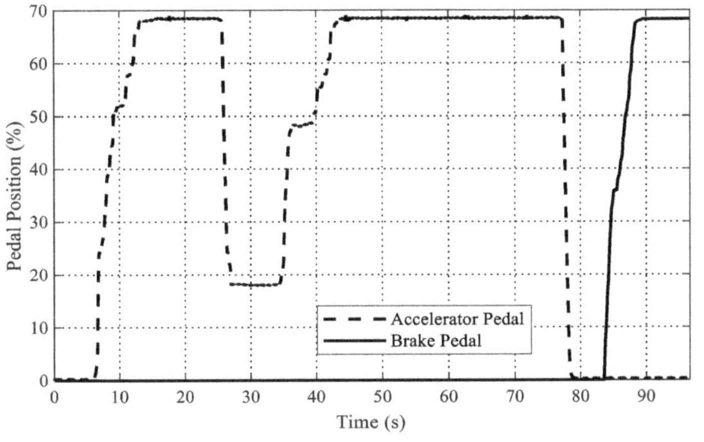

Figure 11. Accelerator and Brake Pedal Position for Scenario 1.

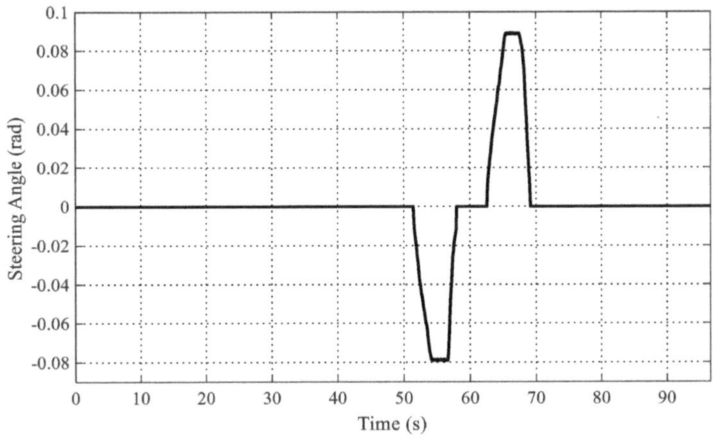

Figure 12. Steering Wheel Angle for Scenario 1.

The response of the car to the driver's command in terms of wheel and vehicle velocities, slip-ratios at each wheel, and position are shown in Figures 13–15, respectively.

In the velocity waveforms presented in Figure 13, the good quality road is reflected by the closely matching of wheel velocities and that of the vehicle. At the beginning of the test, the velocities rise following the increase of accelerator pedal position. This growing trend is flattened when the driver keeps the pedal at lower level, thus the car moves at a nearly constant speed. In the next phase of the acceleration, the vehicle reaches to around 18 m/s (65 km/h) when the accelerator pedal is kept at 68%. The velocities at wheels precisely equal to each other and also almost equal to that of the vehicle from the beginning to t = 52 s thanks to high adhesive coefficient of the road. From this moment, the driver starts cornering the vehicle which causes the differences between the velocities, as shown in two zoomed figures. The Zoom 1 shows the detailed difference at the first cornering and the Zoom 2 illustrates that of the second turning duration. At t = 78 s, the vehicle is slowed down, because the driver starts pushing the brake pedal. If the pedal is more pushed, the vehicle velocity is more decelerated, as can be seen.

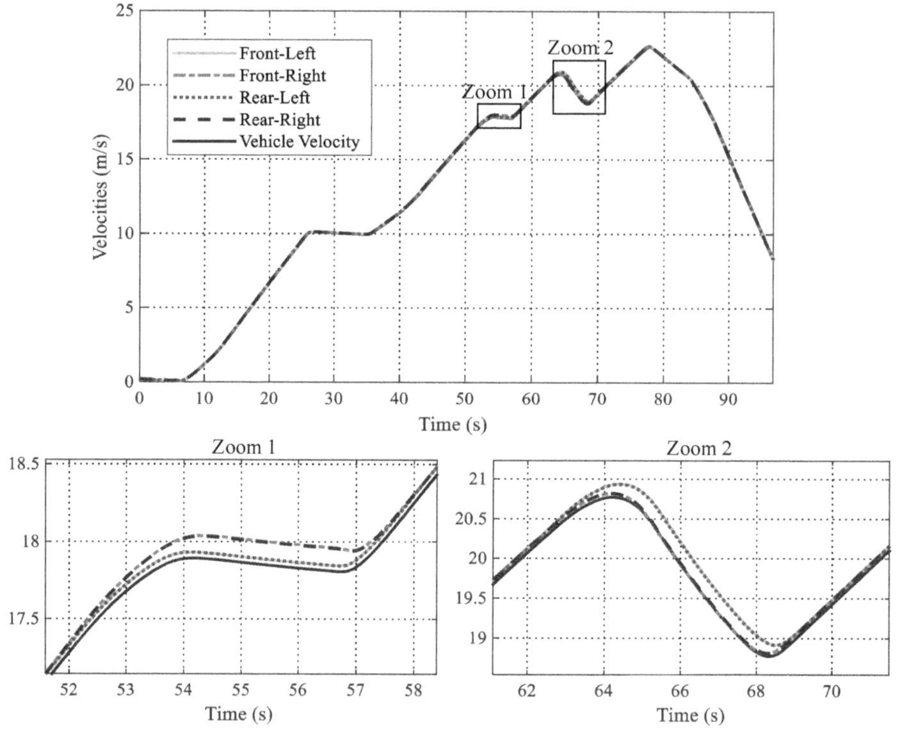

Figure 13. Velocities of Wheels and Vehicle for Scenario 1.

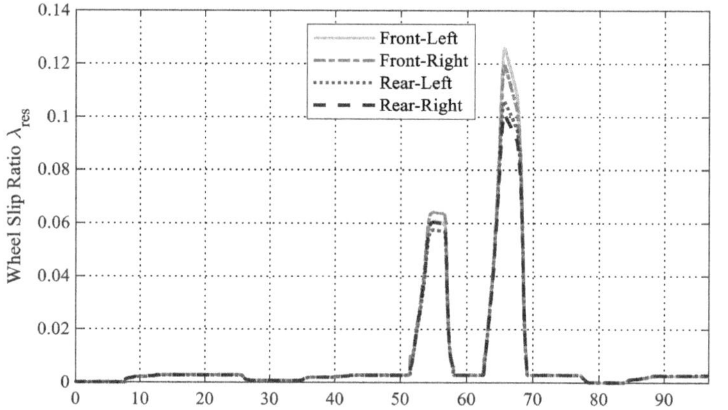

Figure 14. Slip-ratio at Wheels for Scenario 1.

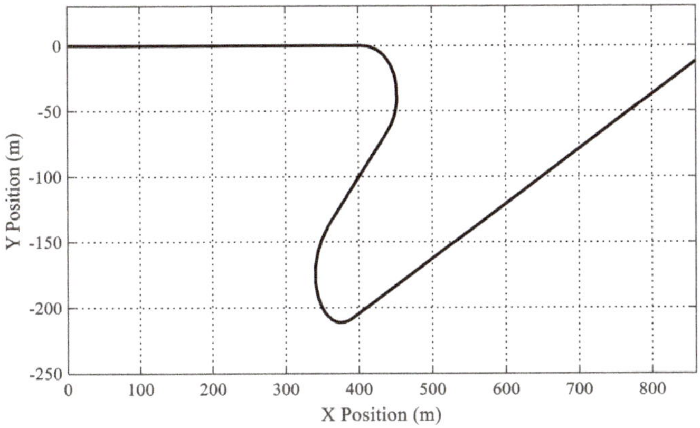

Figure 15. Position of the Vehicle for Scenario 1.

Figure 14 describes the slip-ratios λ_{res} of wheels, which is the remarkable feature of the proposed system. In the duration of straight movement of the vehicle, the slip-ratios are very small, due to the negligible deviation between wheel and vehicle velocities. The significant change of the slip-ratio appears when the car is being cornered. The λ_{res} reaches to maximum of 0.062 in the first cornering and up to 0.123 in the second one. These values are still considered in the safe region of slip-ratio (under 0.2). However, if referring to the conventional slip-ratio calculation (7) in the previous subsection, which only takes into account the difference between wheel velocity ωR_{eff} and longitudinal velocity v_x, the slip-raito cannot obtain such big values. Two zoomed figures in Figure 13 are the evidences for this remark, as the difference among velocities exists, but is not noteworthy. Thanks to the aforementioned calculation presented in Table 2, the total slip-ratio is composed of longitudinal and lateral values. Hence, the significant amount of λ_{res} in Figure 14 are mainly caused by lateral slip-ratio λ_s. In the other words, the vehicle seems to be drifted a bit when cornering at relatively high speed (18 m/s at the first time and 21 m/s at the second time).

Figure 15 illustrates the position of the vehicle on two-dimensional (2-D) plane. The trajectory acts in accordance with the driver's command that is shown in Figures 11 and 12. The car goes straight and then is turned right and left following the steering wheel reference and eventually continues straight to the end of the test. Generally, running on the good road condition, the vehicle can safely obey what the driver wants in terms of secured states of the car.

4.2.2. Scenario 2 and 3 Results

Scenarios 2 and 3 will be carried out in the same test, since they are both used for evaluation on slippery road. In these test cases, the accelerator pedal is kept constant at value of 68% while the brake one is free throughout the entire evaluation, as depicted in Figure 16. Additionally, the steering wheel is held at zero in order to imply that the car is expected going straight for the whole test.

The velocities of wheels and of the vehicle increase, obeying the given accelerator pedal position, as depicted on Figure 17. The vehicle is accelerated to about 11m/s before getting into slippery road. From this moment, evaluation scenarios are respectively started.

Scenario 2 begins at $t \approx 22$ s; when the front-right wheel enters the low adhesive surface and, immediately, its velocity increases very fast. This wheel gets to the maximum speed of 14.3 m/s and leaves the slippery road; following that, the rear-right wheel starts running on this surface. With the same manner, its velocity also rises dramatically to approximated 14.7 m/s before returning to the good road condition. For both right wheels, after leaving the sliding surface, their velocities reduce sharply to that of other wheels and synchronously with the speed of the vehicle. On the contrary, both left wheels still work properly on the good road condition, showing by their velocity trend did not nearly change during the test. The vehicle velocity also has the same behavior to these wheels when its velocity keeps increasing all the time.

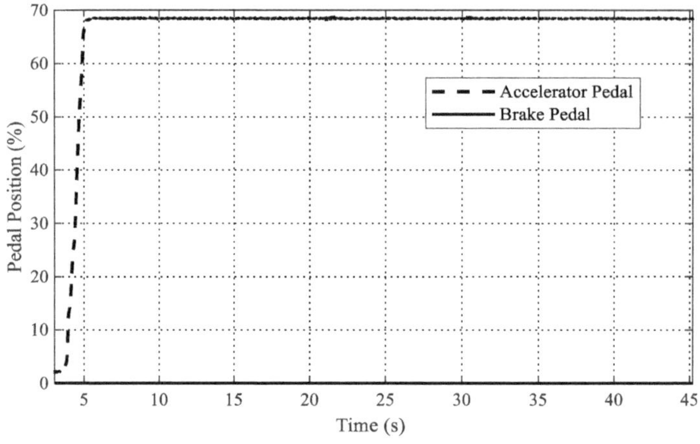

Figure 16. Accelerator and Brake Pedal Position for Scenario 2 and 3.

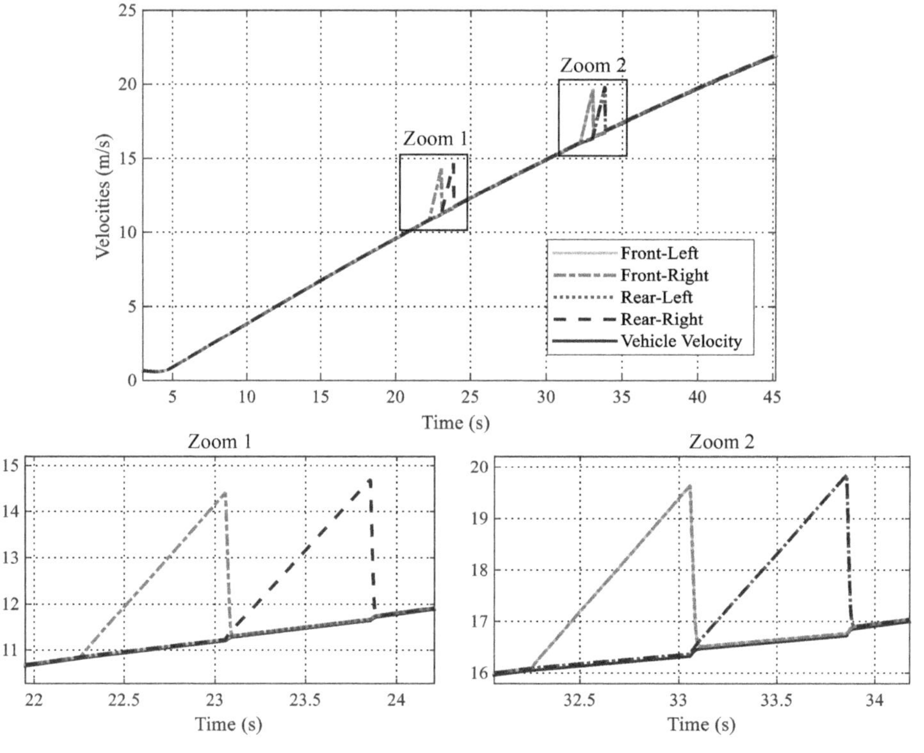

Figure 17. Velocity of Wheels and Vehicle for Scenario 2 and 3. The Zoom 1 is associated with Scenario 2 and the Zoom 2 is for Scenario 3.

When the vehicle reaches to 16m/s of speed ($t \approx 32$ s), once again, it gets in the sliding surface, which is Scenario 3. Different from the previous test, in this case, the two front wheels simultaneously enter the low adhesive road. The friction force is weakened, while motor torques are still remained, which results in the unexpected increase of wheel velocities. The increase in the speed of wheels only stops and reduces dramatically when they get out of the sliding surface. The same behavior happens for both rear wheels. The maximum velocity of four wheels in the test case is nearly 20 m/s. Interestingly, because there are still two wheel running on the high adhesion road while others are moving on bad condition, the vehicle velocity retained its tendency. This is the remarkable advantage of a all-wheel-driven (AWD) vehicle.

The safety aspect of the vehicle in these scenarios can be evaluated by slip-ratio, as illustrated in Figure 18. Empirically, the secured working region of the vehicle is $\lambda_{res} < 0.2$. If the slip-ratio is larger than this number, then the vehicle will be considered to be slipping on the road, i.e., the wheel spins without adhesion on the road surface. This circumstance must be avoided, because the car may loose control easily and this is the duty of all motion controllers, e.g., Traction Control System (TCS) of Toyota or Dynamic Stability Control (DSC) of BWM. Without control, the slip-ratio of the studied vehicle gets over the value of 0.2, as depicted in the figure. Dangerously, these ratios are in the increasing tendency and, if the slippery surface is longer, then the slip-ratios will reach a much higher value. However, the behavior of the slip-ratio at wheels complies with the manner of working situation, which is the key expectation of the proposed system.

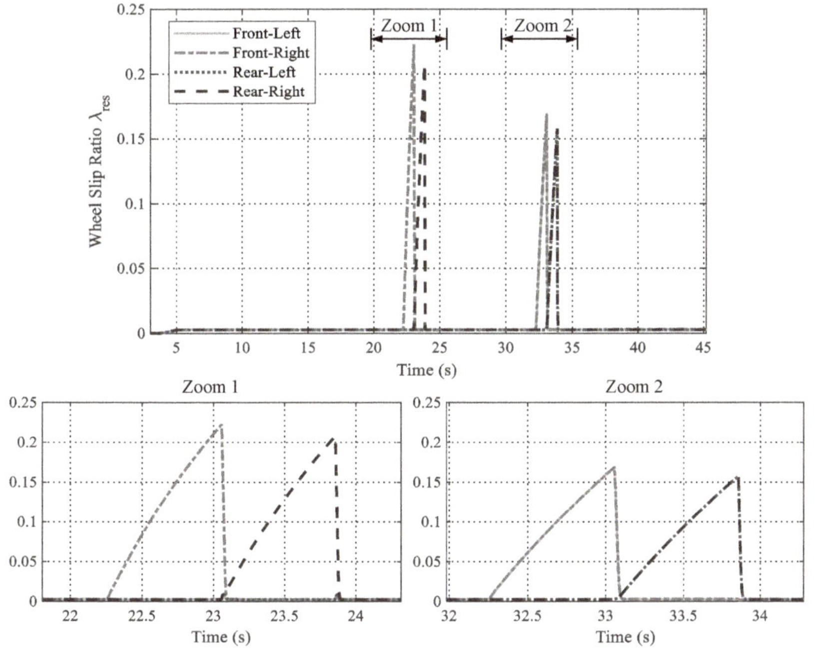

Figure 18. Slip-ratio at Wheels for Scenario 2 and 3. The Zoom 1 is associated with Scenario 2 and the Zoom 2 is for Scenario 3.

The position trajectory of the CoG of the vehicle depicted in Figure 19 also shows the serious situation, especially in the Scenario 2. The vehicle's lateral position is changed suddenly when the side wheels come over the low μ surface. Let us focus on the instant when the front-right wheel enters the slippery surface that causes the adhesion force to be reduced at this wheel. Because the summed force of two left wheels still remains, the difference between the forces on the right and left side of the vehicle generates torque in the clockwise direction. Consequently, the vehicle rotates and moves to the right. When the front-right wheel leaves the low friction surface with high rotational speed and returns to the adhesive road, it causes a "strike force" applied to the right side. This jerks the vehicle to the left in a very small duration. This phenomenon is repeated when the rear-right wheel runs on the low adhesive road. Although the side displacement is small (under 0.5 m) and the speed of the vehicle is relatively low, this unexpected change may still cause a dangerous situation to the car and the driver. For Scenario 3, the vehicle's trajectory is on the straight line, because there is no lateral force or different friction forces between the left and right sides of the vehicle.

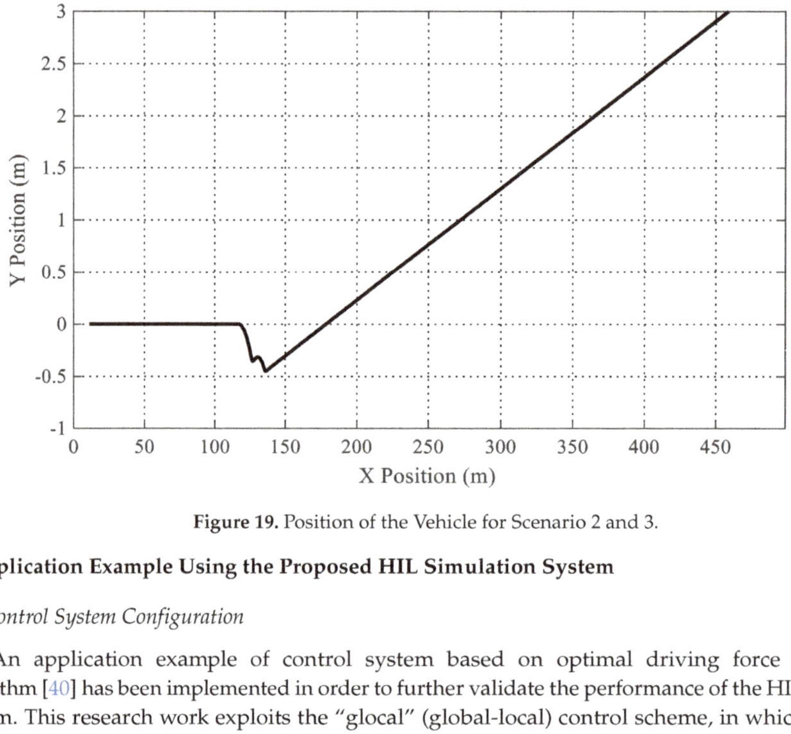

Figure 19. Position of the Vehicle for Scenario 2 and 3.

5. Application Example Using the Proposed HIL Simulation System

5.1. Control System Configuration

An application example of control system based on optimal driving force distribution algorithm [40] has been implemented in order to further validate the performance of the HIL simulation system. This research work exploits the "glocal" (global-local) control scheme, in which the global controller is responsible for generating distributed reference driving forces F_{xi}^* to guarantee the coordination among the motions of wheels and of the vehicle, the local controllers create reference motor torques T_{di}^* that are based on the generated force commands in order to ensure the secured slip-ratios at the wheels of the vehicle. Figure 20 shows the configuration of the example control system.

The input of the control system is the driving force F_{dr}, i.e., driver's command; thus, the accelerator pedal signal should be redefined instead of aforementioned T_{dr}. As mentioned about the design procedure earlier, the whole control system of this application will be deployed on the eZdsp F28335 Control Card. In Figure 20, one local controller for the front-left wheel is described. Other controllers for other wheels are designed identically. The reference motor torques that are generated from the control algorithm are multiplexed and transmitted to the vehicle through the CAN bus. Whereas, the measured signals, including wheel rotational speeds and longitudinal acceleration of the car, are connected to ADC ports of the control card, as they are analogous. On the vehicle model side, instead of using sensors as real experiments do, such signals can be output from DAC ports of the DS1103 Controller Board, which is one of the advantages of the proposed HIL simulation system.

The detail distribution algorithm and control design can be found in [40]. For the driving force observer as well as the slip-ratio estimator, one can find them in [41], as the development of motion controller is not in the scope of this paper.

In this section, the example control scheme will be validated only with Scenarios 2 and 3, because these test cases can effectively highlight the dominant features of the algorithm.

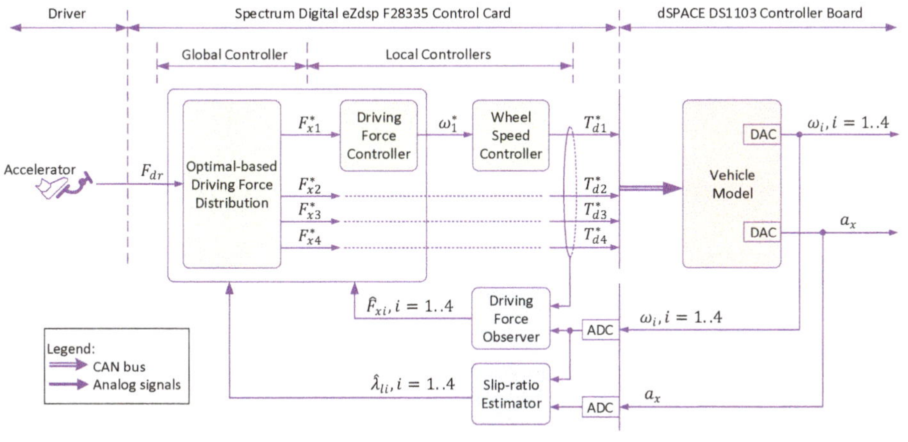

Figure 20. Configuration of Driving Force Distribution and Control for EV with Four In-Wheel Motors, adapted with implementation on the proposed HIL simulation system.

5.2. Results

Figures 21–24 illustrated the validation results of the example control system running on the proposed HIL simulation system.

Let us focus on the reference motor torque that is shown in Figure 21. From the driving force command, four torque values are generated, which are quite stable at the beginning of the test. However, taking a closer look to the waveform within this duration, the controller distributes more torque to the rear wheels than to the front ones. This actually makes sense, because, referring to (16), the normal force at rear wheels increase when accelerating. Hence, more wheel torque can exploit more friction force, leading to better acceleration, which also can be seen in Figure 22 for longitudinal friction force.

When $t \approx 23$ s, Scenario 2 is carried out, which leads to the small difference between in velocity of the right-front wheel and the vehicle, as shown in the Zoom 1 of Figure 23. Remarkably, when traveling on the low μ condition, the friction force decreases sharply, the control system immediately reduces motor torque of the respective wheel, i.e., the right-front one in this case, in order to keep the slip-ratio at a properly small value, as shown in Figure 24. Simultaneously, the torque at the rear-right wheel is increased in order to compensate the reduction in friction force of the sliding wheel. Consequently, the car is kept going safely in a secured state. The same control behavior is applied within the remained Scenario 2 duration, in which the front-right wheel torque is boosted to pull the right side of the vehicle when the rear-right one is running on low friction surface.

Scenario 3 is taken part in the test when $t \approx 33$ s. When both front wheels are traveling on the sliding road, the motor torques are decreased, following the reduction of the friction forces and, thus, keeping the slip-ratio at a small value (around 0.02, as depicted in the Zoom 2 of the Figure 24). The controller compensates for the loss in the total driving force by reinforcing torque at both rear wheels. This leads to the sum of all longitudinal forces remaining the reference value. The same response is made by the controller when the rear wheels are running on the low adhesive surface as the final section of Scenario 3.

For the whole validation, Figure 24 shows the advantage of the control scheme in the safety aspect, in which the slip-ratios are kept under the maximum value of about 0.09, much smaller than those of uncontrolled schema results. Moreover, the wheel speed is controlled very fast; therefore, the unexpected sliding duration is significantly small and the vehicle can overcome the bad road condition smoothly.

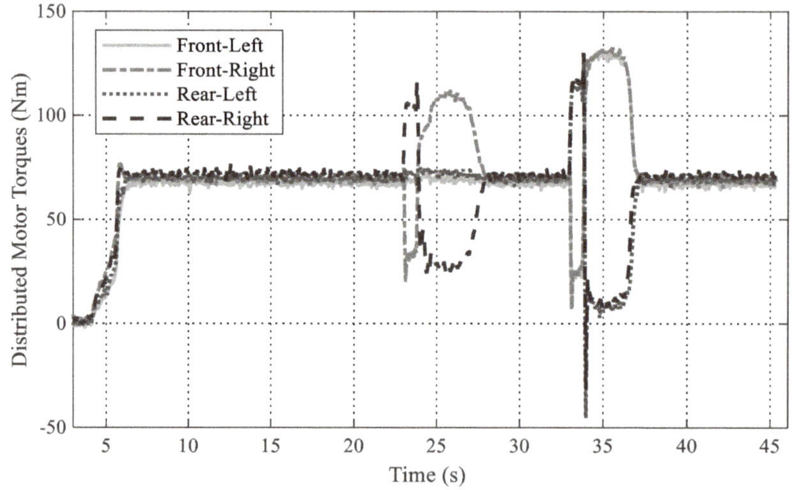

Figure 21. Distributed Motor Torque Generated by the Control System.

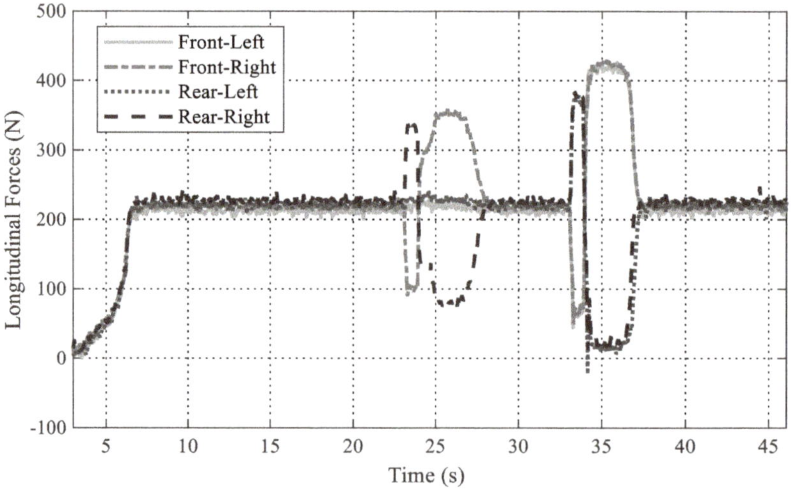

Figure 22. Longitudinal Driving Forces Acted on each Wheels.

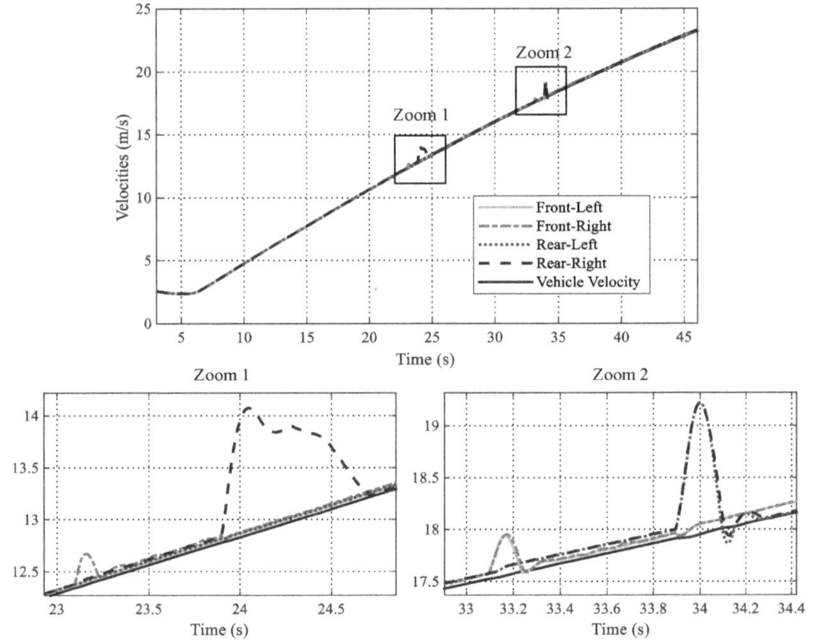

Figure 23. Wheel and Vehicle Velocities. The Zoom 1 is associated with Scenario 2 and the Zoom 2 is for Scenario 3.

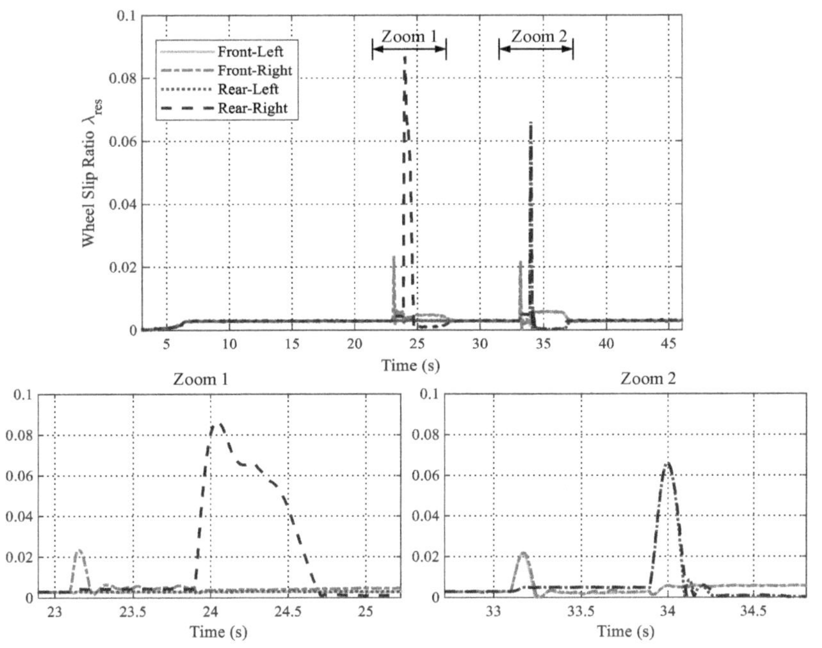

Figure 24. Slip-ratio at each Wheels. The Zoom 1 is associated with Scenario 2 and the Zoom 2 is for Scenario 3.

6. Conclusions

In this paper, an experimental platform that is based on signal HIL simulation system for 4-IWM EVs has been proposed. The HIL system is derived from detailed dynamic models in longitudinal, lateral, and yaw directions, which are frequently used in motion control of EVs. The term "detailed" is shown in adhesion coefficients calculation, slip-ratio computation, and tire force models, which are further complex than the conventional ones. Thanks to these features, the specific aspects in behavior of the vehicle is emphasized and allow for better control performance assessment. The advantages of the proposed signal HIL simulation system are illustrated through various evaluation plans with several road conditions and different driving scenarios. The capability and application of the HIL system are also confirmed by examining a representative control strategy in optimal driving force distribution for four wheels of the studied vehicle. Being developed from DS1103 and F28335 digital signal controllers that are very common research platforms, the proposed system is cost-effective, safe, open, and can be used widely.

In the future, this work on signal HIL simulation system can be extended in both directions: (*i*) applying this system to other vehicle control problems and (*ii*) expanding the system model. Autonomous driving is an example of the former direction, in which adaptive cruise control or automatic lane-changing can exploit the position model of this HIL platform. On the other hand, the proposed system can be enlarged by adding models of energy storage devices or specific electric motors. By that, the system can be used for research on energy management and electric drives that illustrate the open and expandable features of the proposed experimental platform.

Author Contributions: Conceptualization and methodology T.V.-D., M.C.T., B.-H.N. and J.P.F.T.; software, T.V.-D.; validation, T.V.-D. and B.-H.N.; formal analysis, T.V.-D. and B.-H.N.; investigation, T.V.-D. and B.-H.N.; resources, M.C.T. and J.P.F.T.; data curation, T.V.-D.; writing—original draft preparation, T.V.-D.; writing—review and editing, T.V.-D., M.C.T., B.-H.N. and J.P.F.T.; visualization, T.V.-D. and B.-H.N.; supervision, M.C.T. and J.P.F.T.; project administration, M.C.T.; funding acquisition, T.V.-D. and M.C.T. All authors have read and agreed to the published version of the manuscript.

Funding: This research is funded by the Hanoi University of Science and Technology (HUST) under project number T2020-TT-003.

Acknowledgments: We thank Nguyen Tien Dung and Phan Duy Tot from CTI Laboratory for Electric Vehicles, Hanoi University of Science and Technology for their contributions in system debugging, circuit design and their supports when implementing experiments.

Conflicts of Interest: The authors declare no conflict of interest.

References

1. Weiss, M.; Cloos, K.C.; Helmers, E. Energy efficiency trade-offs in small to large electric vehicles. *Environ. Sci. Eur.* **2020**, *32*. [CrossRef]
2. Helmers, E.; Marx, P. Electric cars: Technical characteristics and environmental impacts. *Environ. Sci. Eur.* **2012**, *24*, 1–15. [CrossRef]
3. Karki, A.; Phuyal, S.; Tuladhar, D.; Basnet, S.; Shrestha, B.P. Status of Pure Electric Vehicle Power Train Technology and Future Prospects. *Appl. Syst. Innov.* **2020**, *3*, 35. [CrossRef]
4. Yamamoto, M.; Kakisaka, T.; Imaoka, J. Technical trend of power electronics systems for automotive applications. *Jpn. J. Appl. Phys.* **2020**, *59*. [CrossRef]
5. Chau, K.T. *Electric Vehicle Machine and Drives—Design, Analysis and Application*; John Wiley & Sons: Singapore, 2015.
6. Hori, Y. Motion control of electric vehicles and prospects of supercapacitors. *IEEJ Trans. Electr. Electron. Eng.* **2009**, *4*, 231–239. [CrossRef]
7. Hussain, S.; Ahmed, M.A.; Kim, Y.C. Efficient Power Management Algorithm Based on Fuzzy Logic Inference for Electric Vehicles Parking Lot. *IEEE Access* **2019**, *7*, 65467–65485. [CrossRef]
8. Balaska, H.; Ladaci, S.; Schulte, H.; Djouambi, A. Adaptive cruise control system for an electric vehicle using a fractional order model reference adaptive strategy. *IFAC-PapersOnLine* **2019**, *52*, 194–199. [CrossRef]

9. Ivanov, V.; Savitski, D.; Shyrokau, B. A Survey of Traction Control and Antilock Braking Systems of Full Electric Vehicles with Individually Controlled Electric Motors. *IEEE Trans. Veh. Technol.* **2015**, *64*, 3878–3896. [CrossRef]
10. He, H.; Peng, J.; Xiong, R.; Fan, H. An acceleration slip regulation strategy for four-wheel drive electric vehicles based on sliding mode control. *Energies* **2014**, *7*, 3748–3763. [CrossRef]
11. Lin, C.; Xu, Z. Wheel torque distribution of four-wheel-drive electric vehicles based on multi-objective optimization. *Energies* **2015**, *8*, 3815–3831. [CrossRef]
12. Tian, J.; Tong, J.; Luo, S. Differential Steering Control of Four-Wheel Independent-Drive Electric Vehicles. *Energies* **2018**, *11*, 2892. [CrossRef]
13. Park, J.; Jang, I.G.; Hwang, S.H. Torque distribution algorithm for an independently driven electric vehicle using a fuzzy control method: Driving stability and efficiency. *Energies* **2018**, *11*, 3479. [CrossRef]
14. Thanh, V.D.; Ta, C.M. A Universal Dynamic and Kinematic Model of Vehicles. In Proceedings of the 2015 IEEE Vehicle Power and Propulsion Conference (VPPC), Montreal, QC, Canada, 19–22 October 2015; pp. 1–6. [CrossRef]
15. Baek, S.Y.; Kim, Y.S.; Kim, W.S.; Baek, S.M.; Kim, Y.J. Development and Verification of a Simulation Model for 120 kW Class Electric AWD (All-Wheel-Drive) Tractor during Driving Operation. *Energies* **2020**, *3*, 2422. [CrossRef]
16. Pappalardo, C.M.; Lombardi, N.; Guida, D. A model-based system engineering approach for the virtual prototyping of an electric vehicle of class l7. *Eng. Lett.* **2020**, *28*, 215–234.
17. Mechanical Simulation. Introduction to CarSim. Available online: https://www.carsim.com/downloads/pdf/ (accessed on 31 October 2020).
18. TESIS-DYNAware. veDYNA Entry—The Perfect Entry Point to Vehicle Dynamics Simulation. Available online: https://www.tesis-dynaware.com/ (accessed on 31 October 2020).
19. Bouscayrol, A. Hardware-in-the-loop simulation. In *Industrial Electronics Handbook*, 2nd ed.; Taylor and Francis: Chicago, IL, USA, 2011; Chapter M35.
20. Messier, P.; Nguyễn, B.H.; LeBel, F.A.; Trovão, J.P.F. Disturbance observer-based state-of-charge estimation for Li-ion battery used in light electric vehicles. *J. Energy Storage* **2020**, *27*, 101144. [CrossRef]
21. Kermani, S.; Trigui, R.; Delprat, S.; Jeanneret, B.; Guerra, T.M. PHIL implementation of energy management optimization for a parallel HEV on a predefined route. *IEEE Trans. Veh. Technol.* **2011**, *60*, 782–792. [CrossRef]
22. Nguyen, B.H.; German, R.; Trovao, J.P.F.; Bouscayrol, A. Real-time energy management of battery/supercapacitor electric vehicles based on an adaptation of pontryagin's minimum principle. *IEEE Trans. Veh. Technol.* **2019**, *68*, 203–212. [CrossRef]
23. Mayet, C.; Delarue, P.; Bouscayrol, A.; Chattot, E. Hardware-In-the-Loop Simulation of Traction Power Supply for Power Flows Analysis of Multitrain Subway Lines. *IEEE Trans. Veh. Technol.* **2017**, *66*, 5564–5571. [CrossRef]
24. Yuan, Y.; Zhang, J.; Li, Y.; Li, C. A Novel Regenerative Electrohydraulic Brake System: Development and Hardware-in-Loop Tests. *IEEE Trans. Veh. Technol.* **2018**, *67*, 11440–11452. [CrossRef]
25. Zhang, H.; Zhang, Y.; Yin, C. Hardware-in-the-Loop Simulation of Robust Mode Transition Control for a Series-Parallel Hybrid Electric Vehicle. *IEEE Trans. Veh. Technol.* **2016**, *65*, 1059–1069. [CrossRef]
26. Reitinger, J.; Balda, P.; Schlegel, M. Steam turbine hardware in the loop simulation. In Proceedings of the 2017 21st International Conference on Process Control, PC 2017, Štrbské Pleso, Slovakia, 6–9 June 2017; pp. 380–385. [CrossRef]
27. Yang, J.; Konno, A.; Abiko, S.; Uchiyama, M. Hardware-in-the-loop simulation of massive-payload manipulation on orbit. *Robomech J.* **2018**, *5*. [CrossRef]
28. Wu, J.; Cheng, Y.; Srivastava, A.K.; Schulz, N.N.; Ginn, H.L. Hardware in the Loop test for power system modeling and simulation. In Proceedings of the 2006 IEEE PES Power Systems Conference and Exposition, PSCE 2006, Atlanta, GA, USA, 29 October–1 November 2006; pp. 1892–1897. [CrossRef]
29. Magalhães, Z.; Murilo, A.; Lopes, R.V. Development and evaluation with MIL and HIL simulations of a LQR-based upper-level electronic stability control. *J. Braz. Soc. Mech. Sci. Eng.* **2019**, *41*, 327. [CrossRef]
30. Vo-Duy, T.; C. Ta, M. A signal hardware-in-the-loop model for electric vehicles. *Robomech J.* **2016**, *3*, 29. [CrossRef]

31. Nie, P.; Wu, Y.; Liang, X. Design of HIL Test System for VCU of Pure Electric Vehicle. In Proceedings of the 2017 2nd International Conference on Materials Science, Machinery and Energy Engineering (MSMEE 2017), Dalian, China, 13–14 May 2017; Volume 123, pp. 867–871. [CrossRef]
32. Xia, J.L.; Diao, Q.; Sun, W.; Xing, Z.; Yuan, X.; Dong, T. Development of low cost hardware-in-the-loop test system and a case study for electric vehicle controller. In Proceedings of the 2016 International Conference on Applied System Innovation, (ICASI), Okinawa, Japan, 26–30 May 2016; pp. 4–7. [CrossRef]
33. Ramaswamy, D.; McGee, R.; Sivashankar, S.; Deshpande, A.; Allen, J.; Rzemien, K.; Stuart, W. *A Case Study in Hardware-In-The-Loop Testing: Development of An ECU for a Hybrid Electric Vehicle*; SAE Technical Papers; SAE International: Warrendale, PA, US, 2004. doi:10.4271/2004-01-0303. [CrossRef]
34. Yan, Y.; Liu, C.; Ma, X.; Zhang, Y. Hardware-in-loop real-time simulation of electrical vehicle using multi-simulation platform based on data fusion approach. *Int. J. Distrib. Sens. Netw.* **2019**, *15*. [CrossRef]
35. Nam, K.H. *AC Motor Control and Electrical Vehicle Applications*, 2nd ed.; Taylor & Francis: Boca Raton, FL, USA, 2019.
36. Kiencke, U.; Nielsen, L. *Automotive Control System*; Springer: Berlin/Heidelberg, Germany, 2005.
37. Hans B. Pacejka. *Tire and Vehicle Dynamics*, 3rd ed.; Butterworth-Heinemann: Oxford, UK, 2012.
38. Rajamani, R. *Vehicle Dynamics and Control*, 2nd ed.; Springer: New York, NY, USA, 2012.
39. Jazar, R.N. *Vehicle Dynamics: Theory and Application*; Springer: New York, NY, USA, 2009.
40. Wang, Y.; Fujimoto, H.; Hara, S. Driving Force Distribution and Control for EV With Four In-Wheel Motors: A Case Study of Acceleration on Split-Friction Surfaces. *IEEE Trans. Ind. Electron.* **2017**, *64*, 3380–3388. [CrossRef]
41. Maeda, K.; Fujimoto, H.; Hori, Y. Four-wheel driving-force distribution method based on driving stiffness and slip ratio estimation for electric vehicle with in-wheel motors. In Proceedings of the 2012 IEEE Vehicle Power and Propulsion Conference, VPPC 2012, Seoul, Korea, 9–12 October 2012; pp. 1286–1291. [CrossRef]

energies

MDPI

Article

Longitudinal Modelling and Control of In-Wheel-Motor Electric Vehicles as Multi-Agent Systems

Binh-Minh Nguyen [1,*], Hung Van Nguyen [2], Minh Ta-Cao [2,3] and Michihiro Kawanishi [1]

1 Department of Advanced Science and Technology, Toyota Technological Institute, 2-12-1 Hisakata, Tempaku, Nagoya 468-8511, Japan; kawa@toyota-ti.ac.jp
2 Department of Electrical Engineering, Ha Noi University of Science and Technology, 01 Dai Co Viet Street, Hanoi 100000, Vietnam; ng.v.hungbkhn@gmail.com (H.V.N.); minh.tacao@hust.edu.vn (M.T.-C.)
3 Department of Electrical Engineering, Université de Sherbrooke, 2500, boul. de l'Université, Sherbrooke, QC J1K 2R1, Canada
* Correspondence: binhminh@toyota-ti.ac.jp

Received: 8 September 2020; Accepted: 14 October 2020; Published: 18 October 2020

Abstract: This paper deals with longitudinal motion control of electric vehicles (EVs) driven by in-wheel-motors (IWMs). It shows that the IWM-EV is fundamentally a multi-agent system with physical interaction. Three ways to model the IWM-EV are proposed, and each is applicable to certain control objectives. Firstly, a nonlinear model with hierarchical structure is established, and it can be used for passivity-based motion control. Secondly, a linearized model with rank-1 interconnection matrix is presented for stability analysis. Thirdly, a time-varying state-space model is proposed for optimal control using linear quadratic regulator (LQR). The proposed modellings contribute the new understanding of IWM-EV dynamics from the view point of multi-agent-system theory. By choosing the suitable control theory for each model, the complexity level of system design is maintained constant, no matter what the number of IWMs installed to the vehicle body. The effectiveness of three models and their design approaches are discussed by several examples with Matlab/Carsim co-simulator.

Keywords: electric vehicle; in-wheel-motor; motion control; anti-slip; slip-ratio; multi-agent-system

---

1. Introduction

In-wheel-motors (IWMs) have been shown a novel actuator for motion control system of electric vehicles (EVs) [1]. A remarkable merit of IWM-EV is the ability to precisely generate the driving/braking torques with fast response at individual wheels. Based on this merit, various motion control strategies have been developed in three dimensions. Through literature review, motion control of electric vehicle can be categorized by several sub-fields as follows:

(i) Lateral motion control: By differentiating the torques distributed to the left-side and right-side wheels, a yaw-moment is generated about the vertical axis of the vehicle body. This allows us to stabilize the lateral motion of the vehicle via direct-yaw-moment control [2]. Some important issues of this sub-field are sides-slip angle estimation [2,3], side-slip angle and yaw-rate control [4], and the integration of direct-yaw-moment control with active-steering control [5].

(ii) Roll motion control: By allocating the motor torques suitably, it is also possible to generate the roll moment about the longitudinal axis. This moment can be used to control the roll angle and roll rate for improving the comfort of the driver [6].

(iii) Pitch motion control: By distributing the different torques to the front-wheels and rear-wheels, a pitch moment is generated about the lateral axis. This moment could be used to improve the pitch motion, as presented in [7–9].

(iv) Longitudinal motion control: This is the objective of this paper. Longitudinal motion control (LMC) of EV can be summarized by three main groups in Table 1, namely, (I) slip ratio control [10–17], (II) anti-slip control [18–22], and (III) driving force control [23–26].

Table 1. Review of longitudinal motion control methods for EVs.

Group	Strategy	Consideration of Physical Interaction (YES/NO)	References
(I) Slip ratio control	Sliding mode control	NO	[10–13]
	PI control by linearizing the slip dynamics	NO	[14–16]
	Optimal control by hierarchical LQR	YES	[17]
(II) Anti-slip control	Zero-slip-model following control	NO	[18,19]
	Maximum transmissible torque estimation	NO	[20,21]
	Direct wheel velocity control	NO	[22]
(III) Driving force control	Direct driving force control	NO	[23]
	Driving force control based on wheel velocity control and virtual variable control	NO	[24–26]

In recent years, an uptrend has been observed in the number of IWM actuators installed to the vehicle body. For instance, France Army developed the armored vehicle DPE 6 × 6 which is driven by six independent motors [27]. In Japan, Keio University introduced the prototype of KAZ-EV with eight IWMs [28]. Thanks to the development of wireless IWM technology [29], powerful EVs driven by more than ten IWMs might be soon realized. With respect to this uptrend, IWM-EV should be treated as a multi-agent system in which each local agent is a locally controlled wheel. The local agent cooperates with each other through the vehicle body to generate the global motion of the vehicle body as well as its local motion.

From the above discussion, LMC of IWM-EVs becomes an actual challenge. Fortunately, it is possible to apply the state-of-the-art of multi-agent system theories. Among them, the global (global/local) framework proposed by Hara group has been shown an effective way to model, analyze, and design multi-agent dynamical systems [30]. Several successful examples of glocal control are gene-regulatory network [31], platoon car [32,33], and manufacturing systems [34]. We recently developed a scheme in the glocal theory family, namely, hierarchical linear quadratic regulator (H-LQR), and applied it to optimal control of slip ratio for IWM-EVs in the acceleration mode [17]. The key of H-LQR is to utilize some special features in the structure of the multi-agent system to reduce the design cost. From the view point of glocal control framework, two critical issues still remain in the research field of LMC for IWM-EVs.

Firstly, almost the previous studies (including the works mentioned in Table 1) neglected the physical interaction between the wheel dynamics. For instance, in [22], Suzuki et al. introduced an anti-slip control by directly controlling the wheel velocity. The wheel velocity controller $C_w(s)$ is merely designed by stable pole-placement to the closed-loop of $C_w(s)$ and $P_w(s)$ which represent the local wheel dynamics. In [25], Maeda et al. proposed the driving force control system for the four-IWM-EV. However, the system is just the combination of four decoupled wheel-velocity-based force control loops. In [11], Nam et al. independently design for each wheel a sliding-mode slip ratio controller. However, the design is based on the single wheel dynamics model without any relevant to the others. Actually, the physical interaction exists as the wheels are installed to the same car body. As clarified by a glocal control tool, namely "generalized frequency variable theory" [35], and its application to EV [36], neglecting physical interaction might reduce the system performance. Even if each wheel's control loop is locally stable, this does not automatically guarantee the stability of the IWM-EV system as a whole. Therefore, it is essential to model the IWM-EV as an interconnected system of multi-agents to properly design the LMC.

Secondly, following the uptrend in the number of IWM actuators, analyzing and designing the LMC systems become more and more complex. Increasing the actuator numbers means increasing the order of the system, and hence the implementation effort and the computational cost. Even if we can model the LCM system with physical interactions, how should we select the suitable control theory to reduce the burden of design and analysis?

To tackle the aforementioned issues, proper modelling is necessary. In other words, the model should capture the essential characteristics of the plant in the physical world. Looking at the literature of EV-related-study, this philosophy of modelling has been pointed out by several recent works on battery management and control [37–40]. For instance, in [37], a predictive control algorithm of charging process was proposed by utilizing a thermoelectric model which accurately captures the battery dynamics. Recently, a Gaussian process regression models have been used to address the aging phenomenon of batteries in [39].

To provide the proper models for LMC of IWM-EVs, this paper is based on the glocal framework. The strategy of this paper is described in Figure 1 which includes three spaces. In the physical world, the motion of the IWM-EV is governed by the physical equations. From the view point of multi-agent systems, the physical model is mapped to the control world by three design models (DMs). Each DM is suitable for a certain objective of longitudinal motion control. Moreover, each DMs is associated with a control theory developed in the glocal framework, such that the complexity of system design and analysis is independent of the number of IWM actuators. Three proposed DMs are briefly explained as follows.

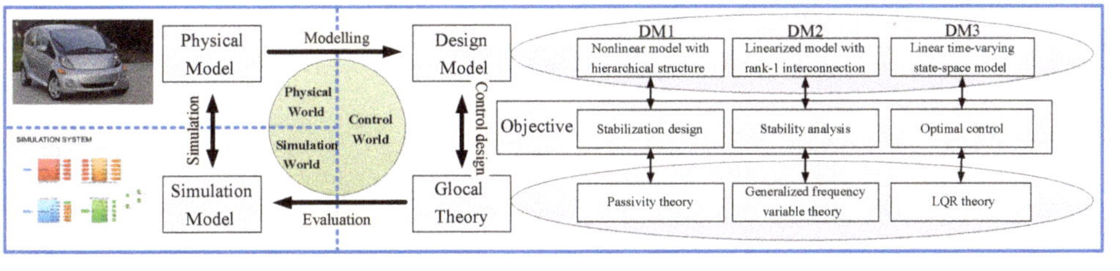

Figure 1. Discussion strategy in this paper: Design procedure of electric vehicles driven by in-wheel-motors (IWM-EV) motion control.

DM-1 (nonlinear model with hierarchical structure): DM-1 consists of the vehicle body dynamics V in the upper-layer, and a bunch of local wheel dynamics $\{W_i\}$ in the lower-layer. The two layers interconnect with each other through the aggregation and distribution channels. This expression is quite useful for designing LMC by passivity theory [41]. The advantage of using DM-1 and the passivity theory is that we can design the control system with less mathematical effort and prove the system stability rigorously. In this paper, a passivity based anti-slip control system is proposed to demonstrate the effectiveness of DM-1.

DM-2 (linearized model with rank-1 interconnection matrix): DM-2 is established by linearizing the nonlinear physical equations of IWM-EV about a given operating point, and it is useful for stability analysis. As an example, DM-2 is applied to wheel velocity control with work-load minimization. Thanks to the rank-1 interconnection matrix and the generalized frequency variable (GFV) theory [35], the complexity of stability analysis is considerably reduced. The stability analysis can be performed by a limited set of inequalities established from the GFV transfer function and the eigenvalues of the interconnection matrix. DM-2 and the GFV makes up the useful tools to design and analyze any existing LMC methods.

DM-3 (time-varying state-space model): DM-3 is proposed for capturing the slip ratio dynamics and driving force dynamics of IWM-EV in the state space expression. Hence, this model can be used for applying the optimal control strategies and updating the optimal control gains in real-time operation. In our recent study [17], DM-3 was established in the acceleration mode. To complete the state space modelling for LMC, this paper investigates the DM-3 in the deceleration mode. We theoretically show that global optimality can be achieved by a hierarchical LQR (H-LQR) algorithm in which we only need to solve the local Riccati equation. Applying the proposed H-LQR, the computational cost is effectively reduced.

The remainder of this paper is organized as follows. Section 2 presents the physical equations that describing the longitudinal motion of IWM-EV. To clarify the motivation of this paper, Section 3 discusses the disadvantages of the traditional LMC methods by numerical simulations. Then, DM-1, DM-2, and DM-3 are investigated in Sections 4–6, respectively. Each of the proposed DMs is established under reasonable assumptions, following by a motion control example. Section 7 presents the simulation results of the motion control examples in Sections 4–6. The simulations were conducted using a Matlab/Casim co-simulator which is based on an experimental electric vehicle developed by our research group. Finally, the concluding remarks are given in Section 8.

2. Longitudinal Dynamics of Vehicle

This paper examines the longitudinal motion of IWM-EV illustrated in Figure 2 of which the number N of IWMs is an even number. Let m be the vehicle mass. Under the assumption that the wheels are homogeneous, J_w and r are the inertia and radius of the wheel, respectively. T_i, F_i, and Z_i are the motor torque, the driving force, and the vertical force acting at the ith wheel, respectively. The local motion of the ith wheel is represented by the rotational velocity ω_i. The global motion of the vehicle body is represented by the longitudinal velocity v_x. F_{air} is the air resistance acting on the vehicle body. The total influence of gravity and rolling resistance is represented by F_ξ.

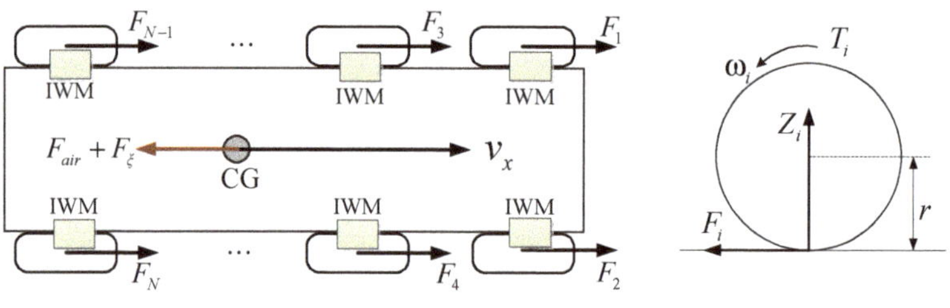

Figure 2. Longitudinal motion of IWM-EV: (**a**) Vehicle body; (**b**) Local wheel.

The motion of the vehicle body is given by

$$m\dot{v}_x = \sum_{i=1}^{N} F_i - F_{air} - F_\xi. \tag{1}$$

Neglecting the wind velocity, the air resistance is expressed as

$$F_{air} = \rho v_x |v_x|, \tag{2}$$

where the coefficient ρ is calculated from the density of air, the cross-sectional area of the vehicle in the air, and the drag coefficient which depends on the shape of the vehicle [42].

The rotational motion of the ith wheel is given as

$$J_w \dot{\omega}_i = T_i - rF_i. \tag{3}$$

Let ε be a small number to avoid division-by-zero, the slip ratio of the ith wheel is defined as

$$\lambda_i = \frac{r\omega_i - v_x}{\max\{r\omega_i, v_x, \varepsilon\}}. \tag{4}$$

The nonlinear relationship between the driving force and the slip ratio can be described by several empirical functions. The most popular one is the magic formula proposed by Pacejka [43]. This formula is given by

$$f_i(\lambda_i) = A_i \sin\Big\{B_i \tan^{-1}\Big[C_i\lambda_i - D_i\Big(C_i\lambda_i - \tan^{-1}(C_i\lambda_i)\Big)\Big]\Big\}, \tag{5}$$

where A_i, B_i, C_i, and D_i are the shape factors, with $A_i = \mu_i Z_i$ where μ_i is the friction coefficient. Note that the friction coefficient and the shape factors are all identifiable from experimental data. Now, the driving force is expressed as

$$F_i(\lambda_i) = \begin{cases} f_i(\lambda_i) \text{ if } \lambda_i \geq 0 \\ -f_i(-\lambda_i) \text{ if } \lambda_i < 0 \end{cases}. \tag{6}$$

The typical curve of the driving force is shown in Figure 3. The driving force gradually increases as the slip ratio increases from 0 to an optimal value λ^* which depends on the road condition. Since the slip ratio exceeds λ^*, the driving force is decreased to a minimum value as the slip ratio approach to 1. Investigating the magic formula, we suggest the following practical remark.

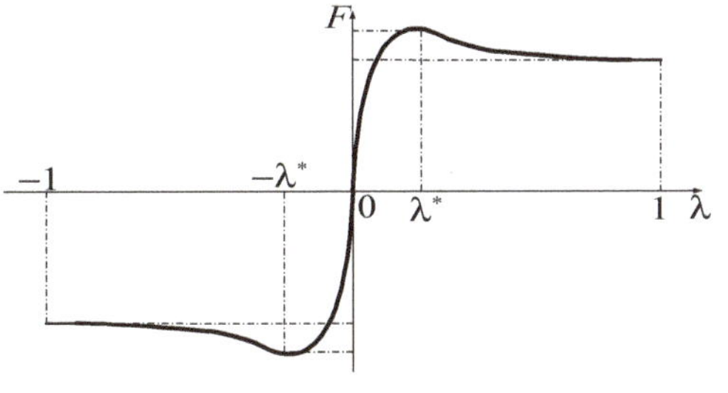

Figure 3. Nonlinear relationship between the driving force and the slip ratio.

Remark 1: *The inequality $\lambda_i f_i(\lambda_i) \geq 0$ holds for any value of slip ratio λ_i in $(-1, 1)$.*

3. Motivating Example

In our recent study [17], we examined the slip ratio control method proposed in [14] by both theoretical discussion and numerical simulations. According to [14], it is possible to linearize the dynamics from the motor torque T_i to the slip ratio λ_i. Then, a PI controller can be designed by placing the stable poles to the local subsystem. However, the performance of the overall system should be determined by both local subsystem's transfer function and a matrix that representing physical interaction [35]. Due to the change of operating points, the poles of the overall system might move towards the imaginary axis with increasing imaginary parts. This phenomenon might degrade the system performance by several scenarios. For instance, oscillated phenomena can be observed when the vehicle runs at high velocity. In addition, the load transfer can introduce large gap between the slip ratios of the front-wheels and rear-wheels. Moreover, the traditional slip ratio control system is highly sensitive to the non-flatness of the road surfaces.

This paper neglects to re-simulate the aforementioned slip ratio control method. Instead, we examine another LMC method, namely, anti-slip control by wheel velocity control [22]. To this end, this paper utilizes Carsim—a standard software for simulating the vehicle dynamics [44]. By using Carsim, it is convenient to perform the control algorithms under different environmental conditions, including the dangerous conditions. The simulation model is built based on the four-wheel experimental electric vehicle developed from the Mitsubishi's i-MiEV prototype. It was an outcome of our research project KC.03.08/11-15 supported by the Ministry of Science and Technology of Vietnam. The main specification of the vehicle is summarized in Appendix A.

To establish the simulation model, we modify the parameters of a hatchback car provided in the Library of Carsim. Since the Carsim only provide the vehicle of gasoline type, we constructed the electrical drive system in Matlab/Simulink. The electrical drive system consists of the battery model, the inverter models, the in-wheel-motor models of interior permanent magnet type, together with the inverter control units and motor control units. To resemble the behavior of the actual drive system, the inverter model imitates the electrical circuit of the experimental inverter. The output torques of the drive system model in Matlab/Simulink are sent to Carsim through the imports namely IMP_MYUSM_*. The notation * represents the IWM position (L1, L2, R1, and R2, where L denote "left" and R denotes "right").

3.1. Anti-Slip Control Based on Wheel Velocity Control

Following the idea of the original work [22], we generalize the block diagram of the control system as in Figure 4. Here, $\mathbf{1}_N$ is the all-one column-vector of size N, and $\mathbf{T} = [T_1 \ldots T_N]^{\mathrm{T}}$, $\boldsymbol{\omega} = [\omega_1 \ldots \omega_N]^{\mathrm{T}}$. The scalar signal T_{cmd} is the driving command given by the driver or the upper motion control layer. In addition, vector $\mathbf{k} = [k_1 \ldots k_N]^{\mathrm{T}}$ is to distribute the driving command to the local wheel. The distribution ratio k_i are commonly selected such that $0 < k_i < 1$ and $k_1 + k_2 + \ldots + k_N = 1$. We assume that the vehicle operates on the acceleration mode when $r\omega_i > v_x$, and let λ^* be the desired slip ratio. From (4), the reference velocity can be calculated from the wheel velocity v_x via the gain

$$\chi = \frac{1}{r(1-\lambda^*)}. \tag{7}$$

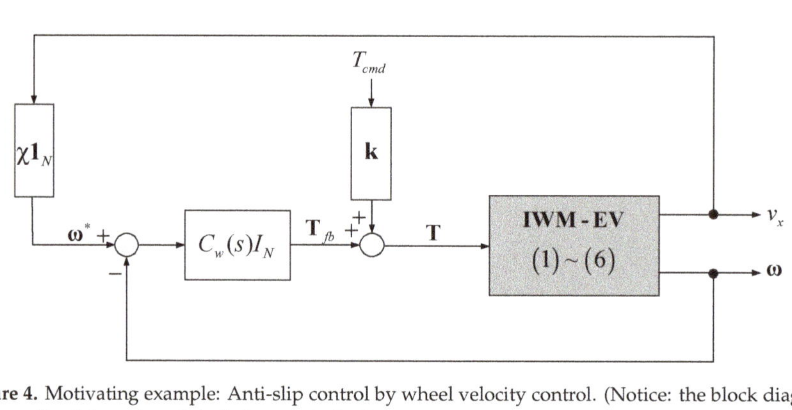

Figure 4. Motivating example: Anti-slip control by wheel velocity control. (Notice: the block diagram is generalized from the original diagram in [22]).

Each wheel is provided a velocity controller $C_w(s)$. Let $P_w(s) = 1/(J_w s)$ be transfer function that characterizes the local dynamics of the wheel. To design $C_w(s)$, the traditional way is to assign the stable poles to the closed loop transfer function

$$P_{cl}(s) = \frac{P_w(s)C_w(s)}{1 + P_w(s)C_w(s)}. \tag{8}$$

For instance, let κ_1 and κ_2 be the stable poles of $P_{cl}(s)$, $C_w(s)$ becomes a PI controller with the P gain and I gain calculated as

$$C_w(s) = \frac{K_p s + K_i}{s}, \; K_p = -J_w(\kappa_1 + \kappa_2), \; K_i = J_w \kappa_1 \kappa_2. \tag{9}$$

3.2. Simulation and Discussion

To evaluate the above design strategy, a test is conducted as follows. A constant driving command T_{cmd} is equally distributed to four wheels such that the EV accelerates from the initial position on the high friction surface (μ = 0.8). It enters the low friction surface (μ = 0.2) at about 4 s. Two test cases are performed, and their results are summarized in Figure 5.

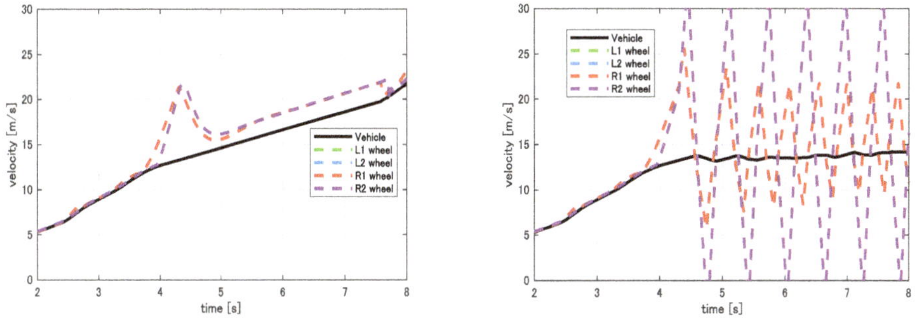

(**a**) Case A (Locally stable & Globally stable) (**b**) Case B (Locally stable & Globally unstable)

Figure 5. Motivating example: anti-slip control by wheel velocity control.

Case A: Two stable poles of $P_{cl}(s)$ are placed at $\kappa_{1,2} = -10 \pm 2j$ (Figure 5a). The wheel velocities are maintained closed to the vehicle velocity. This means the anti-slip function is successfully attained in this case.

Case B: Two stable poles of $P_{cl}(s)$ are placed at $\kappa_{1,2} = -100 \pm 0j$ (Figure 5b). The overall system becomes unstable when the vehicle operates on the low-friction surface.

The simulation results in this sub-section suggest that local stabilization does not guarantee global stability of the control system. By using only the local wheel dynamics to design the controller, we cannot explain such performance of the vehicle in Case B. Therefore, it is necessary to model and design the motion control system of IWM-EV as a whole. Three models that effectively capture the physical interaction will be presented in the following three sections.

4. DM-1: Nonlinear Model with Hierarchical Structure

This section proposes a way to directly model the IWM-EV system using its nonlinear equations. To design the control system, our strategy is based on the passivity theory which has been widely utilized in various fields of control engineering for three decades. As an example, DM-1 is utilized to design an anti-slip control system. Notice that, it is possible to neglect the disturbance F_{ξ} in the theoretical discussion.

4.1. Hierarchical Model of IWM-EV

Based on (3)–(6), the model of the local agent W_i is shown as in Figure 6a. It includes two inputs $\{v_x, T_i\}$ and two outputs $\{F_i, \omega_i\}$. The vehicle body dynamics V is obtained from (1) and (2). V has the input $1_N^T\mathbf{F}$ which is the aggregation of the local driving force, and the output v_x which is distributed to the local W_i. Combining $\{\mathrm{W}_i\}$ in the lower-layer and V in the upper-layer, the DM-1 is established as in Figure 6b with the hierarchical structure.

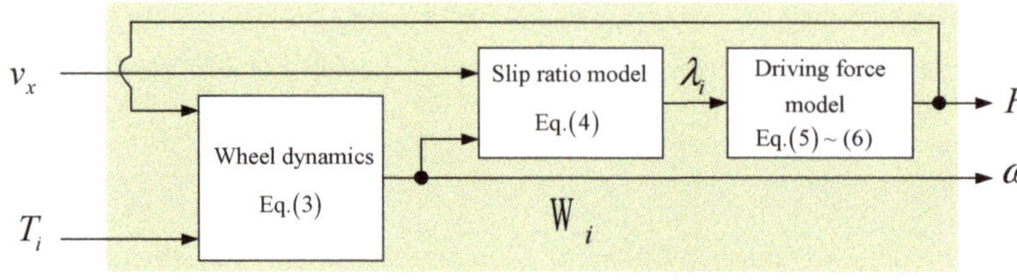

(**a**) Model of a single wheel dynamics.

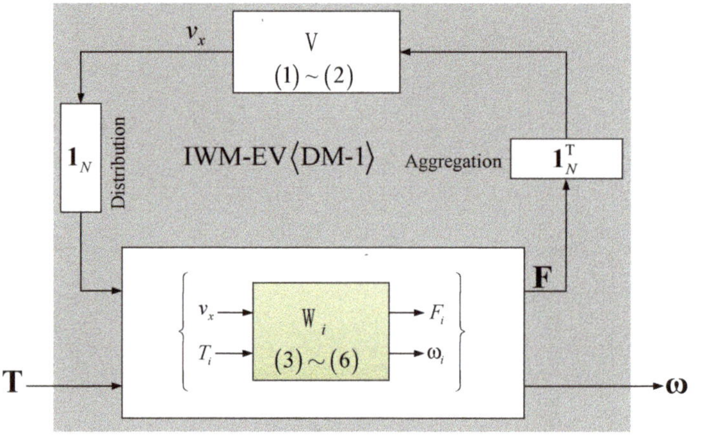

(**b**) Hierarchical model of IWM-EV dynamics.

Figure 6. Block diagram of the DM-1.

4.2. Passivity Theory and Its Application to DM-1

Consider a system given by a state space equation

$$\begin{cases} \dot{x} = f(x,u) \\ y = h(x,u) \end{cases}, \tag{10}$$

with the input vector $u(t) \in \mathbb{R}^p$, output vector $y \in \mathbb{R}^p$, and the state vector $x(t) \in \mathbb{R}^n$.

Definition 1 [41]: *The system (10) is said to be passive from input u to output y if there exists a positive semidefinite function S:* $\mathbb{R}^n \to \mathbb{R}_+$*, called storage function, such that the inequality* $\dot{S} \leq y^T u$ *holds for all* $x \in \mathbb{R}^n$ *and* $u \in \mathbb{R}^p$*. In addition,* (10) *is input strictly passive (ISP) if* $\dot{S} \leq y^T u - \delta_u ||u||^2$ *for some* $\delta_u > 0$*, and output strictly passive (OSP) if* $\dot{S} \leq y^T u - \delta_y ||y||^2$ *for some* $\delta_y > 0$*.*

Proposition 1: *The IWM-EV system in Figure 6b is passive from the input* $\mathbf{T}$ *to the output* $\boldsymbol{\omega}$*.*

Proof of Proposition 1: Define the storage function

$$S = \frac{1}{2} m v_x^2 + \sum_{i=1}^{N} \left(\frac{1}{2} J_w \omega_i^2 \right), \tag{11}$$

then with respect to Remark 1, it is transparent that the inequality $\dot{S} \leq \boldsymbol{\omega}^T \mathbf{T}$ holds true. The detailed calculation is neglected for the sake of paper length.

Passivity-based-control (PBC) was introduced three decades ago to the adaptive control problem of rigid robot [45]. It has been widely applied in various fields of control engineering, especially in multi-agent systems [46]. Following the fundamental passivity theorems [41], if the plant is passive, system stability can be guaranteed by connecting the plant with a strictly passive feedback controller. Motivated by the Proposition 1, there exists many ways to design motion control of IWM-EV using PBC [47]. A design example of anti-slip control will be examined in the next subsection.

4.3. Example of DM-1: Passivty Based Anti-Slip Control of IWM-EV

This sub-section demonstrates anti-slip control as an application of the DM-1 and passivity theory. The significance and necessity of anti-slip control can be explained from two points of view as follows.

From a safety point of view: As discussed by Hori et al. in [14], not only the driving force but also the side force strongly depends on the slip ratio. The side force becomes smaller as the slip ratio becomes bigger. If the slip ratio increases considerably due to the change of road condition, the side force becomes drastically smaller. This causes serious problems, including drift-out in front-wheel-drive vehicles, spin in rear-wheel-drive vehicles, and drift-out with rotation in four-wheel-drive vehicles. Such situations are extremely dangerous for the drivers.

From an energy point of view: For simplicity in discussion, we consider a single-wheel vehicle model. Following the Proposition 1, its energy storage function is

$$S = \frac{1}{2}mv_x^2 + \frac{1}{2}J_w\omega^2, \tag{12}$$

With respect to the definition of slip ratio expressed in (4), we obtain

$$S = \begin{cases} \frac{1}{2}\left[m + \frac{J_w}{r^2(1-\lambda)^2}\right]v_x^2 \text{ if } r\omega > v_x \\ \frac{1}{2}\left[m + \frac{J_w(1+\lambda)^2}{r^2}\right]v_x^2 \text{ if } r\omega < v_x \end{cases}, \tag{13}$$

In the acceleration mode ($r\omega > v_x$), a sharp increasing of λ means a sharp increasing of S. Thus, to attain a given velocity pattern v_x in the acceleration mode, it is required to use more electric energy from the battery. Next, we examine the deceleration mode ($r\omega < v_x$). In this case the side of λ is negative. A drastic decreasing of λ (or increasing of $-\lambda$) results in a drastic decreasing of S. Thus, to attain a given velocity pattern v_x in the deceleration mode, only small amount of energy can be regenerated back to the battery. In summary, the driving range of electric vehicles would be considerably reduced if we wasted energy by neglecting anti-slip control.

The proposed control system is shown in Figure 7 where the scalar T_{cmd} is the driving command given by the driver. The command distributed to each W_i is $T_{r,i} = k_i \times T_{cmd}$ where the set of distribution ratio satisfies $k_i > 0$ for all i from 1 to N, and $k_1 + k_2 + ... + k_N = 1$. Consequently, vectors $\mathbf{T}_r$ and $\mathbf{k}$ are defined as $\mathbf{T}_r = [T_{r,1}\ T_{r,2}\ ...\ T_{r,N}]^T$ and $\mathbf{k} = [k_1\ k_2\ ...\ k_N]^T$, respectively. Note that vector $\mathbf{k}$ can be updated in real-time to attain the additional control objective, such as yaw moment control or minimization of energy consumption. Let K_a and K_ω be the control gains. Considering the acceleration mode, a possible option of the local anti-slip control law $C_{l,i}$ is given by

$$T_{l,i} = -K_a(r\omega_i - v_x)sign(\omega_i)sign(r\omega_i - v_x) - K_\omega\omega_i. \tag{14}$$

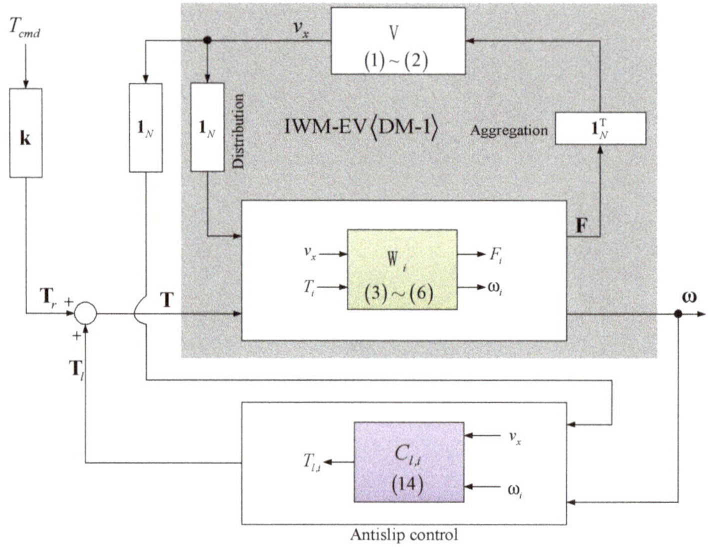

Figure 7. Expression of anti-slip control system of IWM-EV using DM-1.

The stability of the control system is guaranteed by the following theoretical result.

Proposition 2: *Considering the acceleration mode, the anti-slip control system in* Figure 7 *with the local controller $C_{l,i}$ given by (14) is OSP from $\mathbf{T}_r$ to $\boldsymbol{\omega}$ if the control gains are selected such that $K_a > 0$ and $K_\omega > 0$.*

Proof of Proposition 2: With respect to the storage function (11), the control law (14), and the dynamics of W_i described by (3)–(6), we have

$$\begin{aligned}\boldsymbol{\omega}^T\mathbf{T}_r - \dot{S} = &\sum_{i=1}^{N}(\max\{r\omega_i, v_x, \varepsilon\}\lambda_i f_i(\lambda_i)) + (\rho|v_x|)v_x^2 \\ &+\sum_{i=1}^{N} K_a((r\omega_i - v_x)sign(r\omega_i - v_x))(\omega_i sign(\omega_i)) + \sum_{i=1}^{N} K_\omega\omega_i^2\end{aligned} \tag{15}$$

There exist four terms on the right-hand side of (15). It is clear from Remark 1 that the first term is non-negative. Similarly, the second term is non-negative. With the positive control gains, we can see that $\boldsymbol{\omega}^T\mathbf{T}_r - \dot{S} \geq \sum_{i=1}^{N} K_\omega\omega_i^2 = K_\omega(\boldsymbol{\omega}^T\boldsymbol{\omega})$ holds true. In other words, $\dot{S} \leq \boldsymbol{\omega}^T\mathbf{T}_r - K_\omega(\boldsymbol{\omega}^T\boldsymbol{\omega})$. Following the Definition 1, the system in Figure 7 is shown to be OSP. This completes the proof.

5. DM-2: Linearized Model with Rank-1 Interconnection Matrix

The previous section focuses on the design problem using DM-1. Thanks to the passivity theory, we can rigorously prove the stability of the control system even if it is nonlinear and quite complex. However, passivity theory limits the class of controllers that we can select. Therefore, it is also essential to consider the problem of stability analysis. In other words, given a motion control system, how should we evaluate its stability? This section introduces a design model which can reduce the complexity level of the analysis problem. As an example, it is applied to a wheel velocity control system of IWM-EV. Note that, the air resistance and the other disturbances are neglected in the discussion in this section. For the sake of paper length, we only examine the vehicle motion in the acceleration mode, such that max$\{r\omega_i,$

$v_x, \varepsilon\} = r\omega_i$. We focus our stability analysis to the operating point O = $\{\omega_{o,1}, \omega_{o,2}, \ldots, \omega_{o,N}, v_{xo}\}$ with the following assumption.

Assumption 1: *(i) The vehicle operates on the uniform road condition and the friction coefficients of all wheels are the same. (ii) The slip ratio is small, and the driving force is linearized as $F_i \approx S_n\lambda_i$ where S_n is the nominal driving stiffness. At O, the values of max$\{r\omega_{o,i}, v_{xo}, \varepsilon\}$ are almost the same. In other words, we might approximate*

$$\kappa_n \approx \frac{1}{\max\{r\omega_{o,i}, v_{xo}, \varepsilon\}} \; \forall i \in [1, N]. \tag{16}$$

5.1. Linearized Model of IWM-EV

Under the Assumption 1, the driving force can be expressed as

$$F_i = \widetilde{S}_n \eta_i, \tag{17}$$

where $\widetilde{S}_n \approx \kappa_n S_n$, and $\eta_i = r\omega_i - v_x$. Based on (17), the linearized model of IWM-EV (DM-2) is established as in Figure 8 where $\boldsymbol{\eta} = [\eta_1 \ldots \eta_N]^{\mathrm{T}}$ and

$$A = -1_N 1_N^T, \tag{18}$$

$$P_g(s) = \frac{1}{ms}, P_w(s) = \frac{1}{J_w s}. \tag{19}$$

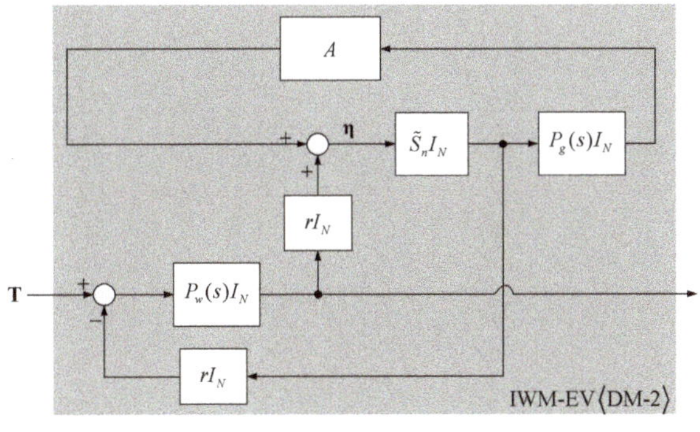

Figure 8. DM-2: Linearized model of IWM-EV about an operating point.

Remark 2: *The interconnection matrix A is a matrix of rank-1 with N-1 zero eigenvalues, and only one non-zero eigenvalue which equals to -N.*

5.2. Example of DM-2: Wheel Velocity Control System and Its Stability Condition

This section considers a wheel velocity control system for distributing the driving command T_{cmd} with a set of distribution ratios $\{k_i\}$. From (3), the reference of wheel acceleration is calculated as [48]

$$\dot{\omega}_i^* = \frac{1}{J_w} k_i T_{cmd} - \frac{r}{J_w} \hat{F}_i, \tag{20}$$

where $\hat{F}_i$ is the obtained by the driving force observer (DFO) as follows

$$\hat{F}_i = Q(s) \cdot \left(\frac{1}{r}T_i - \frac{J_w}{r} \cdot s \cdot \omega_i\right), \tag{21}$$

where the low-pass-filter is expressed as $Q(s) = 1/(\tau_f s + 1)$ with the small time constant τ_f. From (19)–(21) and the DM-2, block diagram of the wheel velocity control system is shown in Figure 9 where $C_w(s)$ is the wheel velocity controller. Notice that we neglect to discuss the algorithms to optimize in this section. Such algorithms can be founded in several other works on range extension control of IWM-EV [48–50]. The main goal of this section is to provide a suitable model for stability analysis. About the operating point O, the stability of the system in Figure 9 can be discussed from the system in Figure 10 where the transfer function from **y** to **z** is diagonalized by

$$H(s) = \frac{\widetilde{S}_n\big(1 + \widetilde{C}_w(s)P_w(s)\big)P_g(s)}{1 + \widetilde{C}_w(s)P_w(s) + \widetilde{S}_n r^2 P_w(s)}, \ \widetilde{C}_w(s) = \frac{C_w(s)(1 - Q(s))}{1 + C_w(s)P_w(s)Q(s)}. \tag{22}$$

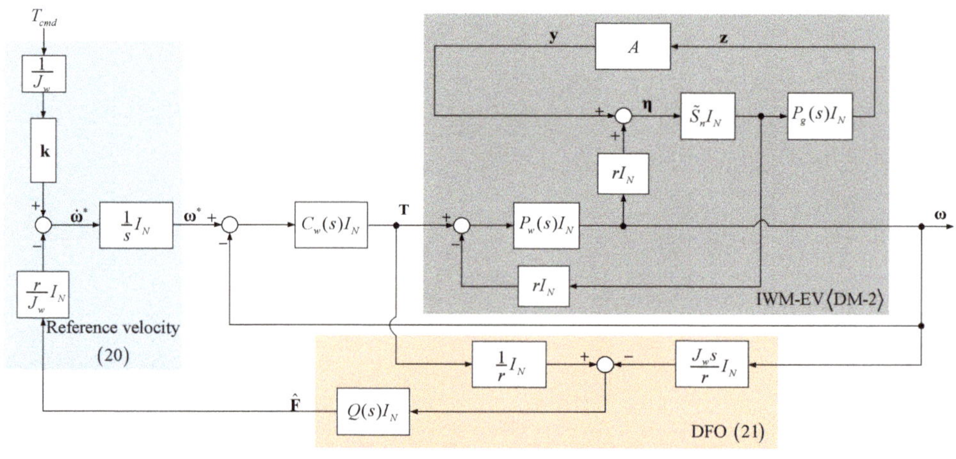

Figure 9. Expression of wheel velocity control system using DM-2.

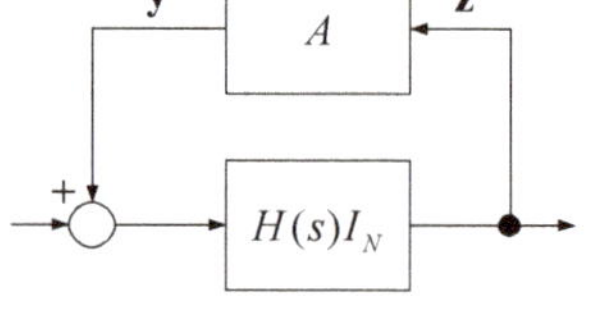

Figure 10. Representation of wheel velocity control system using DM-2 as a multi-agent-system.

The system in Figure 10 is a multi-agent-system of N agent $H(s)$, interconnecting via the matrix A. The system is stable at O if and only if

$$v_i H(s) - 1 \neq 0 \forall\, s \in \mathbb{C}_+, \tag{23}$$

for all eigenvalues v_i of matrix A. We notice that the zero eigenvalues certainly satisfy (23). Following the generalized frequency variable theory (GFV) [35] with respect to Remark 2, (23) can be checked by the following condition.

Stability condition 1: *The wheel velocity control system is stable at the operating point O if the point (-N, 0) is located in the stable domain Ω_+^c defined as $\Omega_+ := \phi(C_+)$, $\Omega_+^c := C\backslash\Omega_+$ where the GFV is $\phi(s) = H^{-1}(s)$.*

5.3. Stability Test

To demonstrate the Stability condition 1, this section considers the PI controller

$$C_w(s) = \frac{K_P s + K_I}{s}. \tag{24}$$

Substitute (24) to (22), the transfer function of the GFV is derived as

$$\phi(s) = \frac{1}{H(s)} = \frac{s^4 + a_3 s^3 + a_2 s^2 + a_1 s + a_0}{b_3 s^3 + b_2 s^2 + b_1 s + b_0}, \tag{25}$$

where the set of $\{a_i, b_i\}$ is expressed in the Appendix B. Define the polynomial

$$pl = s^4 + a_3 s^3 + a_2 s^2 + a_1 s + a_0 - (x + jy)\left(b_3 s^3 + b_2 s^2 + b_1 s + b_0\right) \equiv s^4 + \sum_{i=1}^{4} (p_i + jq_i) s^{4-i}. \tag{26}$$

Following the Theorem 1 in [35], a set of $\{D_k > 0\}$ that defining the stable domain is obtained as

$$\begin{cases} D_1 = \det\left|p_1\right| \\ D_k = \det\begin{bmatrix} F(\mathbf{p}_k) & -F(\mathbf{q}_k)R \\ UF(\mathbf{q}_k) & F(\mathbf{p}_{k-1}) \end{bmatrix} (k = 2, \ldots, 4) \end{cases}, \tag{27}$$

where

$$\mathbf{p}_k = \begin{bmatrix} 1 & p_1 & p_2 & \ldots & p_{2k-1} \end{bmatrix}^T, \mathbf{q}_k = \begin{bmatrix} 0 & 0 & q_1 & q_2 & \ldots & q_{2k-2} \end{bmatrix}^T,$$

$$R = \begin{bmatrix} 0 \\ I \end{bmatrix} \in \mathbb{R}^{k\times(k-1)},\ U = \begin{bmatrix} I & 0 \end{bmatrix} \in \mathbb{R}^{(k-1)\times k},\ F(x) = \begin{bmatrix} x_2 & x_4 & x_6 & \ldots & x_{2k} \\ x_1 & x_3 & x_5 & \ddots & \vdots \\ 0 & x_2 & x_4 & \ddots & x_{k+3} \\ \vdots & \ddots & \ddots & \ddots & x_{k+2} \\ 0 & \ldots & 0 & x_{k-1} & x_{k+1} \end{bmatrix}.$$

Notice: $p_i = q_i = 0$ if $i > 4$.

Based on the GFV theory [35], the Stability condition 1 can be tested as follows.

Stability test 1: *We consider an operating point $O = \{\omega_{o,1}, \omega_{o,2}, \ldots, \omega_{o,N}, v_{xo}\}$. We establish the linearized model DM-2 under the Assumption 1. We calculate the GFV using (25), and then establish the set of $\{D_k\}$ using (26) and (27). The wheel velocity control system in* Figure 9 *is stable at O if*

$$D_k(x = -N, y = 0) > 0\ \forall k \in \{1, 2, 3, 4\}. \tag{28}$$

The complexity level of the Stability test 1 is independent of the number of IWM actuators. At each operating point, the test is performed by a limited set of inequalities (28).

5.4. Discussion: Possibility of Robust Stability Test

By using Assumption 1, the LMC system of electric vehicle is finally modelled as a homogeneous multi-agent system as in Figure 10. However, in actual operation, the friction coefficients of all wheel might be not the same. In addition, the friction coefficients might vary due to the change of road condition. Considering this scenario, we can express the LMC system as a multi-agent system with perturbation as in Figure 11a. The perturbation matrix is $\mathbf{\Delta}(s) = diag\{\Delta_i(s)\}$. We can assume that each local perturbation satisfies $\|\Delta_i(s)\|_\infty \leq \xi$ where the volume ξ of perturbation represents the heterogeneous level between the agents and the uncertainties due to the change of road condition. With respect to the structure of matrix A given by (18), there exists the unitary matrix U that diagonalizes matrix A. Thus, we can transform the system in Figure 11a to the equivalent system in Figure 11b where $\{\lambda_i\}$ is the set of eigenvalues of matrix A. The transfer function from $\overline{w}$ to $\overline{z}$ is diagonalized by N transfer functions

$$\overline{H}_i(s) = \frac{\lambda_i H(s)}{1 - \lambda_i H(s)}. \tag{29}$$

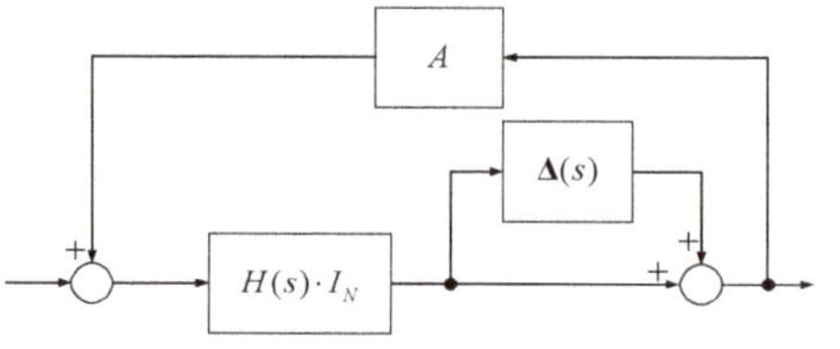

(**a**) Original perturbed system.

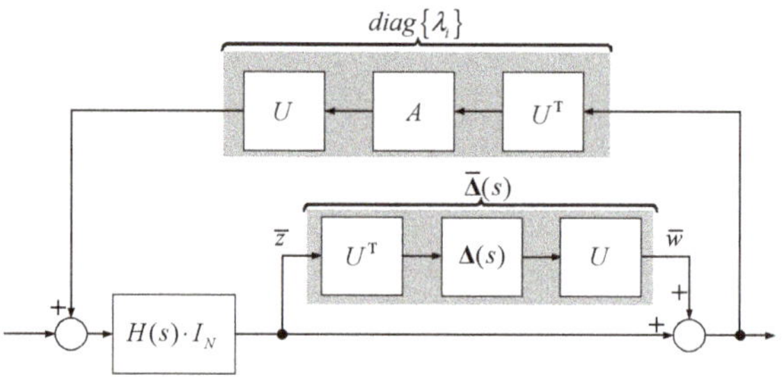

(**b**) Equivalent perturbed system.

Figure 11. Vehicle control system as a multi-agent-system with perturbation.

As $\|\Delta_i(s)\|_\infty \leq \xi$, it follows that $\left\|\mathbf{\Delta}(s)\right\|_\infty \leq \xi$, and therefore $\left\|\bar{\mathbf{\Delta}}(s)\right\|_\infty \leq \xi$ since U is a unitary matrix. Applying the small gain theorem, the perturbed system is robust stable if

$$\left\|\frac{\lambda_i H(s)}{1 - \lambda_i H(s)}\right\|_\infty \leq \frac{1}{\xi} \forall \lambda_i \in \sigma(A). \tag{30}$$

The zero eigenvalues certainly satisfy (30). With respect to Remark 2, the robust stability condition is reduced to

$$\left\|\frac{NH_n(s)}{1 + NH_n(s)}\right\|_\infty \leq \frac{1}{\xi}. \tag{31}$$

In summary, the DM-2 is still applicable to deal with the uncertainties. In this situation, we have to design for each wheel a local controller such that a triad of conditions are satisfied: (i) the stability condition of the nominal system given by (28); (ii) the model matching condition given by the norm inequality $\|\Delta_i(s)\|_\infty \leq \xi$; and (iii) the robust stability condition given by (31).

Another way to deal with robust stability is to utilize "sector bounded nonlinearity" of the tire-force characteristics (Figure 12). We can directly apply the Circle and Popov criteria which was presented in Khalil's famous book on nonlinear control [41]. Fortunately, the special structure of matrix A could be utilized to reduce the complexity of "Circle and Popov criteria". We let this discussion for future study, as the goal of this paper is only to discuss several ways to model the vehicle dynamics such that longitudinal motion control can be designed properly.

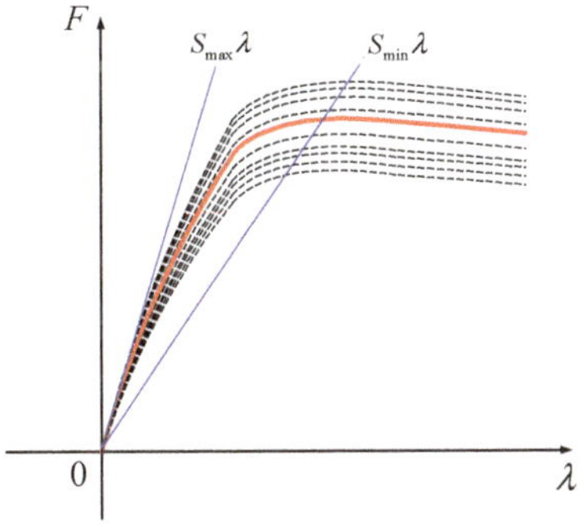

Figure 12. Sector bounded nonlinearities to capture the uncertainty of tire-force characteristics.

6. DM-3: Time-Varying State-Space Model

In Section 4, the DM-1 is established based on the "physical equations" of the IWM-EV. Section 5 proposes the DM-2 expressed by "transfer functions and interconnection matrix". This section shows that the dynamics of IWM-EV can also be captured by a "state-space model". We examine the vehicle motion in the deceleration mode, such that $\max\{r\omega_i, v_x, \varepsilon\} = v_x$. Notice that it is possible to neglect the unknown disturbances of F_{air} and F_ξ in the expression of this model.

6.1. State-Space Model of IWM-EV in the Deceleration Mode

Calculating the derivative of the slip ratio (4) with respect to (1) and (3), we have

$$\dot{\lambda}_i = -\frac{r^2}{J_w v_x} F_i - \frac{\dot{v}_x}{v_x}\lambda_i + \frac{r}{J_w v_x} T_i - \frac{1}{m v_x}\sum_{j=1}^{N} F_j. \tag{32}$$

For the sake of estimation and control design, it is possible to use the first order dynamics model [51] of the driving force represented by

$$\tau_i \dot{F}_i + F_i = S_i \lambda_i, \tag{33}$$

where τ_i is the relaxation time constant and S_i is the driving stiffness which is a parameter for capturing the linear region of the Pacejka model. Here, S_i can be identified from IWM's torque and on-board sensor [52]. Selecting the slip ratio and driving force as state variables, and considering wheel velocities as time-varying parameter, from (30) and (31), the dynamical equation of the ith agent is expressed as follows

$$\dot{\mathbf{x}}_i(t) = \mathbf{A}_{l,i}(t)\mathbf{x}_i(t) + \mathbf{B}_{l,i}\mathbf{u}_i(t) + \sum_{j=1}^{N} \mathbf{A}_{g,ij}(t)\mathbf{x}_j(t), \tag{34}$$

where

$$\mathbf{x}_i(t) = \begin{bmatrix} F_i(t) & \lambda_i(t) \end{bmatrix}^T, \mathbf{u}_i(t) = T_i(t), \tag{35}$$

$$\mathbf{A}_{l,i}(t) = \begin{bmatrix} \frac{-1}{\tau_i} & \frac{S_i}{\tau_i} \\ \frac{-r^2}{J_w v_x} & \frac{-\dot{v}_x}{v_x} \end{bmatrix}, \mathbf{B}_{l,i}(t) = \begin{bmatrix} 0 \\ \frac{r}{J_w v_x} \end{bmatrix}, \mathbf{A}_{g,ij}(t) = \begin{bmatrix} 0 & 0 \\ -\frac{1}{m v_x} & 0 \end{bmatrix}. \tag{36}$$

Define $\mathbf{x}(t) = \begin{bmatrix} \mathbf{x}_1^T(t) & \dots & \mathbf{x}_N^T(t) \end{bmatrix}^T$ and $\mathbf{u}(t) = \begin{bmatrix} \mathbf{u}_1^T(t) & \dots & \mathbf{u}_N^T(t) \end{bmatrix}^T$, the DM-3 is expressed as

$$\dot{\mathbf{x}}(t) = (\mathbf{A}_L(t) + \mathbf{A}_G(t))\mathbf{x}(t) + \mathbf{B}_L(t)\mathbf{u}(t), \tag{37}$$

where $\mathbf{A}_L(t) = diag\{\mathbf{A}_{l,i}(t)\}, \mathbf{B}_L(t) = diag\{\mathbf{B}_{l,i}(t)\}, \mathbf{A}_G(t) = \{\mathbf{A}_{g,ij}(t)\}$. The full block matrix $\mathbf{A}_G(t)$ represents the physical interaction between the local agents.

The acceleration-mode-version of DM-3 when $\max\{r\omega_i, v_x, \varepsilon\} = r\omega_i$ can be found in our recent study [17]. DM-3 can be used for controlling either driving force or slip ratio. To this end, we proposed in [17] a hierarchically decentralized LQR algorithm which was verified by real-time experiments using an EV driven by two IWMs.

6.2. Example of DM-3: Slip Ratio Control in the Deceleration Mode

To demonstrate the practical application of DM-3, we investigate the slip ratio control in the deceleration mode. The following assumptions are considered for the sake of simplicity in presenting our main idea.

Assumption 2: *All the state variables and time-varying parameters of the system (34) or their appropriate estimates are available.*

Assumption 3: *The parameters $\{\tau_i, S_i\}$ are not much different from wheel to wheel, and we can represent them by the nominal parameters $\{\tau_n, S_n\}$.*

Assumption 2 is reasonable since the driving force can be estimated accurately using DFO presented in the previous section. From DFO and the estimated/calculated slip ratio, the driving stiffness can be identified off-line or on-line [52]. On the other hand, the vehicle velocity and wheel slip ratio can be estimated using on-board sensor, such as GPS receiver, accelerometer, and wheel encoders. Several methods for estimating slip ratio can be found in [17,53]. We neglect to focus on state estimation and parameter identification, since it is not the main goal of this paper.

The parameters $\{\tau_i, S_i\}$ depend on the road condition at each wheel and load transfer. This means the parameters $\{\tau_i, S_i\}$ are not always the same in real-time operation. The Assumption 3 is reasonable if the wheels are symmetrically distributed in the vehicle body, and the local road conditions of the wheels are quite uniform. Since we implement the feedback control using DM-3, the parameter difference among the local agents can be compensated.

For controlling the slip ratios such that they follow the reference value λ_i^*, the tracking errors are introduced as the augmented states. In the following, time stamping "t" is eliminated for simplifying the expression. The augmented dynamical model of the ith agent is given as

$$\dot{\bar{\mathbf{x}}}_i = \bar{\mathbf{A}}_1\bar{\mathbf{x}}_i + \bar{\mathbf{B}}_1\mathbf{u}_i + \sum_{j=1}^{N}\bar{\mathbf{A}}_2\bar{\mathbf{x}}_j + \bar{\mathbf{G}}_\lambda\lambda_i^*, \tag{38}$$

where $\bar{\mathbf{x}}_i = \begin{bmatrix} F_i & \lambda_i & e_i \end{bmatrix}^T$, $e_i = \int\left(\lambda_i - \lambda_i^*\right)dt$ and

$$\bar{\mathbf{A}}_1 = \begin{bmatrix} -\frac{1}{\tau_n} & \frac{S_n}{\tau_n} & 0 \\ -\frac{r^2}{J_w v_x} & -\frac{-\dot{v}_x}{v_x} & 0 \\ 0 & 1 & 0 \end{bmatrix}, \bar{\mathbf{B}}_1 = \begin{bmatrix} 0 \\ \frac{r}{J_w v_x} \\ 0 \end{bmatrix}, \bar{\mathbf{A}}_2 = \begin{bmatrix} 0 & 0 & 0 \\ -\frac{1}{m v_x} & 0 & 0 \\ 0 & 0 & 0 \end{bmatrix}, \bar{\mathbf{G}}_\lambda = \begin{bmatrix} 0 \\ 0 \\ -1 \end{bmatrix}.$$

State-space modelling of the total system is expressed as follows using Kronecker product $\otimes$:

$$\dot{\bar{\mathbf{x}}} = \bar{\mathbf{A}}\bar{\mathbf{x}} + \bar{\mathbf{B}}\mathbf{u} + \bar{\mathbf{G}}\boldsymbol{\lambda}^*, \tag{39}$$

where $\bar{\mathbf{x}} = \begin{bmatrix} \bar{\mathbf{x}}_1^T & \dots & \bar{\mathbf{x}}_N^T \end{bmatrix}^T, \boldsymbol{\lambda}^* = \begin{bmatrix} \lambda_1^* & \dots & \lambda_N^* \end{bmatrix}^T$ and

$$\bar{\mathbf{A}} = \mathbf{I}_N \otimes \bar{\mathbf{A}}_1 + \boldsymbol{\Gamma}_N \otimes \bar{\mathbf{A}}_2, \boldsymbol{\Gamma}_N = 1_N 1_N^T, \bar{\mathbf{B}} = \mathbf{I}_N \otimes \bar{\mathbf{B}}_1, \bar{\mathbf{G}} = \mathbf{I}_N \otimes \bar{\mathbf{G}}_\lambda.$$

The LQR controller of the augmented system (39) can be obtained by minimizing the quadratic cost function

$$\Im = \int_{t_o}^{t_f} \bar{\mathbf{x}}^T \mathbf{Q} \bar{\mathbf{x}} dt + \int_{t_o}^{t_f} \mathbf{u}^T \mathbf{R} \mathbf{u} dt + \mathbf{x}_f^T \mathbf{H} \mathbf{x}_f, \tag{40}$$

where t_0 is the initial time, t_f is the finished time. $\mathbf{Q}$ and $\mathbf{S}$ are symmetric positive semidefinite matrices, and $\mathbf{R}$ is strictly symmetric positive definite matrix. $\mathbf{Q}$, $\mathbf{H}$, and $\mathbf{R}$ are used to penalize the transient state deviation, the final state, and the control effort, respectively. By solving the Riccati equation

$$\dot{\mathbf{P}} + \mathbf{P}\bar{\mathbf{A}} + \bar{\mathbf{A}}^T \mathbf{P} - \mathbf{P}\bar{\mathbf{B}}\mathbf{R}^{-1}\bar{\mathbf{B}}^T \mathbf{P} + \mathbf{Q} = 0, \tag{41}$$

that satisfies the boundary condition

$$\mathbf{P}_f \triangleq \mathbf{P}\left(t_f\right) = \mathbf{H}, \tag{42}$$

the control gain is obtained as

$$\mathbf{K} = -\mathbf{R}^{-1}\bar{\mathbf{B}}^T \mathbf{P}, \tag{43}$$

Equation (41) with condition (42) can be solved using the Hamiltonian established from the system model [54]. To reduce the computational effort, especially when N is a big number, we recently proposed in [17] that the feedback gain can be calculated as

$$\mathbf{K} = \mathbf{I}_N \otimes \mathbf{K}_1 + \mathbf{\Gamma}_N \otimes \mathbf{K}_{g1} + \mathbf{\Psi}_N \otimes \mathbf{K}_{g2}, \tag{44}$$

by appropriate choices of the weighting matrices. As can be seen in Figure 13, the proposed LQR controller has the hierarchical structure. In the lower layer, the local feedback gain $\mathbf{K}_1$ is used to improve the local control performance of slip ratio. The upper-layer has two term, $\mathbf{K}_{g1}$ and $\mathbf{K}_{g2}$. $\mathbf{K}_{g1}$ is to deal with the physical interconnection represented by $\mathbf{\Gamma}_N$. On the other hand, $\mathbf{K}_{g2}$ is introduced to improve some other global performances, such as the consensus of slip ratios. Since $\mathbf{K}_1$ is obtained by solving the Riccati equation of a local system, the computation cost is independent of the number of IWM actuators. See Appendix C for the details of the hierarchical LQR algorithm.

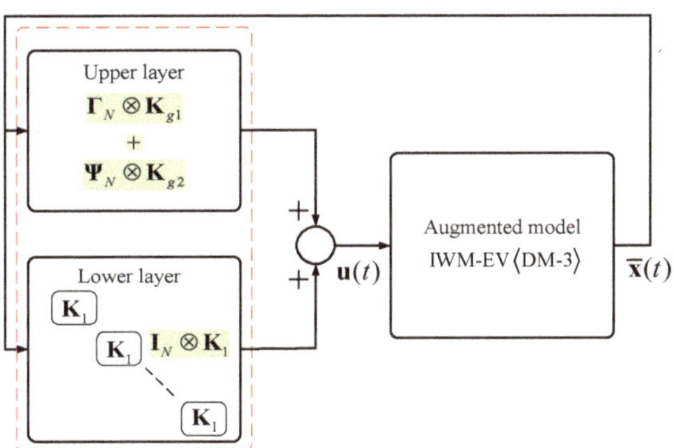

Figure 13. Structure of the hierarchical decentralized LQR using DM-3.

Notice 1: *Similar to Assumption 1, Assumption 3 is just to limit our discussion to homogenous multi-agent systems. Motivated by a recent work of Nguyen and Hara [33], we will develop the H-LQR version for heterogeneous multi-agent systems with physical interaction in our future study.*

7. Evaluation of the Proposed Design Models by Carsim/Matlab Co-Simulator

Using the Carsim/Matlab co-simulator introduced in Section 3, this section is to demonstrate the effectiveness of the proposed design models.

7.1. Evaluation of Passivity Based Anti-Slip Control Using DM-1

To evaluate the proposed anti-slip control strategy, a test is conducted as follows. As shown in Figure 14a, the EV accelerates from the initial position on the high friction surface (μ = 0.8). It enters the low friction surface (μ = 0.2) at about 5.7 s. The total driving command T_{cmd} is expressed in Figure 14b. The distribution vector **k** is updated to minimize the total workload of all wheels [55].

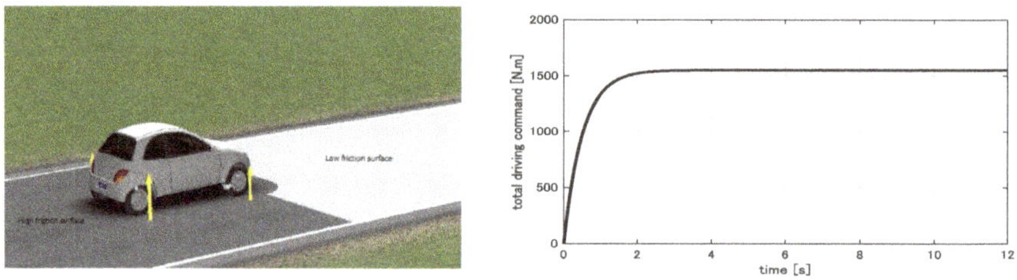

(**a**) Road surface of the test (**b**) Total driving command

Figure 14. Simulation setting for evaluating passivity-based anti-slip control using DM-1.

Case 1 (Without anti-slip control: Figure 15): The motor torque of each wheel is only distributed from the total driving command, i.e., $T_i = k_i T_{cmd}$ where k_i is the distribution ratio of the ith wheel. Consequently, the slip occurs. Especially, the rear-wheel velocities increase considerably as the vehicle enters the low friction surface.

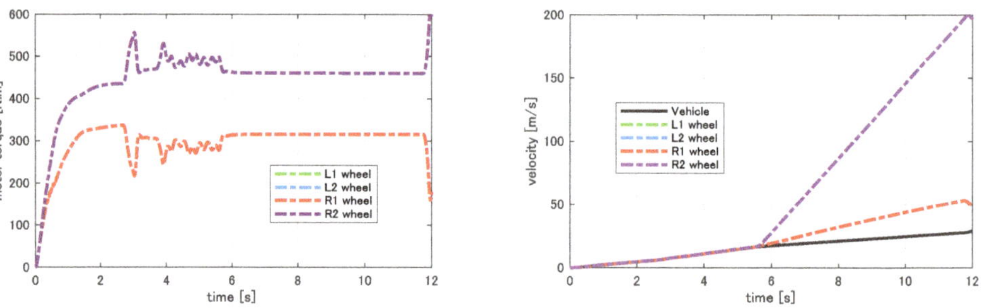

(**a**) Motor torques (**b**) Velocities

Figure 15. Simulation results of Case 1: without anti-slip control.

Case 2 (With passivity-based anti-slip control: Figure 16): The anti-slip control law is implemented based on passivity and the DM-1. By trial-and-error, we select the control gains K_a = 100 and K_ω = 0.0001. Notice that a small positive value of K_ω is enough to assure the strict passivity of the control system. We should not select a big value of K_ω which certainly degrades the acceleration performance of the vehicle. In contrast to Case 1, the motor torques are generated properly on different road surfaces. Thus, the wheel velocities are always closed to the vehicle velocity.

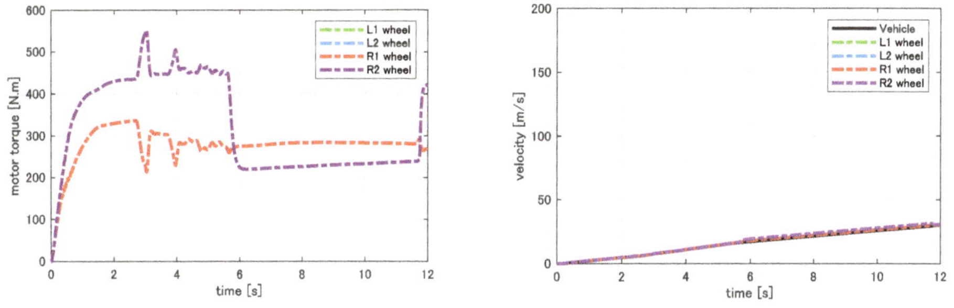

(**a**) Motor torques (**b**) Velocities

Figure 16. Simulation results of Case 2: with passivity-based anti-slip control.

The above test cases clarify the merit of the proposed anti-slip strategy using DM-1. Although the physical system of the EV is quite complex and nonlinear, it can be modelled hierarchically with the aggregation and distribution. Based on this model, we can analyze the passivity of the EV dynamics, and design the anti-slip control law that rigorously guarantees the stability of the overall system. Thanks to passivity theory, we design the system with less mathematical and computational effort. Using the DM-1, the energy storage function of the EV is shown to be $S = \frac{1}{2}mv_x^2 + \sum_{i=1}^{N}\left(\frac{1}{2}J_w\omega_i^2\right)$. Given a desired velocity pattern of v_x, the energy consumption of the EV can be reduced if the longitudinal velocities of the wheels $\{r\omega_i\}$ are closed to v_x. Therefore, the proposed anti-slip control is not only for improving the safety, but also extending the use of the energy source for the EVs.

7.2. Evaluation of GFV Based Stability Analysis Using DM-2

To demonstrate the GFV based stability analysis using DM-2, we conducted wheel velocity control using the system in Figure 9. The EV operates on the high friction surface with μ = 0.85, as shown in Figure 17a. The total driving command T_{cmd} is expressed in Figure 17b. Similar to the previous sub-section, the distribution vector **k** is updated to minimize the total workload of all wheels. The driving force observer is designed with the time constant τ_f = 0.033 [s]. This paper examines the operating points with v_{xo} = 10 (m/s). Considering the high friction surface, it is reasonable to assume that a small slip ratio of 0.05 is maintained constantly when the vehicle velocity changes between $v_{xo} - \Delta v$ and $v_{xo} + \Delta v$ where Δv = 2 (m/s). We select the wheel velocity controller with $\{K_p$ =52.8; K_i = 528.0$\}$. It is transparent that this controller stabilizes the local transfer function $H(s)$ about the operating points. Applying the Stability test 1, the stability regions at the operating points of v_{xo}, $v_{xo} - \Delta v$ and $v_{xo} + \Delta v$ are plotted as the shaded regions in Figure 18a–c, respectively. It is transparent that the points (−4, 0) and (0,0) are always placed in such shaded regions. This means the wheel velocity control system in Figure 9 is stable at the aforementioned operating points.

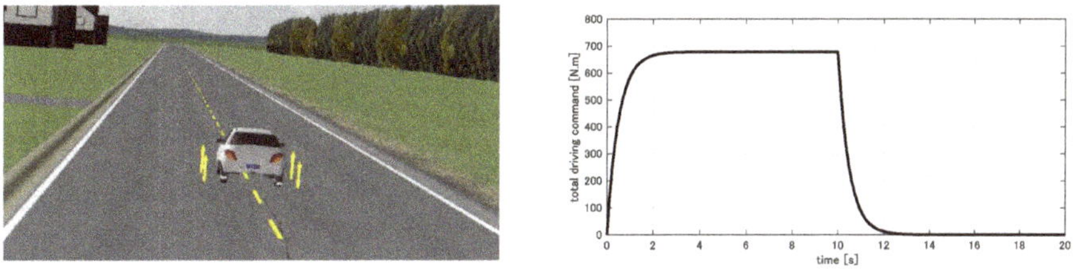

(**a**) Road surface of the test (**b**) Total driving command

Figure 17. Simulation setting for evaluating wheel velocity control using DM-2.

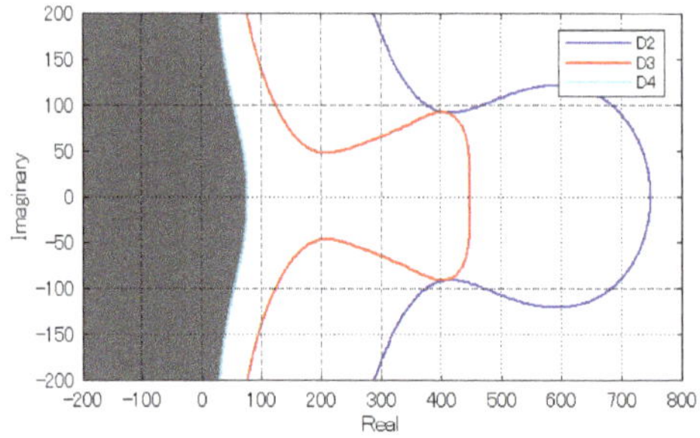

(**a**) Stable domain plotted at v_{xo}

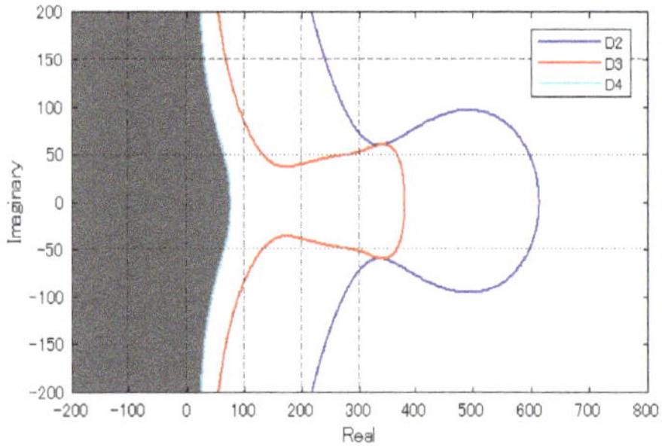

(**b**) Stable domain plotted at $v_{xo} - \Delta v$

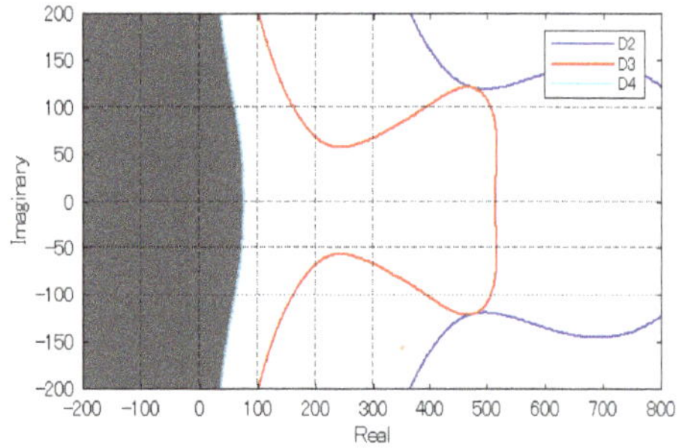

(**c**) Stable domain plotted at $v_{xo} + \Delta v$

Figure 18. Stable domain given by the generalized frequency variable (GFV) (shaded region). Notice: the curve $D_1(x,y) = 0$ is a vertical line. It is not plotted since it exceeds the range of the figures.

The above example of stability test unveils a merit of GFV approach using DM-2. Thanks to the rank-1 physical interaction expressed by DM-2 and the GFV theory, the complexity of the stability test is independent of the number of IWM actuators. We can repeat the above test to assure the stability of the control system at different operating points with the vehicle velocity from zero to v_{xo}. This statement is confirmed again by the simulation results summarized in Figure 19. In this simulation, the vehicle is accelerated until 10 s. From 10 s, the total driving command becomes zero, and the vehicle is gradually decelerated due to friction.

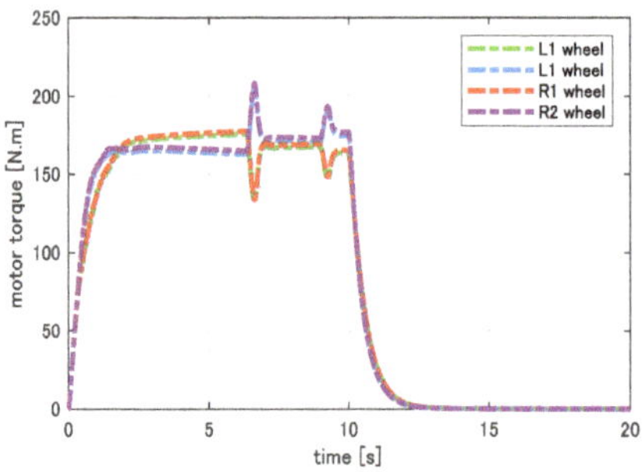

(**a**) Motor torques

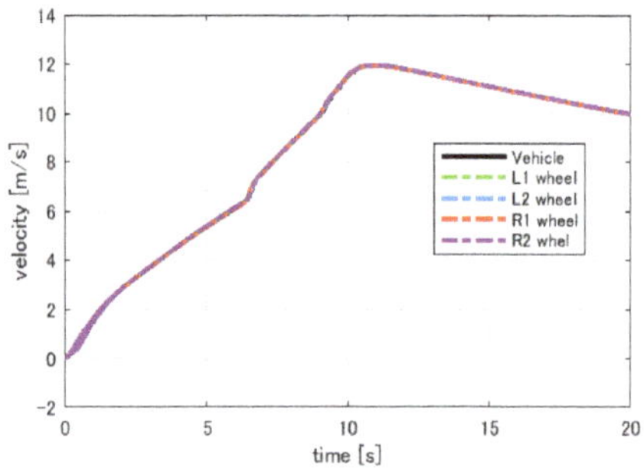

(**b**) Velocities

Figure 19. Simulation results of wheel velocity control using DM-2.

In Appendix D, we present several results of the real-time experiments of the wheel velocity control system in Figure 9.

7.3. Evaluation of LQR-Based Slip-Ratio Control Using DM-3

Using the DM-3, we proposed the LQR algorithm for controlling slip-ratio of IWM-EV [17]. This algorithm is verified in the acceleration mode. We also compared the performance of the LQR algorithm with the others, such as sliding mode algorithm and model-following algorithm in [17]. This comparison presented the merits of the proposed LQR: (i) As a hierarchically decentralized algorithm, its computational cost is considerably reduced in comparison with the conventional centralized LQR. (ii) Unlike the completely decentralized control algorithms, the physical interaction between the local wheels are properly treated. This gives the degree of freedom to improve the consensus of local wheels' slip ratios. In addition, this can reduce the vibration of slip ratios introduced by the non-flat surfaces. (iii) The optimal control gains can be updated at every control period by the measurements of vehicle motion (vehicle velocity, wheel velocities, driving forces).

For the sake of paper space, we neglect to repeat the comparison between the control approaches in this study. Instead, we only demonstrate the performance of the proposed algorithm in deceleration mode (Figure 20a). A total driving command is given as in Figure 20b. This command is distributed to the local wheels to minimize the total workload. The vehicle runs on high friction surface ($\mu = 0.8$) from the beginning until 12 s. From 12 s, the vehicle starts to decelerate on the low friction surface ($\mu = 0.2$). The proposed LQR algorithm is applied to maintain a slip ratio of $\lambda^* = -0.1$ during the deceleration period. To this end, we design the LQR algorithm with

$$\mathbf{Q}_1 = diag\left\{1\times10^{-4}, 2\times10^{2}, 4\times10^{3}\right\},\ \mathbf{R}_1 = 4\times10^{-4},\ \mathbf{R}_{g1} = 1\times10^{-1},\ \mathbf{R}_{g2} = 1.$$

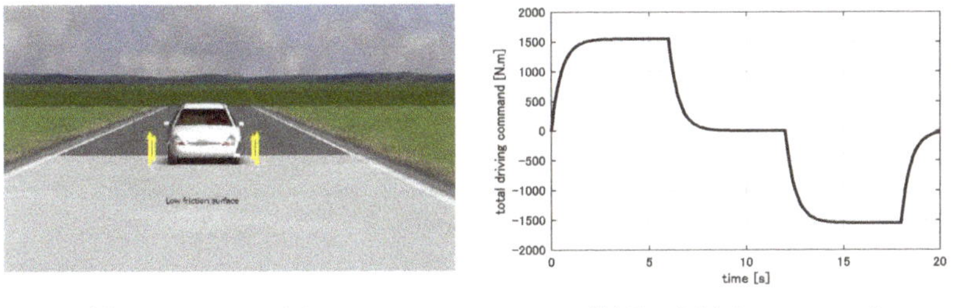

(**a**) Road surface of the test (**b**) Total driving command

Figure 20. Simulation setting for evaluating LQR-based slip-ratio control using DM-3.

To compensate the influence of load transfer between the front and rear wheels during the deceleration period, we select the positive semidefinite matrix $\mathbf{\Psi}_N$ as follows with $\varphi_{front} = \varphi_{rear} = 1$.

$$\mathbf{\Psi}_4^{front-rear} = \begin{bmatrix} 1 & -1 \\ -1 & 1 \end{bmatrix} \otimes \begin{bmatrix} \varphi_{front} & 0 \\ 0 & \varphi_{rear} \end{bmatrix}.$$

Simulation results of two test cases are summarized as follows.

Case 1 (without slip-ratio control: Figure 21a): The total driving command T_{cmd} in Figure 18b is directly distributed to each wheel. Consequently, the vehicle suffers severe slip during the deceleration periods. The wheel velocities are reduced too fast in comparison with the vehicle velocity. Especially, the wheels reverse their rotation after 14 s.

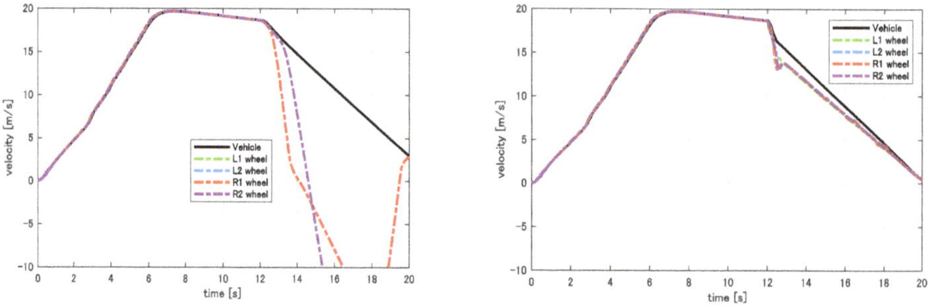

(**a**) Case 1: without slip-ratio control (**b**) Case 2: LQR-based slip-ratio control

Figure 21. Simulation results of slip-ratio control in deceleration mode.

Case 2 (with LQR-based slip-ratio control: Figure 21b): The motor torque of each wheel is the summary of two signals: the command distributed from T_{cmd} expressed in Figure 18b; and the output of the LQR-based slip-ratio controller. Thanks to the proposed method, the wheel velocities are slightly smaller than the vehicle velocity during the deceleration period. This means the safety and controllability of the vehicle are successfully maintained. In addition, the efficiency of electric energy utilization is also improved by maintaining a small slip ratio.

8. Conclusions

This paper suggests that IWM-EV should be treated as a special type of multi-agent-system with physical interaction between the local agents or the locally controlled wheels. Three ways to model the IWM-EV are proposed. Each is shown to be suitable for certain control objectives and certain system design approach. From a multi-agent system point of view, the proposed design models, namely DM-1,2,3, contribute some new understandings on IWM-EV dynamics.

Firstly, the DM-1 points out that the IWM-EV naturally has the hierarchical structure: the vehicle body dynamics in the upper-layer, and the local wheel dynamics in the lower-layer. The two layers are connected via the aggregation and distribution channels. Although DM-1 is nonlinear and complex, it is quite helpful for applying passivity theory to clarify the energy storage function of IWM-EV. Passivity theory, therefore, is a possible candidate for improving the safety and optimizing the use of electricity in IWM-EVs.

Secondly, the DM-2 shows another interesting characteristic of IWM-EV. This is the rank-1 physical interaction. Based on DM-2, IWM-EV motion control system can be modelled by a local transfer function $H(s)$ and the interaction matrix. This model is quite useful for applying generalized frequency variable to analyze the stability of the control system. The complexity of stability analysis is independent of the number of IWM actuators.

Last but not least, the DM-3 is a state-space representation of IWM-EV. It is a suitable choice for applying optimal control strategies, such as LQR control of the slip-ratios and driving forces. In addition to attaining the local control objective at each wheel, DM-3 can be utilized to address the additional global control objectives.

This paper demonstrates several control design examples using the proposed DMs. The effectiveness of the DMs and their design approaches are evaluated using Carsim/Matlab co-simulator. For the sake of simplicity, DM-2 and DM-3 are established under the assumptions that the local agents are homogeneous (Assumptions 1 and 3). To overcome this real limitation, in future study we will employ the multi-agent theories to deal with the heterogeneous of the local wheels.

This paper only focuses on the dynamical models for vehicle motion control. Therefore, we do not consider the models of other parts of the EV systems, such as the motor model, the inverter model, and the energy source model. Motivated by several recent works on battery modelling [37–40], we will also develop the full model of the EVs using Matlab-Carsim co-simulator. Such simulation models will be utilized to integrate motion control, motor control, and optimal energy management for IWM-EVs.

Author Contributions: B.-M.N. designed the study and wrote the manuscript. H.V.N. contributed to the simulation model. M.T.-C. and M.K. provided critical feedback for revising the manuscript. All the authors discussed the results and contributed to the final version of the manuscript. All authors have read and agreed to the published version of the manuscript.

Funding: This study is supported in part by the Toyota Technological Institute (Aichi, Japan).

Acknowledgments: We would like to thank Shinji Hara (Tokyo Institute of Technology) who gave kind advice in the theoretical background of this paper. We also thank Hori-Fujimoto Lab (The University of Tokyo) for the experimental electric vehicle presented in Appendix A.

Conflicts of Interest: The authors declare no conflict of interest.

Appendix A

Table A1 summarizes some main parameters of the simulation model which is based on the electric vehicle prototype shown in Figure A1.

Table A1. Specification of the vehicle model.

Vehicle	
Vehicle mass	$m = 1080$ (kg)
Radius of wheel	$r = 0.285$ (m)
Wheel moment of inertia	$J_w = 1.25$ (kg.m^2)
Cross-sectional area of vehicle in the air	$A_F = 2.37$ (m^2)
Drag coefficient	$C_d = 0.35$
Distance between front axle and rear axle	$L = 2.55$ (m)
Distance from center of gravity to front axle	$L_f = 1.45$ (m)
Distance from center of gravity to rear axle	$L_r = 1.10$ (m)
Height of the center of gravity	$H_g = 0.356$ (m)
Number of wheels	$N = 4$

Figure A1. Electric vehicle for establishing the simulation model.

Appendix B

The GFV $\phi(s) = \frac{s^4+a_3s^3+a_2s^2+a_1s+a_0}{b_3s^3+b_2s^2+b_1s+b_0}$ is calculated with

$$a_3 = \frac{mJ_w\left(J_w + \tau_f K_P + \widetilde{S}_n r^2 \tau_f\right)}{m\tau_f J_w^2}, a_2 = \frac{mJ_w\left(K_P + \tau_f K_I + \widetilde{S}_n r^2\right)}{m\tau_f J_w^2}, a_1 = \frac{m\left(J_w K_I + \widetilde{S}_n r^2 K_P\right)}{m\tau_f J_w^2}, a_0 = \frac{m\widetilde{S}_n r^2 K_I}{m\tau_f J_w^2},$$

$$b_3 = \frac{\widetilde{S}_n \tau_f J_w^2}{m\tau_f J_w^2}, b_2 = \frac{\widetilde{S}_n J_w\left(J_w + \tau_f K_P\right)}{m\tau_f J_w^2}, b_1 = \frac{\widetilde{S}_n J_w\left(K_P + \tau_f K_I\right)}{m\tau_f J_w^2}, a_0 = \frac{\widetilde{S}_n J_w K_I}{m\tau_f J_w^2},$$

Appendix C

To get an optimal feedback gain of the form (44), the weighting matrices are selected as

$$\mathbf{Q} = \mathbf{I}_N \otimes \mathbf{Q}_1 + \mathbf{\Gamma}_N \otimes \mathbf{Q}_{g1} + \mathbf{\Psi}_N \otimes \mathbf{Q}_{g2},\ \mathbf{R}^{-1} = \mathbf{I}_N \otimes \mathbf{R}_1^{-1} + \mathbf{\Gamma}_N \otimes \mathbf{R}_{g1}^{-1} + \mathbf{\Psi}_N \otimes \mathbf{R}_{g2}^{-1},\ \mathbf{H} = \mathbf{I}_N \otimes \mathbf{H}_1, \quad \text{(A1)}$$

where the matrices ($\mathbf{Q}_1$, $\mathbf{R}_1$, $\mathbf{H}_1$) are set for the lower layer, ($\mathbf{Q}_{g1}$, $\mathbf{R}_{g1}$) and ($\mathbf{Q}_{g2}$, $\mathbf{R}_{g2}$) are set for the upper layer. The control gains are obtained by the three-step procedure as follows

Step 1 (Lower-layer):

Select the weighting matrices ($\mathbf{Q}_1$, $\mathbf{R}_1$, $\mathbf{S}_1$) such that $\mathbf{Q}_1$ and $\mathbf{S}_1$ are positive semidefinite, and $\mathbf{R}_1$ is strictly positive definite. Then, solve the differential Riccati equation

$$\dot{\mathbf{P}}_1 + \mathbf{P}_1\bar{\mathbf{A}}_1 + \bar{\mathbf{A}}_1^T\mathbf{P}_1 - \mathbf{P}_1\bar{\mathbf{B}}_1\mathbf{R}_1^{-1}\bar{\mathbf{B}}_1^T\mathbf{P}_1 + \mathbf{Q}_1 = 0, \quad \text{(A2)}$$

with the boundary condition $\mathbf{P}_1\left(t_f\right) = \mathbf{P}_{1f} = \mathbf{H}_1$.

Step 2 (Upper-layer):

Select the positive semidefinite matrix $\mathbf{\Psi}_N$ and the positive definite matrices $\mathbf{R}_{g1}$ and $\mathbf{R}_{g2}$ to obtain the weighting matrix $\mathbf{R}$ as in (45). Then, the two matrices $\mathbf{Q}_{g1}$ and $\mathbf{Q}_{g2}$ are set as following such that $\mathbf{Q}_{g1}$ is positive semidefinite:

$$\mathbf{Q}_{g1} = \mathbf{P}_1\bar{\mathbf{B}}_1\mathbf{R}_{g1}^{-1}\bar{\mathbf{B}}_1^T\mathbf{P}_1 - \mathbf{P}_1\bar{\mathbf{A}}_2 - \bar{\mathbf{A}}_2^T\mathbf{P}_1,\ \mathbf{Q}_{g2} = \mathbf{P}_1\bar{\mathbf{B}}_1\mathbf{R}_{g2}^{-1}\bar{\mathbf{B}}_1^T\mathbf{P}_1. \quad \text{(A3)}$$

Step 3 (Feedback gain):

Finally, the feedback gains are calculated as

$$\mathbf{K}_1 = -\mathbf{R}_1^{-1}\bar{\mathbf{B}}_1^T\mathbf{P}_1,\ \mathbf{K}_{g1} = -\mathbf{R}_{g1}^{-1}\bar{\mathbf{B}}_1^T\mathbf{P}_1,\ \mathbf{K}_{g2} = -\mathbf{R}_{g2}^{-1}\bar{\mathbf{B}}_1^T\mathbf{P}_1. \quad \text{(A4)}$$

The above procedure is a kind of partial inverse LQR method that provides the hierarchical control structure in Figure 13, since the choices of $\mathbf{Q}_{g1}$ and $\mathbf{Q}_{g2}$ are not completely free but restricted as in (45). The optimality of the feedback gains is shown by Proposition 1 in [17].

Appendix D

In this paper, we present the general design models for the electric vehicle driven by N IWM actuators. All the simulations are performed by using a 4-wheel-model. We actually implemented several motion control approaches to a real electric vehicle, such as lateral motion control in [3], and longitudinal motion control in [17]. It is the electric vehicle COMS shown in Figure A2. It has four wheels but only two rear wheels are driven by IWMs of IPMSM type with the maximum power of 2 kW. The motor drives and inverters are provided by Myway. The vehicle is equipped with RTK-GPS receiver produced by Hemisphere, the gyroscopes, and the accelerometers. Through the fusion of GPS receiver, gyroscopes, and accelerometers, the vehicle velocity can be obtained. The encoders are installed at the wheels to measure the rotational speed. The heart of the system is a RT-Linux computer which processes the control algorithm and stores the experimental data.

In Figure A3, we demonstrate an experimental result of the wheel velocity control system in Figure 9. This was an autonomous driving test in which the vehicle is required to accelerate to the velocity of 5 m/s and then, maintaining this velocity constantly. To this end, the driving command is given by the upper-layer motion control layer. To design the PI controller for controlling the wheel velocity, we apply the Stability condition 8 in case $N = 2$. Thanks to this condition, we can stabilize the overall system and smoothly attain both the global motion and local motions of the vehicle.

Figure A2. Electric vehicle COMS (Hori-Fujimoto Lab, Kashiwa Campus, The University of Tokyo).

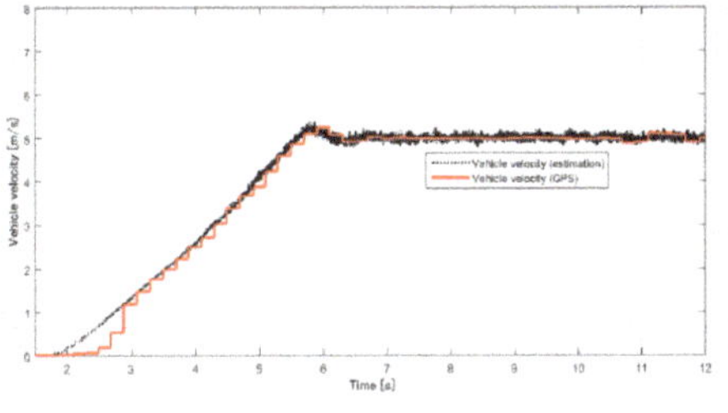

(**a**) Global motion: vehicle velocity

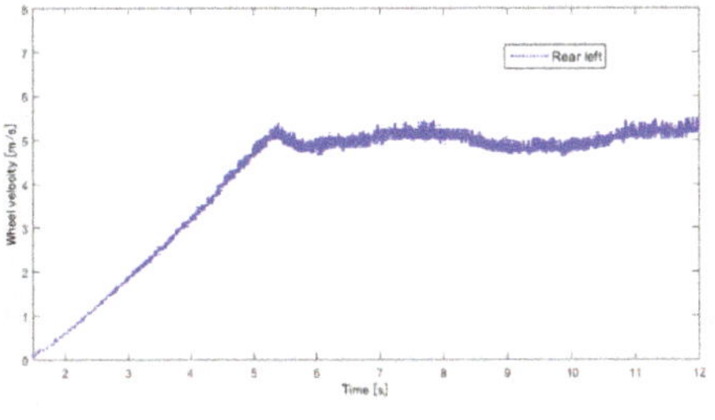

(**b**) Local motion: rear-left wheel.

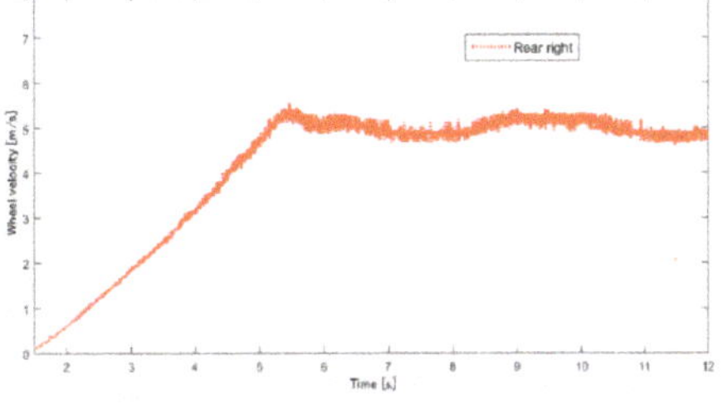

(**c**) Local motion: rear-right wheel.

Figure A3. Experiments of wheel velocity control using electric vehicle COMS.

References

1. Hori, Y. Future vehicle driven by electricity and control-Research on 4 wheel motored UOT March II. *IEEE Trans. Ind. Electron.* **2004**, *51*, 954–962. [CrossRef]
2. Geng, C.; Mostefai, L.; Denaï, M.A.; Hori, Y. Direct yaw-moment control of an in-wheel-motored electric vehicle based on body slip angle fuzzy observer. *IEEE Trans. Ind. Electron.* **2009**, *56*, 1411–1419. [CrossRef]
3. Nguyen, B.-M.; Wang, Y.; Fujimoto, H.; Hori, Y. Lateral stability control of electric vehicle based on disturbance accommodating Kalman filter using the integration of single antenna GPS receiver and yaw rate sensor. *J. Electr. Eng. Technol.* **2013**, *8*, 899–910. [CrossRef]

4. Tota, A.; Lenzo, B.; Lu, Q.; Sorniotti, A.; Gruber, P.; Fallah, S.; Velardocchia, M.; Galvagno, E.; De Smet, J. On the experimental analysis of integral sliding modes for yaw rate and sideslip control of an electric vehicle with multiple motors. *Int. J. Automot. Technol.* **2018**, *19*, 811–823. [CrossRef]
5. Nam, K.; Fujimoto, H.; Hori, Y. Lateral stability control of in-wheel-motor-driven electric vehicles based on sideslip angle estimation using lateral tire force sensors. *IEEE Trans. Veh. Technol.* **2012**, *61*, 1972–1985. [CrossRef]
6. Kawashima, K.; Uchida, T.; Hori, Y. Rolling stability control based on electronic stability program for in-wheel-motor electric vehicle. *World Electr. Veh. J.* **2009**, *3*, 34–41. [CrossRef]
7. Fujimoto, H.; Sato, S. Pitching control method based on quick torque response for electric vehicle. In Proceedings of the 2010 International Power Electronics Conference-ECCE ASIA, Sapporo, Japan, 21–24 June 2010; Institute of Electrical and Electronics Engineers (IEEE): New York, NY, USA, 2010; pp. 801–806.
8. Kanchwala, H.; Wideberg, J. Pitch reduction and traction enhancement of an EV by real-time brake biasing and in-wheel motor torque control. *Int. J. Veh. Syst. Model. Test.* **2016**, *11*, 165. [CrossRef]
9. Yu, Y.; Xiong, L.; Yu, Z.; Yang, X.; Hou, Y.; Leng, B. Model-based pitch control for distributed drive electric vehicle. *SAE Tech. Pap. Ser.* **2019**. [CrossRef]
10. He, H.; Peng, J.; Xiong, R.; Fan, H. An acceleration slip regulation strategy for four-wheel drive electric vehicles based on sliding mode control. *Energies* **2014**, *7*, 3748–3763. [CrossRef]
11. Nam, K.; Hori, Y.; Lee, C.-Y. Wheel slip control for improving traction-ability and energy efficiency of a personal electric vehicle. *Energies* **2015**, *8*, 6820–6840. [CrossRef]
12. Savitski, D.; Ivanov, V.; Augsburg, K.; Emmei, T.; Fuse, H.; Fujimoto, H.; Fridman, L.M. Wheel slip control for the electric vehicle with in-wheel motors: Variable structure and sliding mode methods. *IEEE Trans. Ind. Electron.* **2020**, *67*, 8535–8544. [CrossRef]
13. He, L.; Ye, W.; He, Z.; Song, K.; Shi, Q. A combining sliding mode control approach for electric motor anti-lock braking system of battery electric vehicle. *Control. Eng. Pr.* **2020**, *102*, 104520. [CrossRef]
14. Hori, Y.; Toyoda, Y.; Tsuruoka, Y. Traction control of electric vehicle: Basic experimental results using the test EV "UOT electric march". *IEEE Trans. Ind. Appl.* **1998**, *34*, 1131–1138. [CrossRef]
15. Ivanov, V.; Savitski, D.; Augsburg, K.; Barber, P.; Knauder, B.; Zehetner, J. Wheel slip control for all-wheel drive electric vehicle with compensation of road disturbances. *J. Terramechanics* **2015**, *61*, 1–10. [CrossRef]
16. De Pinto, S.; Chatzikomis, C.; Sorniotti, A.; Mantriota, G. Comparison of traction controllers for electric vehicles with on-board drivetrains. *IEEE Trans. Veh. Technol.* **2017**, *66*, 6715–6727. [CrossRef]
17. Nguyen, B.-M.; Hara, S.; Fujimoto, H.; Hori, Y. Slip control for IWM vehicles based on hierarchical LQR. *Control. Eng. Pr.* **2019**, *93*, 104179. [CrossRef]
18. Li, L.; Kodama, S.; Hori, Y. Design of anti-slip controller for an electric vehicle with an adhesion status analyzer based on the ev simulator. *Asian J. Control.* **2008**, *8*, 261–267. [CrossRef]
19. Fujimoto, H.; Saito, T.; Noguchi, T. Motion stabilization control of electric vehicle under snowy conditions based on yaw-moment observer. In Proceedings of the 8th IEEE International Workshop on Advanced Motion Control, Kawasaki, Japan, 28 March 2004; AMC '04. Institute of Electrical and Electronics Engineers (IEEE): New York, NY, USA, 2004.
20. Yin, D.; Oh, S.; Hori, Y. A novel traction control for ev based on maximum transmissible torque estimation. *IEEE Trans. Ind. Electron.* **2009**, *56*, 2086–2094. [CrossRef]
21. Chiang, W.-P.; Yin, D.; Omae, M.; Shimizu, H. Integrated slip-based torque control of antilock braking system for in-wheel motor electric vehicle. *IEEJ J. Ind. Appl.* **2014**, *3*, 318–327. [CrossRef]
22. Suzuki, T.; Fujimoto, H. Slip ratio estimation and regenerative brake control without detection of vehicle velocity and acceleration for electric vehicle at urgent brake-turning. In Proceedings of the 2010 11th IEEE International Workshop on Advanced Motion Control (AMC), Nagaoka, Japan, 21–24 March 2010; pp. 273–278. [CrossRef]
23. Amada, J.; Fujimoto, H. Torque based direct driving force control method with driving stiffness estimation for electric vehicle with in-wheel motor. In Proceedings of the IECON 2012–38th Annual Conference on IEEE Industrial Electronics Society, Montreal, QC, Canada, 25–28 October 2012; Institute of Electrical and Electronics Engineers (IEEE): New York, NY, USA, 2012; pp. 4904–4909.
24. Maeda, K.; Fujimoto, H.; Hori, Y. Four-wheel driving-force distribution method for instantaneous or split slippery roads for electric vehicle. *Automatika* **2013**, *54*, 103–113. [CrossRef]

25. Maeda, K.; Fujimoto, H.; Hori, Y. Driving force control of electric vehicles with estimation of slip ratio limitation considering tire side slip. *Trans. Soc. Instrum. Control. Eng.* **2014**, *50*, 259–265. [CrossRef]
26. Fuse, H.; Fujimoto, H. Fundamental study on driving force control method for independent-four-wheel-drive electric vehicle considering tire slip angle. In Proceedings of the IECON 2018-44th Annual Conference of the IEEE Industrial Electronics Society, Washington, DC, USA, 20–23 October 2018; Institute of Electrical and Electronics Engineers (IEEE): New York, NY, USA, 2018; pp. 2062–2067.
27. Boulon, L.; Hissel, D.; Bouscayrol, A.; Pape, O.; Péra, M.-C. Simulation model of a military HEV with a highly redundant architecture. *IEEE Trans. Veh. Technol.* **2010**, *59*, 2654–2663. [CrossRef]
28. Shimizu, H. Multi-purpose electric vehicle "Kaz.". *IATSS Res.* **2001**, *25*, 96–97. [CrossRef]
29. Sato, M.; Yamamoto, G.; Gunji, D.; Imura, T.; Fujimoto, H. Development of wireless in-wheel motor using magnetic resonance coupling. *IEEE Trans. Power Electron.* **2015**, *31*, 5270–5278. [CrossRef]
30. Hara, S.; Imura, J.-I.; Tsumura, K.; Ishizaki, T.; Sadamoto, T. Glocal (global/local) control synthesis for hierarchical networked systems. In Proceedings of the 2015 IEEE Conference on Control Applications (CCA), Sydney, Australia, 21–23 September 2015; Institute of Electrical and Electronics Engineers (IEEE): New York, NY, USA, 2015; pp. 107–112.
31. Kim, T.-H.; Hori, Y.; Hara, S. Robust stability analysis of gene–protein regulatory networks with cyclic activation–repression interconnections. *Syst. Control. Lett.* **2011**, *60*, 373–382. [CrossRef]
32. Khaisongkram, W.; Hara, S. Performance analysis of decentralized cooperative driving under non-symmetric bidirectional information architecture. In Proceedings of the 2011 IEEE International Conference on Control Applications (CCA), Denver, CO, USA, 28–30 September 2011; Institute of Electrical and Electronics Engineers (IEEE): New York, NY, USA, 2010; pp. 2035–2040.
33. Nguyen, D.-H.; Hara, S. Hierarchical decentralized controller synthesis for heterogeneous multi-agent dynamical systems by LQR. *SICE J. Control. Meas. Syst. Integr.* **2015**, *8*, 295–302. [CrossRef]
34. Tsumura, K.; Nguyen, B.M.; Wakayama, H.; Hara, S. Distributed stabilization by probability control for deterministic-stochastic large scale systems: Dissipativity approach. In Proceedings of the IEEE Control Decision Conference, Jeju Island, Korea, 14–18 December 2020.
35. Hara, S.; Tanaka, H.; Iwasaki, T. Stability analysis of systems with generalized frequency variables. *IEEE Trans. Autom. Control.* **2013**, *59*, 313–326. [CrossRef]
36. Nguyen, B.M.; Hara, S.; Fujimoto, H. Stability analysis of tire force distribution for multi-actuator electric vehicles using generalized frequency variable. In Proceedings of the 2016 IEEE Conference on Control Applications (CCA), Buenos Aires, Argentina, 19–22 September 2016; Institute of Electrical and Electronics Engineers (IEEE): New York, NY, USA, 2016; pp. 91–96.
37. Liu, K.; Li, K.; Zhang, C. Constrained generalized predictive control of battery charging process based on a coupled thermoeletric model. *J. Power Sources* **2017**, *347*, 145–158. [CrossRef]
38. Ouyang, Q.; Wang, Z.; Liu, K.; Xu, G.; Li, Y. Optimal charging control for lithium-ion battery packs: A distributed average tracking approach. *IEEE Trans. Ind. Informatics* **2019**, *16*, 3430–3438. [CrossRef]
39. Liu, K.; Hu, X.; Wei, Z.; Li, Y.; Jiang, Y. Modified gaussian process regression models for cyclic capacity prediction of lithium-ion batteries. *IEEE Trans. Transp. Electrification* **2019**, *5*, 1225–1236. [CrossRef]
40. Liu, K.; Li, Y.; Hu, X.; Lucu, M.; Widanage, W.D. Gaussian process regression with automatic relevance determination kernel for calendar aging prediction of lithium-ion batteries. *IEEE Trans. Ind. Inform.* **2019**, *16*, 3767–3777. [CrossRef]
41. Khalil, H.K. *Nonlinear Systems*, 3rd ed.; Prentice Hall: Upper Saddle River, NJ, USA, 2002.
42. Rajamani, R. *Vehicle Dynamics and Control*; Springer: Berlin, Germany, 2012.
43. Pacejka, H.B. *Tyre and Vehicle Dynamics*; Elsevier BV: Amsterdam, Netherlands, 2006.
44. Mechanical Simulation Homepage. Available online: https://www.carsim.com/ (accessed on 20 August 2020).
45. Ortega, R.; Spong, M.W. Adaptive control of robot manipulators: A tutorial. *Automatica* **1989**, *25*, 877–888.
46. Hatanaka, T.; Chopra, N.; Fujita, M.; Spong, M.W. *Passivity-Based Control. and Estimation in Networked Robotics*; Springer Science and Business Media LLC: Berlin, Germany, 2015.
47. Nguyen, B.-M.; Tsumura, K.; Hara, S. Glocal traction control for in-wheel-motor electric vehicles-A passivity approach. In Proceedings of the 1st Virtual IFAC World Congress, Berlin, Germany, 11–17 July 2020.
48. Wang, Y.; Fujimoto, H.; Hara, S. Torque distribution-based range extension control system for longitudinal motion of electric vehicles by LTI modeling with generalized frequency variable. *IEEE/ASME Trans. Mechatron.* **2015**, *21*, 1. [CrossRef]

49. Harada, S.; Fujimoto, H. Range extension control system for electric vehicles based on optimal-deceleration trajectory and front-rear driving-braking force distribution considering maximization of energy regeneration. In Proceedings of the 2014 IEEE 13th International Workshop on Advanced Motion Control (AMC), Yokohama, Japan, 14–16 March 2014; Institute of Electrical and Electronics Engineers (IEEE): New York, NY, USA, 2014; pp. 173–178.
50. Fujimoto, H.; Harada, S. Model-based range extension control system for electric vehicles with front and rear driving–braking force distributions. *IEEE Trans. Ind. Electron.* **2015**, *62*, 3245–3254. [CrossRef]
51. Rill, G. First oder tire dynamics. In Proceedings of the III European Conference on Computational Mechanics Solids, Structures and Coupled Problems in Engineering, Lisbon, Portugal, 5–8 June 2006.
52. Sado, H.; Sakai, S.; Hori, Y. Road condition estimation for traction control in electric vehicle. In Proceedings of the ISIE '99 IEEE International Symposium on Industrial Electronics (Cat. No.99TH8465), Bled, Slovenia, 12–16 July 1999; Institute of Electrical and Electronics Engineers (IEEE): New York, NY, USA, 2003; Volume 2, pp. 973–978.
53. Fujii, K.; Fujimoto, H. Traction control based on slip ratio estimation without detecting vehicle speed for electric vehicle. In Proceedings of the 2007 Power Conversion Conference, Nagoya, Japan, 2–5 April 2007; Institute of Electrical and Electronics Engineers (IEEE): New York, NY, USA, 2007; pp. 688–693.
54. Wilson, D.I. *Advanced Control. using MATLAB or Stablising the Unstabilisable*; Auckland University of Technology: Auckland, New Zealand, 2015.
55. Suzuki, Y.; Kano, Y.; Abe, M. A study on tyre force distribution controls for full drive-by-wire electric vehicle. *Veh. Syst. Dyn.* **2014**, *52*, 235–250. [CrossRef]

Publisher's Note: MDPI stays neutral with regard to jurisdictional claims in published maps and institutional affiliations.

energies

MDPI

Article

On the Comparison of 2- and 4-Wheel-Drive Electric Vehicle Layouts with Central Motors and Single- and 2-Speed Transmission Systems

Stefano De Pinto [1,2], Pablo Camocardi [1], Christoforos Chatzikomis [1], Aldo Sorniotti [1,*], Francesco Bottiglione [2], Giacomo Mantriota [2] and Pietro Perlo [3]

1 Department of Mechanical Engineering Sciences, University of Surrey, Guildford GU2 7XH, UK; m14931@surrey.ac.uk (S.D.P.); m14930@surrey.ac.uk (P.C.); c.chatzikomis@surrey.ac.uk (C.C.)

2 Dipartimento di Meccanica, Matematica e Management, Politecnico di Bari, 70126 Bari, Italy; francesco.bottiglione@poliba.it (F.B.); giacomo.mantriota@poliba.it (G.M.)

3 Interactive Fully Electrical VehicleS (IFEVS), 10040 La Loggia, Torino, Italy; pietro.perlo@ifevs.com

* Correspondence: a.sorniotti@surrey.ac.uk

Received: 5 June 2020; Accepted: 19 June 2020; Published: 30 June 2020

Abstract: Electric vehicles (EVs) are characterized by a significant variety of possible powertrain configurations, ranging from one to four electric machines, which can have an on-board or in-wheel layout. Multiple models of production EVs have recently been introduced on the market, with 4-wheel-drive (4WD) architectures based on a central motor within each axle, connected to the wheels through a gearbox, a differential, and half-shafts. In parallel, an important body of research and industrial demonstrations have covered the topic of 2-speed transmission systems for EVs, with the target of enhancing longitudinal acceleration and gradeability performance, while increasing the operating efficiency of the electric powertrain. Although several recent studies compare different electric powertrain architectures, to the best of the authors' knowledge the literature misses a comparison between 2-wheel-drive (2WD) and 4WD configurations for the same EV, from the viewpoint of drivability and energy consumption. This paper targets this gap, by assessing 2WD and 4WD powertrain layouts with central motors, for a case study light passenger car for urban mobility, including consideration of the effect of single- and 2-speed transmission systems. An optimization routine is used to calculate the energy-efficient gear state and/or torque distribution for each considered configuration. For the specific EV, the results highlight the favourable trade-off of the single-speed 4WD layout, capable of reducing the energy consumption during driving cycles by approximately 9% with respect to the conventional 2WD layout with single-speed transmission, while providing satisfactory drivability and good gradeability, especially in low tire–road friction conditions.

Keywords: Electric vehicle; 2-wheel-drive; 4-wheel-drive; single- and 2-speed transmissions; optimal torque distribution; optimal gear selection

1. Introduction

Electric vehicles (EVs) are characterized by a variety of potential powertrain architectures, including one to four electric motors [1]. The electric machines can have on-board [2] or in-wheel [3] installations, depending on whether they are part of the sprung or unsprung mass of the EV. While the in-wheel configurations imply the adoption of two motors within the same axle, the on-board configurations can have either a single motor (central motor configuration) or two motors per axle. In the central motor configurations, the electric machine is connected to the wheels through a mechanical transmission, a mechanical differential and half-shafts. Production EVs usually have a 2-wheel-drive (2WD) architecture, with a central motor configuration including a single-speed transmission system [4].

The simplicity of such powertrain layout meets the performance requirements of most EVs, and results in low cost and weight. This solution is enabled by the favourable torque characteristics of electric machines, which, in a first approximation, have a constant torque region up to the base speed, and then a constant power region up to the top speed. However, the automotive industry is exploring alternative EV layouts, with a series of recent 4-wheel-drive (4WD) production EVs with one motor per axle, such as the Tesla Model S (built in Fremont, California, USA), Audi e-tron (built in Brussels, Belgium), and Jaguar I-PACE (built in Graz, Austria) [4]. Such 4WD architectures provide redundancy in case of powertrain failures, and permit the implementation of advanced vehicle dynamics and traction control concepts, through the variation of the front-to-total torque distribution [5,6]. Moreover, the 4WD layouts with one motor per axle do not require the complex—yet achievable through engineering efforts—functional safety analyses (see the relevant standard ISO-26262, [7]) of the configurations with two motors per axle, either on-board or in-wheel [8,9], which can generate undesired direct yaw moments in case of motor drive failures. Nevertheless, in medium to long term, a fair market share can be expected for the configurations with two motors per axle, which: (i) offer the possibility of implementing continuous direct yaw moment control, to enhance active safety [2] and energy efficiency in cornering [10]; and (ii) facilitate advanced individual wheel slip control [11], especially in case of in-wheel motors, which are not affected by the torsional dynamics of the half-shafts. These layouts, covered with a specific set of performance indicators for cornering conditions in the comparison in De Novellis et al. [12], will not be considered in this study, which is focused on central motor arrangements.

In parallel, in the last 10 years, the automotive industry and academia have been evaluating 2-speed transmission systems for electric powertrains with central motor configurations, with the purpose of enhancing the longitudinal acceleration, gradeability, top speed, and energy consumption performance, through the appropriate design and on-line selection of the gear ratios [13–18]. The first examples of electric powertrains with 2-speed transmissions are currently reaching the production phase [19]. The studies on 2-speed transmission systems include gearbox configurations and algorithms for reducing the wheel torque gap and longitudinal acceleration oscillations during gearshifts [17,18], as well as controllers for 4WD EV configurations [20,21], in which: (i) significant operational flexibility is available in terms of number of states and front-to-total wheel torque distributions to obtain, in an energy efficient way, a total wheel torque demand at a given speed; and (ii) the torque-fill effect [22] during the gearshift can be achieved through simplified 2-speed transmission system layouts, e.g., without controlled friction clutches, by modulating the torque profile on the axle not involved in the gearshift [21]. However, to the best of the authors' knowledge, the production controllers for gear selection and/or management of the front-to-total wheel torque distribution in 4WD configurations tend to adopt relatively simple but robust algorithms, which are far from providing optimal performance, especially in terms of energy consumption.

The EV literature includes contributions analysing and optimizing powertrain efficiency, energy consumption along driving cycles, and operating costs, [13,23,24], as well as studies on novel central motor configurations, with comparisons with more conventional layouts. For example, Bottiglione et al. [25] discuss the increased operational flexibility, and thus efficiency, of the electric motor drive, allowed by infinitely variable transmission systems. Holdstock et al. [26] present a four-speed dual motor electric powertrain concept, in which two gears are available for each electric machine, and compare the energy consumption of the novel powertrain with that of single motor configurations coupled with single- and 2-speed transmissions. The dual motor layout is simplified in Kwon et al. [23] and Ruan et al. [27], which consider one gear ratio for each electric motor. Nguyen et al. [28] extend the four-speed dual motor powertrain concept to a case study plug-in parallel hybrid electric bus, i.e., including an internal combustion engine in the powertrain. De Carlo et al. [29] propose a dual motor electric powertrain layout based on an epicyclical gearset, which shows an average ~9% energy efficiency increase along an urban driving cycle, with respect to the corresponding single motor solution. However, the previous 2WD dual motor layouts, although requiring two electric machines, do not bring the traction capability benefits of a 4WD EV architecture, unless they are replicated on both vehicle axles, which would imply the installation of four motors.

Despite the availability of studies comparing different EV powertrain architectures [30], to the best of the authors' knowledge, the literature misses a thorough performance comparison between 2WD and 4WD EV architectures with a single motor per axle, which are the focus of the current industrial interest. Moreover, although on-board electric powertrains are subject to important torsional oscillations [2], which have to be compensated by anti-jerk controllers [31], the available electric powertrain comparisons tend to overlook the drivability implications of the considered layouts. Hence, for a case study EV, this paper aims to address the identified gap through:

- A drivability analysis based on the frequency and time response characteristics of 2WD and 4WD powertrain layouts, excluding and including the effect of the anti-jerk control action;
- A unified optimization routine to find the energy-efficient transmission state and front-to-total torque distribution for each operating condition of the EV, to provide a fair comparison among the considered architectures;
- A comparison of the resulting acceleration, gradeability and energy consumption performance.

The paper is organized as follows: Section 2 discusses the EV layouts and simulation models; Section 3 deals with the drivability assessment; Section 4 presents the unified optimization routine for energy-efficient gear selection and front-to-total wheel torque distribution; Section 5 includes the acceleration, gradeability and energy consumption results, which are followed by the conclusions.

2. Electric Vehicle Layouts and Simulation Models

2.1. Electric Vehicle Layouts

The case study EV platform (Figure 1) is a light urban passenger car prototype developed by IFEVS (Interactive Fully Electrical VehicleS) within the European project PLUS-MOBY, and enhanced during the follow-up project TELL [32]. The considered EV layouts, reported in Figure 2, are:

- Layout L1: a 2WD configuration with a single front 28 kW (peak power value) electric motor coupled with a single-speed transmission. The gear ratio was chosen by the involved industrial companies as a trade-off between gradeability and top speed.
- Layout L2: a 2WD configuration, which differs from L1 for the adoption of a 2-speed transmission. The design is a cost-driven simplification of the 2-speed transmission in [14,18], by excluding the friction and sprag clutches, which are replaced by a dog clutch. During the gearshifts, the synchronization is controlled through the electric motor drive [20,21], at the price, with respect to the more sophisticated original design, of a wheel torque gap. The first gear ratio provides the required longitudinal acceleration and gradeability performance, while the second gear ratio ensures the desired top speed.
- Layout L3: a 4WD configuration with two 14 kW electric motors, connected to the respective axle through single-speed transmissions.
- Layout L4: a 4WD configuration, which differs from L3 for the adoption, within each axle, of the same 2-speed transmission system as in L2. In this EV layout, a torque-fill controller provides seamless gearshifts at low-to-medium wheel torque demands, through the variation of the torque on the axle not involved in the gearshift [20,21].

The EV prototype was initially implemented and experimentally tested in the L4 configuration, see [21]. The efficiency maps of the individual single- and 2-speed powertrains used in the 4WD layouts, based on the data provided by the motor and transmission manufacturers, are in Figure 3, expressed as functions of the total wheel torque and vehicle speed. The maps consider the inverter, electric machine and transmission power losses. In addition to increasing the available torque-speed region, through appropriate gear selection the 2-speed layout significantly expands the operating region where the individual powertrain efficiency exceeds 90%. To provide a fair comparison between the 2WD and 4WD EV layouts, characterized by different sizes of the individual powertrains, the efficiency maps were scaled according to the maximum motor torque.

(a) (b) (c)

Figure 1. The case study 4-wheel-drive (4WD) electric vehicle platform prototype: (**a**) Exterior view; (**b**) The EV during testing on a rolling road facility; (**c**) The 2-speed electric powertrain.

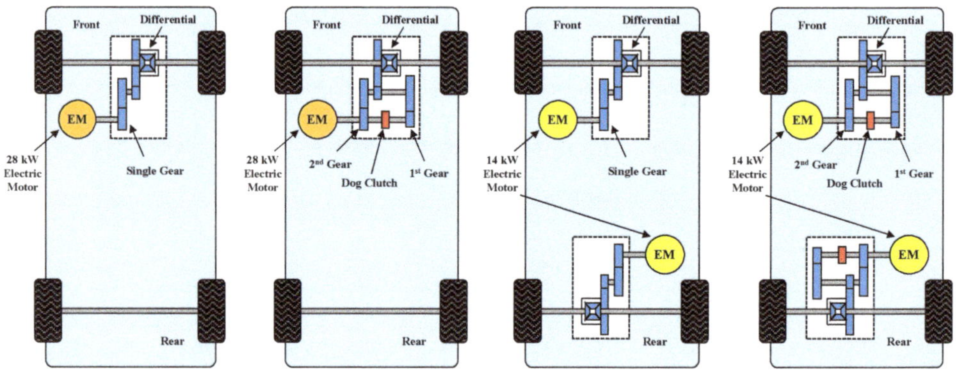

L1 L2 L3 L4

Figure 2. The considered EV layouts, with one (L1 and L2) or two (L3 and L4) electric motors, coupled with single-speed (L1 and L3) or 2-speed (L2 and L4) transmissions.

The main parameters of the four EV architectures are in Table 1, including indication of the EV mass variation associated with the powertrain configurations.

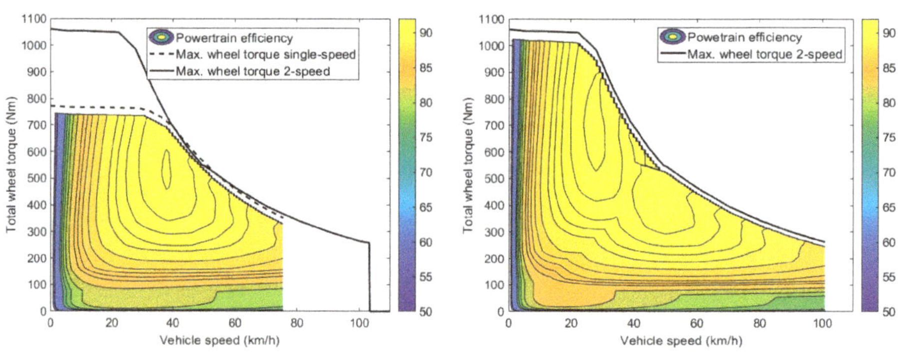

(a) (b)

Figure 3. Efficiency maps of the considered powertrains, including electric motor drive and transmission: (**a**) Single-speed powertrain; (**b**) 2-speed powertrain (with optimal gear selection).

Table 1. Main parameters of the considered EV layouts.

Electric Vehicle Layouts	Vehicle Mass (kg)	Motor Rotor Inertia (kgm^2)	Transmission Mass (kg)	Inverter Mass (kg)	1st Gear Ratio (–)	2nd Gear Ratio (–)	Max. Wheel Torque * (Nm)
L1: single-speed 2WD	820.6	0.0338	12.6	5	8.9	–	772
L2: 2-speed 2WD	831.5	0.0338	23.5	5	12.2	6.76	1060
L3: single-speed 4WD	848.2	0.0181	12.6 × 2	5 × 2	8.9	–	772
L4: 2-speed 4WD	870.0	0.0181	23.5 × 2	5 × 2	12.2	6.76	1060

* The values reported in the table do not account for transmission efficiency.

2.2. Nonlinear Model

A nonlinear forward facing Matlab-Simulink model of the longitudinal vehicle dynamics was implemented to assess the considered EV layouts. This Section presents the main Equations describing the dynamics of the 4WD EV configuration with 2-speed transmissions, i.e., L4 (see Figure 4), since L1-3 can be considered as its subcases. The front and rear axle models include the features responsible for the first order torsional drivetrain dynamics: (i) the half-shafts, modeled as a torsional spring and damper in parallel, and parametrized to include the torsional compliance of the whole driveline; and (ii) the tires, simulated through the Pacejka magic formula (version 5.2) with a relaxation length model [33].

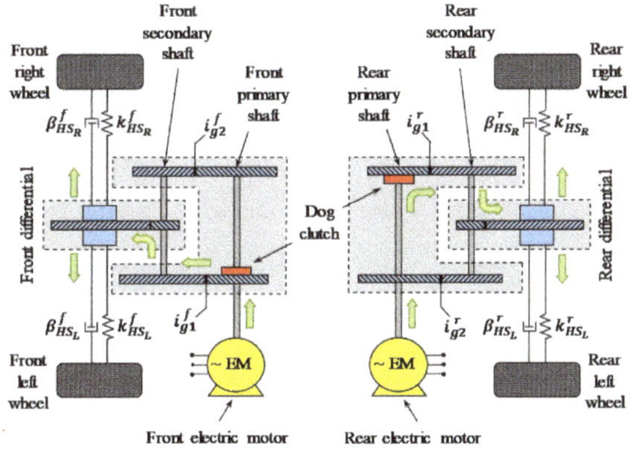

Figure 4. Simplified conceptual schematic of the 2-speed 4WD EV (L4).

Each 2-speed transmission can operate in three conditions, governed by different Equations:

- Engaged first gear, i.e., with the dog clutch engaged such that the primary and secondary shafts rotate according to the first gear ratio;
- Engaged second gear, i.e., with the dog clutch engaging the corresponding gear;
- Disengaged gear: in this condition the drivetrain is characterized by two degrees of freedom, as the electric motor is decoupled from the secondary transmission shaft.

At the EV level, these conditions allow for a total number of eight states, as indicated in Table 2. The Equations are reported for the front axle only, as the front and rear drivetrains are identical. As the assessed scenarios do not include asymmetric tire-road friction conditions on the two EV sides, for simplicity the internal dynamics of the mechanical differential are neglected, and the model considers an equivalent half-shaft and an equivalent tire per axle.

The electric motor torque dynamics are described through the combination of a pure time delay and first order transfer function:

$$\frac{T_{EM}^{f}}{T_{EM,des}^{f}}(s) = \frac{e^{-\tau_{d,EM}^{f}s}}{1+\tau_{EM}^{f}\,s} \tag{1}$$

The parametrization of Equation (1) considered the expected variation of the electrical system dynamics with the motor size, i.e., larger machines are usually characterized by larger values of their time constant. More specifically, in the following analyses, a time constant $\tau_{EM} = 25$ ms was selected for the larger motors of the 2WD EV configurations, and $\tau_{EM} = 12.5$ ms was chosen for the machines of the 4WD layouts.

Table 2. Available gear state combinations for the 2-speed 4WD EV (L4).

State	Front Transmission	Rear Transmission	State	Front Transmission	Rear Transmission
1	1st gear	1st gear	5	Neutral	1st gear
2	1st gear	2nd gear	6	Neutral	2nd gear
3	2nd gear	1st gear	7	1st gear	Neutral
4	2nd gear	2nd gear	8	2nd gear	Neutral

In first gear, the torque balance of the powertrain is given by:

$$\ddot{\theta}^f_{diff} = \frac{1}{J^f_{eq_{g1}}}\left[i^f_{g1}\, i^f_{diff}\eta^f_{tr}\, T^f_{EM} - \left(\beta^f_{HS_L} + \beta^f_{HS_R}\right)\left(\dot{\theta}^f_{diff} - \dot{\theta}^f_w\right) - \left(k^f_{HS_L} + k^f_{HS_R}\right)\left(\theta^f_{diff} - \theta^f_w\right)\right] \quad (2)$$

In Equation (2), the transmission efficiency, η^f_{tr}, can be reversed depending on the direction of the power flow through the coupling, and is expressed through a look-up table, as a function of the input torque, speed, and operating temperature. The equivalent mass moment of inertia of the drivetrain in first gear is:

$$J^f_{eq_{g1}} = J^f_{diff} + 0.5\left(J^f_{HS_R} + J^f_{HS_L}\right) + \left(J^f_{SS} + J^f_{g2}\, {i^f_{g2}}^2\right){i^f_{diff}}^2 + \left(J^f_{EM} + J^f_{PS} + J^f_{g1}\right){i^f_{g1}}^2{i^f_{diff}}^2 \quad (3)$$

In second gear, the system dynamics are expressed by replacing the gear ratio in Equation (2), while the equivalent mass moment of inertia of the drivetrain is:

$$J^f_{eq_{g2}} = J^f_{diff} + 0.5\left(J^f_{HS_L} + J^f_{HS_R}\right) + \left(J^f_{SS} + J^f_{g1}\, {i^f_{g1}}^2\right){i^f_{diff}}^2 + \left(J^f_{EM} + J^f_{PS} + J^f_{g2}\right){i^f_{g2}}^2{i^f_{diff}}^2 \quad (4)$$

In conditions of disengaged gear, the primary and secondary shafts are decoupled, and the drivetrain has two degrees of freedom, i.e., the first one for the rotating parts of the electric motor and transmission system components rigidly connected to the primary shaft, and the second one for the transmission components rotating together with the differential:

$$\begin{cases} \ddot{\theta}^f_{EM} = \frac{T^f_{EM} - T^f_{w\&f}}{J^f_{EM} + J^f_{PS}} \\ \ddot{\theta}^f_{diff} = -\frac{1}{J^f_{eq_{dis}}}\left[\left(\beta^f_{HS_L} + \beta^f_{HS_R}\right)\left(\dot{\theta}^f_{diff} - \dot{\theta}^f_w\right) + \left(k^f_{HS_L} + k^f_{HS_R}\right)\left(\theta^f_{diff} - \theta^f_w\right)\right] \end{cases} \quad (5)$$

where:

$$J^f_{eq_{dis}} = J^f_{diff} + 0.5\left(J^f_{HS_L} + J^f_{HS_R}\right) + \left(J^f_{SS} + J^f_{g1}{i^f_{g1}}^2 + J^f_{g2}{i^f_{g2}}^2\right){i^f_{diff}}^2 \quad (6)$$

The axial position of the dog clutch for gear selection is controlled though an electro-mechanical actuator. The actuator position is modeled through a pure time delay, τ^f_d, and a first order transfer function with time constant τ^f_{act}:

$$\frac{x^f_{act}}{x^f_{des}}(s) = \frac{e^{-\tau^f_d s}}{1 + \tau^f_{act}\, s} \quad (7)$$

The simulation model of L4 includes the gearshift algorithm with torque-fill capability, discussed and experimentally evaluated in De Pinto et al. [21], which modifies the torque demand on the axle not involved in the gearshift to compensate for the torque gap on the other axle.

The wheel dynamics are described by:

$$\ddot{\theta}^f_w = \frac{1}{2J_w + 0.5\left(J^f_{HS_L} + J^f_{HS_R}\right)}\left[\left(\beta^f_{HS_L} + \beta^f_{HS_R}\right)\left(\dot{\theta}^f_{diff} - \dot{\theta}^f_w\right) + \left(k^f_{HS_L} + k^f_{HS_R}\right)\left(\theta^f_{diff} - \theta^f_w\right) - 2T^f_{dw} - 2T^f_{br} - 2T^f_{roll}\right] \quad (8)$$

where T^f_{br} is the braking torque on each front corner, and T^f_{roll} is the rolling resistance torque of the front tires. The delayed tire torque, T^f_{dw}, i.e., the moment corresponding to the actual longitudinal tire forces, is calculated through a relaxation model:

$$T^f_w = \frac{L_{rel}}{\dot{\theta}^f_{w0} R_w}\dot{T}^f_{dw} + T^f_{dw} \quad (9)$$

where T^f_w is the steady-state tire torque, corresponding to the longitudinal force output by the Pacejka magic formula model. The tire relaxation length is expressed as a function of the vertical tire load. Finally, the dynamics of the equivalent EV inertia are described by:

$$mR^2_w\ddot{\theta}_v = 2T^f_{dw} + 2T^r_{dw} - T_{aero} \quad (10)$$

where $\ddot{\theta}_v$ is the angular acceleration of the equivalent EV inertia, and T_{aero} is the equivalent torque corresponding to the aerodynamic drag.

2.3. Linearized Model

On-board electric powertrains are usually characterized by a rather low first natural frequency, because of the relatively large rotating inertias of the motor rotor and wheels, and low torsional stiffness of the half-shafts. Given the limited damping level of the system, torsional vibrations provoked by the electric motor torque transients are common issues that affect drivability and cause longitudinal acceleration oscillations [34]. In this context, 4WD EV configurations and 2-speed transmissions add complexity and can amplify the problem. Hence, a linearized version of the model in Section 2.2 was implemented to obtain the frequency response characteristics of the considered EV layouts in conditions of engaged gear, for torque demand inputs.

The main system nonlinearity is represented by the longitudinal tire force characteristics, which, in the linearized model, are approximated through the following linear function:

$$F \approx F_0 + C_0(\sigma - \sigma_0) = F_0 + C_0/\dot{\theta}_{w0}\left(\dot{\theta}_w\dot{\theta}_{v0}/\dot{\theta}_{w0} - \dot{\theta}_v\right) = F_0 + C_0/\dot{\theta}_{w0}\left[\dot{\theta}_w(1-\sigma_0) - \dot{\theta}_v\right] \quad (11)$$

where the longitudinal slip stiffness at the linearization point, C_0, is calculated through the magic formula model. Equation (11) shows that the tire behaves like a nonlinear damper between the wheel and vehicle inertias, where the damping coefficient is a decreasing function of the wheel speed at the linearization point, $\dot{\theta}_{w0}$, i.e., the torsional drivetrain dynamics change with speed.

From the linearized model, the Bode plot of the following transfer function is considered for each EV configuration:

$$H_{req}(j\omega) = \frac{\ddot{\theta}_v}{T_{req}}(j\omega) \quad (12)$$

which expresses the EV acceleration, $\ddot{\theta}_v$, resulting from a wheel torque demand input, T_{req}. $H_{req}(j\omega)$ characterizes the dynamic response to a driver torque request, and thus allows to evaluate the EV drivability properties. For the 2WD layouts, $H_{req}(j\omega)$ and the respective Bode plot are directly derived from the linearized model formulation. In the 4WD layouts, $H_{req}(j\omega)$ results from the superposition of the EV response to the front and rear torque demands, T^f_{req} and T^r_{req}:

$$T_{req} = T^f_{req} + T^r_{req} \tag{13}$$

From the definition of the front-to-total wheel torque distribution ratio p, it follows that:

$$\begin{aligned} T^f_{req} &= p\,T_{req} \\ T^r_{req} &= (1-p)\,T_{req} \end{aligned} \tag{14}$$

By assuming a constant value of p and the presence of a phase shift ϕ between the front and rear torque demands, $H_{req}(j\omega)$ is calculated as:

$$H_{req}(j\omega) = \frac{\ddot{\theta}_v}{T_{req}}(j\omega) = p\,\frac{\ddot{\theta}_v}{T_{req}}(j\omega)\bigg|_f + (1-p)\,\frac{\ddot{\theta}_v}{T_{req}}(j\omega+\phi)\bigg|_r \tag{15}$$

3. Drivability Analysis

3.1. Frequency Response Characteristics

Figure 5 reports the magnitude plots of $H_{req}(j\omega)$ for the considered EV configurations, for a selection of gear states (see Table 2) and front-to-total wheel torque distributions, under the assumption of using the same half-shafts for all EVs. All cases show an evident resonance peak, at a frequency that decreases when the gear ratio increases, i.e., the resonance frequency is lower in first gear in the 2-speed powertrains. The resonance frequency of the single-speed 4WD configuration is rather high because of the reduced mass moment of inertia of each downsized motor. The effect of the state selection on the resonance frequencies is significant, because of the different values of equivalent mass moment of inertia of the powertrain. In the 4WD layouts, the front-to-total torque distribution does not affect the frequency and magnitude of the resonance, if the two drivetrains are in the same gear ratio. If different gears ratios are selected, the system shows two resonance frequencies, i.e., one per drivetrain, and experiences an important variation of $|H_{req}(j\omega)|$ as a function of p.

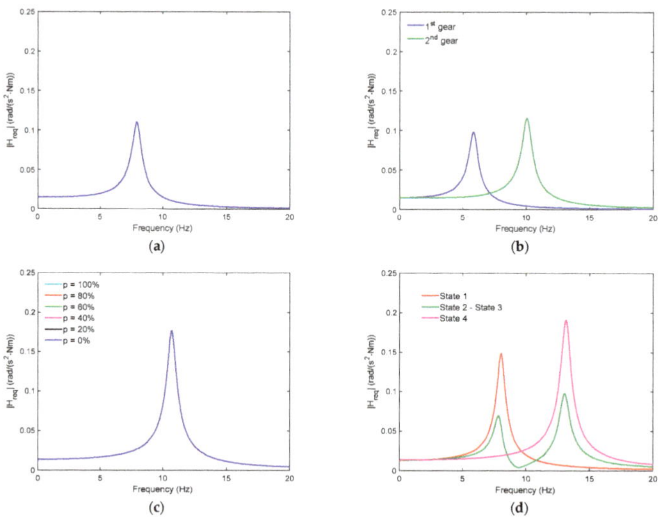

(a) (b) (c) (d)

Figure 5. *Cont.*

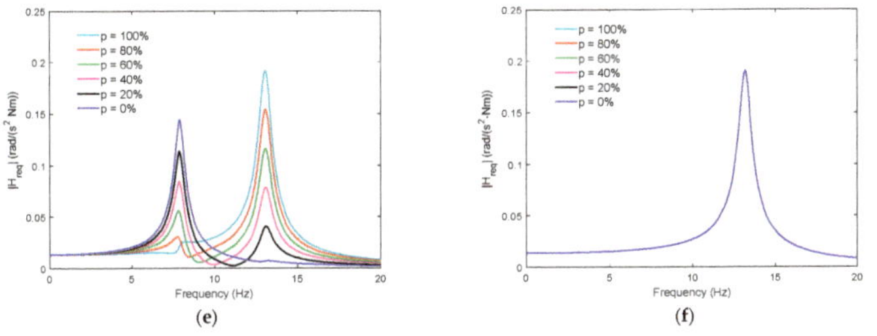

Figure 5. Examples of magnitude plots of $H_{req}(j\omega)$ for the different EV layouts at 20 km/h: (**a**) L1: Single-speed 2WD; (**b**) L2: 2-speed 2WD; (**c**) L3: Single-speed 4WD; (**d**) L4: 2-speed 4WD (with states indicated according to Table 2); (**e**) L4: 2-speed 4WD (second gear on front axle, first gear on rear axle); (**f**) L4: 2-speed 4WD (second gear on front and rear axles).

Figure 6 is a sensitivity analysis on the effect of the phase shift between the front and rear drivetrain torque demands. During real EV operation, input torque phase shifting is inevitable, and can be caused by the variation of the wheel torque distribution decided by the energy management system, the gearshift controller, as well as the interventions of the anti-jerk and traction controllers. The phase shifts affect $|H_{req}(j\omega)|$ especially when different gear ratios are selected on the two axles, in the frequency region between the respective resonances.

The important conclusion is that the 2-speed 4WD configuration (L4) is characterized by a significantly increased complexity of its drivability response in constant gear condition, with respect to all the other layouts, and its frequency response is strongly dependent on the selected state, wheel torque distribution, and torque demand phase shift. Hence, such EV architecture needs a more careful parametrization of the anti-jerk controller, i.e., the controller compensating for the torsional drivetrain dynamics, which must be adaptable to the variety of operating conditions of the EV.

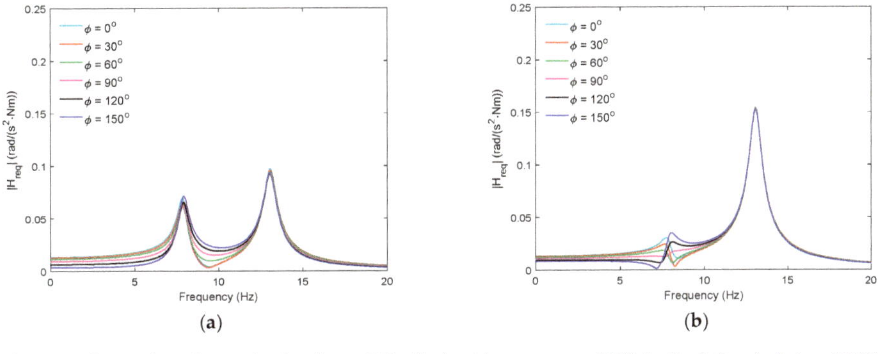

Figure 6. Examples of magnitude plots of $H_{req}(j\omega)$ with constant p (50% in the left subplot and 80% in the right subplot) and different phase shifts ϕ between the front and rear motor torque demands, at 20 km/h, for L4 (2-speed 4WD): (**a**) Second gear on front axle, first gear on rear axle; (**b**) Second gear on front and rear axles.

3.2. Anti-Jerk Controller

The implemented anti-jerk controller, i.e., the controller damping the torsional drivetrain oscillations, is based on a tachometric feedback architecture [2,35], in which, in a first approximation, the corrective anti-jerk torque, $T^{f}_{EM,corr}$, reported for the front powertrain, is proportional to the drivetrain torsion rate (see also Figure 7).

$$T^{f}_{EM,corr} = K\left(\dot{\theta}^{f}_{EM} - \dot{\theta}^{f}_{w} i^{f}_{sel} i^{f}_{diff}\right) \tag{16}$$

A gain scheduling approach as a function of the vehicle speed and selected gear ratio was adopted to obtain rather uniform performance along the variety of possible operating scenarios. Appropriate activation and deactivation conditions were defined, e.g., the anti-jerk controller is not active during the interventions of the traction controller or during gearshift actuation.

The tuning of the anti-jerk controller was performed through simulation-based iterations, for a set of tip-in and tip-out tests, i.e., manoeuvres characterized by abrupt positive and negative variations of the powertrain torque. The tuning objective was to reduce the longitudinal acceleration oscillations, and in particular the first overshoot or undershoot following the torque demand variation, without excessive penalization of powertrain responsiveness.

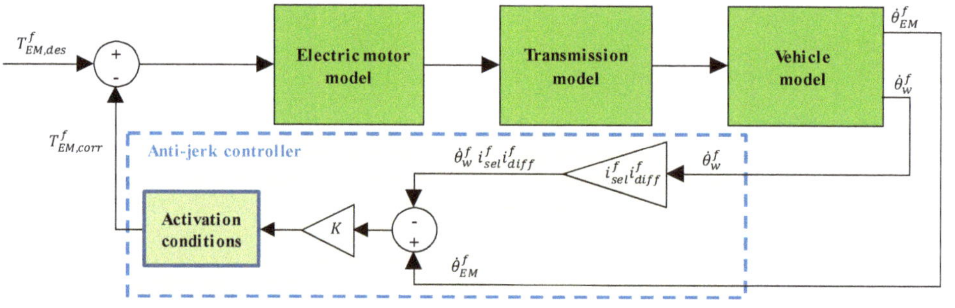

Figure 7. Simplified block diagram of the vehicle system with the implemented anti-jerk controller.

3.3. Drivability Assessment

Figure 8 presents the effect of the anti-jerk controller on the longitudinal acceleration response of the four EV configurations, along a tip-in test from an initial torque level required to keep a constant EV speed of 10 km/h, to the maximum motor torque demand. No gearshift is actuated in the 2-speed configurations. The initial gear is the first one for the layouts with 2-speed transmission systems, which, given the larger gear ratio, show higher average values of the longitudinal vehicle acceleration. The lowest steady-state value of the longitudinal acceleration is associated with the single-speed 4WD layout, and is caused by the smaller gear ratio and marginal mass increase due to the second driven axle.

All configurations without anti-jerk control respond with major oscillations, which severely degrade drivability. The 4WD layouts are more responsive and prone to oscillations, because of the reduced time constant of their electric machines, which increases the excitation level of the torsional drivetrain dynamics. Without anti-jerk controller, in the 2-speed 2WD EV the oscillations are attenuated by the intervention of the traction controller, which reduces the wheel torque demand to prevent wheel spinning. A minor intervention of the traction controller also occurs in the single-speed 2WD case.

The proposed anti-jerk control scheme brings consistent performance improvements in all EV configurations, e.g., with a reduction of the first acceleration peak by more than 1.5 m/s^2 for the 2-speed 4WD EV (L4), and a general compensation of the oscillations in less than 0.3 s after the application of the tip-in torque.

For the test in Figure 8, Table 3 reports the values of typical drivability performance indicators (see also the standards ISO 2631 [36] and ISO 8041 [37]):

- The fourth-power vibration dose value of the longitudinal acceleration, $VDV_{\tilde{a}}$, evaluating the vehicle comfort level:

$$VDV_{\tilde{a}} = \sqrt[4]{\int_{t1}^{t2} \tilde{a}^4 dt} \quad (17)$$

where $\widetilde{a}(t)$ is the difference between the actual EV acceleration profile, $a(t)$, and the acceleration value, $a_{ss} = a(t_2)$, computed at the time t_2 = 1.5 s, in which the acceleration can be considered stabilized to its steady-state value in all controlled configurations:

$$\widetilde{a}(t) = a(t) - a_{ss} \tag{18}$$

t_1 is the time in which $\widetilde{a}$ crosses zero for the first time during the tip-in.

- The root-mean-square value of $\widetilde{a}$, i.e., $RMS_{\widetilde{a}}$:

$$RMS_{\widetilde{a}} = \sqrt{\frac{1}{t_2 - t_1} \int_{t1}^{t2} \widetilde{a}^2 dt} \tag{19}$$

- The difference of the resulting EV speeds at $t = t_2$, without and with anti-jerk controller, i.e., $V_{AJ-off}(t_2)$ and $V_{AJ-on}(t_2)$:

$$\Delta_V = V_{AJ-off}(t_2) - V_{AJ-on}(t_2) \tag{20}$$

which evaluates the degradation of the acceleration performance induced by the controller.

- The response time, t_{resp}, evaluated as the difference between the times in which the longitudinal acceleration profile crosses 90% and 10% of a_{ss} for the first time after the variation of the torque demand request.
- The settling time, t_{sett}, i.e., the time that is required for a to remain within a given range of percentage (2.5%) of a_{ss}, evaluated in the interval $[t_1, t_2]$.
- The peak response time, t_{peak}, evaluated as the difference between the time of the first acceleration peak, and the time corresponding to the first crossing of a_{ss} after the variation of torque demand request.
- The overshoot, OS, expressed in percentage, calculated from the difference between the peak acceleration value and a_{ss}.
- The integral (normalized with time) of the absolute value of the anti-jerk control actions, which evaluates the magnitude of the control effort:

$$IACA = \frac{1}{t_2 - t_1} \int_{t1}^{t2} \left(\left| T^{f}_{EM,corr} \right| + \left| T^{r}_{EM,corr} \right| \right) dt \tag{21}$$

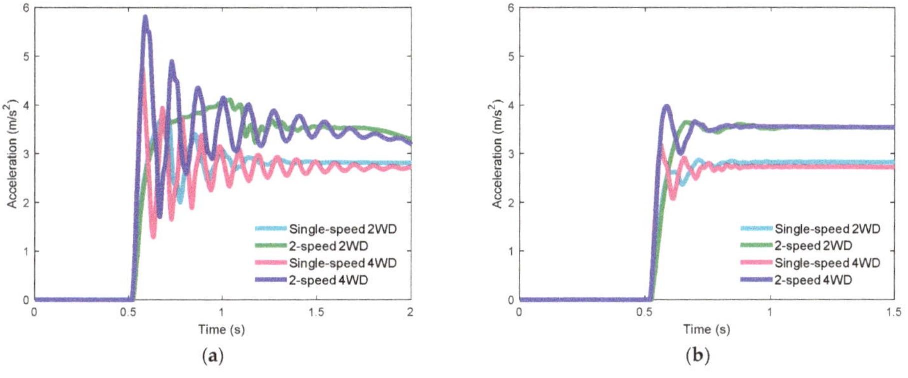

Figure 8. Longitudinal vehicle acceleration profiles for the four EV layouts, during a tip-in from an initial speed of 10 km/h: (**a**) Without anti-jerk controller; (**b**) With anti-jerk controller.

The indicators confirm the qualitative behaviors observed in Figure 8, with the 4WD layouts characterized by higher responsiveness and reduced comfort with respect to the 2WD layouts, and a major performance improvement provided by the anti-jerk controllers in all EV configurations.

Table 3. Drivability performance indicators for the tip-in test of Figure 8.

Layout	Anti-jerk Controller	$VDV_{\tilde{a}_x}$ (m/s$^{1.75}$)	$RMS_{\tilde{a}_x}$ (m/s^2)	Δv (km/h)	t_{resp} (s)	t_{sett} (s)	t_{peak} (s)	OS (%)	$IACA$ (Nm)
L1: Single-speed 2WD	Off	0.491	0.341	0	0.055	0.646	0.150	33.22	0
	On	0.035	0.020	−0.370	0.056	0.259	0.184	1.67	2.588
L2: 2-speed 2WD	Off	0.329	0.279	0	0.096	0.789	0.505	16.29	0
	On	0.056	0.039	−0.423	0.081	0.258	0.132	3.08	3.752
L3: Single-speed 4WD	Off	0.799	0.528	–	0.021	NA	0.048	69.65	0
	On	0.238	0.120	−0.232	0.023	0.252	0.037	16.54	1.675
L4: 2-speed 4WD	Off	1.046	0.701	–	0.027	NA	0.062	68.90	–
	On	0.219	0.120	−0.372	0.032	0.180	0.058	12.45	2.044

In reality, when considering the operation of the EV beyond the specific tip-in test, the simplicity of the adopted 2-speed transmission layout, which is needed to limit the cost of the multiple-speed EV configurations, originates drivability limitations in maneuvers involving gearshifts at high torque demand. This is shown in the second tip-in test of Figure 9 (with a lower torque gradient than the one in Figure 8), in which: (i) the 2-speed 2WD EV performs an upshift; and (ii) the 2-speed 4WD EV is subject to a gearshift from the first to the second gear on the front axle, and then from the first to the second gear on the rear axle. Each gearshift has a duration of approximately 0.8 s. In these conditions, the gearshift torque-fill strategy of the 2-speed 4WD EV [21] is not effective, since the axle not involved in the gearshift is already operating at its torque limit and cannot be used for filling in the torque gap during the gearshift of the other axle. This can originate a decay of the drivability performance in real driving conditions, which shifts the balance of the comparison toward the single-speed solutions.

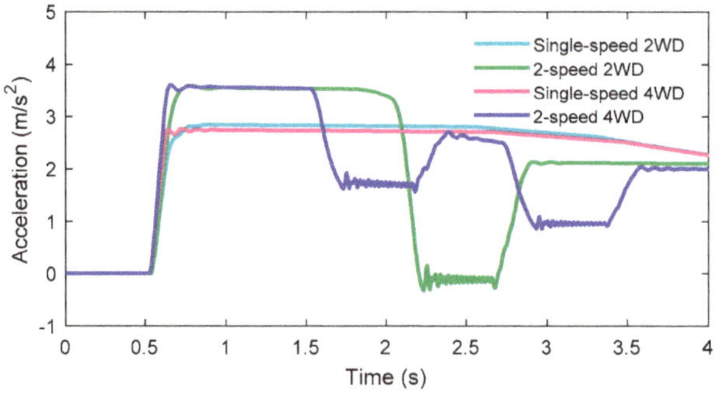

Figure 9. Longitudinal vehicle acceleration profiles for the four EV layouts, during a tip-in test including gearshift actuation for the layouts with 2-speed transmissions.

4. Routine for Selection of Optimal Gear State and Front-to-Total Wheel Torque Distribution

4.1. Structure of the Optimization Routine

With the exception of the single-speed 2WD layout (L1), the considered EV configurations are characterized by actuation redundancy in many operating conditions, i.e., a given wheel torque demand can be achieved through multiple transmission states and/or front-to-total wheel torque distributions. Obviously, the selection of the state and torque distribution has an impact on the EV energy consumption. Therefore, for fairness of comparison, an off-line optimization routine was implemented to calculate the commands that minimize the instantaneous powertrain power loss for each EV speed and wheel torque demand. The optimal actuations are compiled in look-up tables, which are used by the on-line controllers. Because of the relatively limited number of decision variables, a brute-force search algorithm was adopted for the generation of the look-up tables.

The optimization process is summarized in Figure 10. The routine, detailed in De Pinto et al. [21] for L4, defines a grid of operating points, $(V_h, T_{w,j})$, in terms of EV speed and total wheel torque, each of them identified by the indices h and j (see Figure 11a). The limits of the grid depend on the gear ratios and motor torque characteristics of each EV architecture. In braking, the results of the optimization routine are generated up to the maximum regenerative capability of each EV configuration, see the 'Maximum possible regen. torque' curves in Figure 11b. However, in the on-line implementation, the regenerative torque area has been limited as indicated by the 'Maximum actual regen. torque' characteristics. In fact, after a progressive transition from regenerative braking to dissipative braking, at very low speed it is desirable to use only the friction brakes to stop the vehicle, in accordance with the current common practice on production EVs. Note that the case study EV is characterized by a conventional hydraulic braking system, which does not allow the implementation of the most advanced continuous brake blending algorithms.

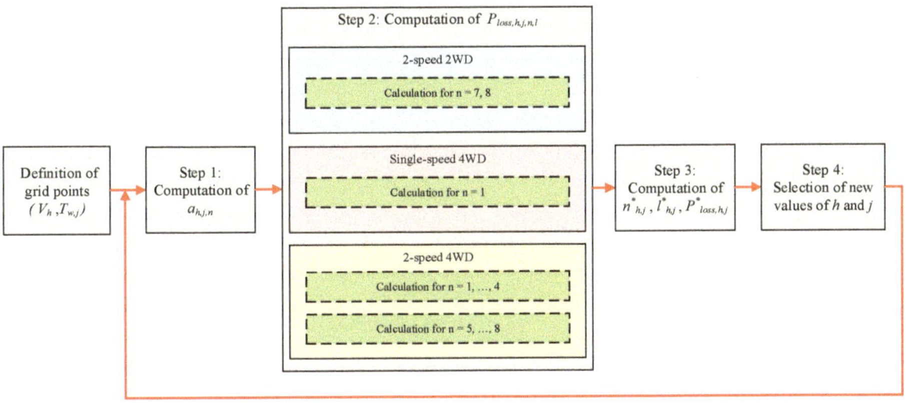

Figure 10. Simplified flow chart of the optimal state and front-to-total wheel torque distribution search routine.

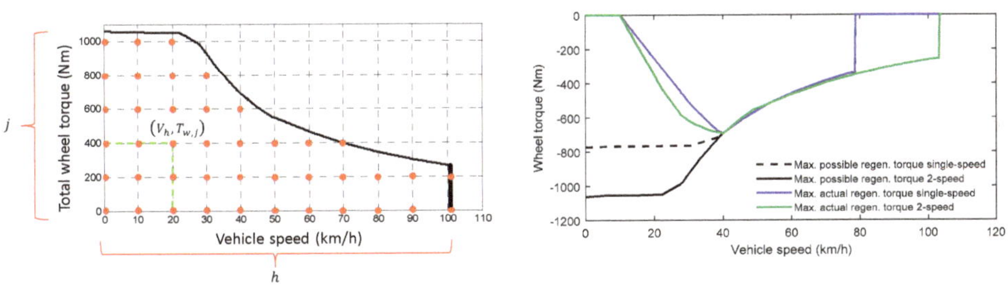

(a) (b)

Figure 11. Features of the implemented optimization routine: (**a**) Examples of grid points; (**b**) Regenerative braking torque limits.

After the definition of the grid, for the selected values of h and j, the optimization routine iterates the following steps:

- Step 1: Computation of the longitudinal vehicle acceleration, $a_{h,j,n}$, for each feasible EV state, n, under the assumption of known road gradient:

$$a_{h,j,n} = \frac{T_{w,j} - F_{aero}(V_h)R_w - F_{roll}(V_h)R_w - mgR_w sin\alpha}{m_{app,n}R_w} \tag{22}$$

where the apparent vehicle mass, $m_{app,n}$, function of n, is:

$$m_{app,n} = m + 4\frac{J_w}{R_w^2} + \frac{J^f_{eq_{g1/g2/dis}}}{R_w^2} + \frac{J^r_{eq_{g1/g2/dis}}}{R_w^2} \tag{23}$$

$J^{f/r}_{eq_{g1/g2/dis}}$ represents the equivalent mass moment of inertia of the rotating parts of the considered powertrain. The feasibility of a state corresponds to its capability of generating the required wheel torque at the assigned speed. An approximate option, available in the routine, is to neglect the inertial effects in Steps 1–2, which simplifies the problem without significant drawbacks, especially if the vehicle is not equipped with an on-line estimator of mass and road gradient.

- Step 2: Computation of the sum, $P_{loss,h,j,n,l}$, of the transmission, electric motor, inverter and rolling resistance power losses at the selected grid point $\left(V_h, T_{w,j}\right)$, for each of its feasible states and—where relevant depending on the state—along a further grid, defined by the index l, covering the feasible range of front-to-total wheel torque distributions. The front-to-total torque distribution corresponds to the inner loop of the routine. In the most general case, corresponding to L4, $P_{loss,h,j,n,l}$ is given by:

$$\begin{aligned} P_{loss,h,j,n,l} &= P^f_{loss,EM+D}\left(T^f_{EM,h,j,n,l}, \dot{\theta}^f_{EM,h,n}\right) + P^r_{loss,EM+D}\left(T^r_{EM,h,j,n,l}, \dot{\theta}^r_{EM,h,n}\right) \\ &+ P^f_{loss,tr}\left(T^f_{tr,h,j,n,l}, \dot{\theta}^f_{EM,h,n}\right) + P^r_{loss,tr}\left(T^r_{tr,h,j,n,l}, \dot{\theta}^r_{EM,h,n}\right) + P^f_{roll}\left(V_h, a_{h,j,n}, n\right) \\ &+ P^r_{roll}\left(V_h, a_{h,j,n}, n\right) \end{aligned} \tag{24}$$

where $P^{f/r}_{loss,EM+D}$, $P^{f/r}_{loss,tr}$ and $P^{f/r}_{roll}$ are provided in the form of maps. The motor drive and transmission power loss maps are functions of the respective input torque and angular speed, calculated for the specific grid point through a backward facing rigid drivetrain model. The tire slip power losses are neglected by the optimization routine.

- Step 3: Selection of the state, $n^*_{h,j}$, and (where applicable) front-to-total torque distribution, $l^*_{h,j}$, which minimize the drivetrain power losses for the operating condition defined by h and j:

$$\left\{n^*_{h,j}, l^*_{h,j}\right\} = \text{argmin} P_{loss,h,j,n,l} \tag{25}$$

$n^*_{h,j}$ and $l^*_{h,j}$ are stored in the look-up tables of the on-line implementation of the controller.

- Step 4: Selection of new values of h and j on the defined grid and repetition the process from Step 1.

4.2. Optimal State Selection for the 2-Speed 2WD Layout (L2)

In this case, the whole required traction or regenerative torque is delivered by one drivetrain, and the undriven axle only produces a rolling resistance torque, unless the friction brakes are activated. Equation (24) is simplified into:

$$P_{loss,h,j,n} = P^f_{loss,EM+D}\left(T^f_{EM,h,j,n}, \dot{\theta}^f_{EM,h,n}\right) + P^f_{loss,tr}\left(T^f_{tr,h,j,n}, \dot{\theta}^f_{EM,h,n}\right) + P^f_{roll}\left(V_h, a_{h,j}, n\right) + P^r_{roll}\left(V_h, a_{h,j}, n\right) \tag{26}$$

The resulting optimal gearshift map for traction conditions is presented in Figure 12a.

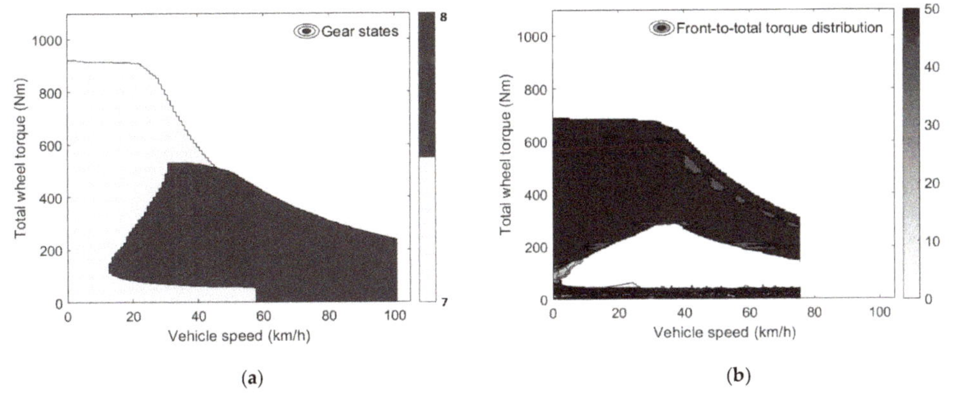

Figure 12. Optimization results: (**a**) Gearshift map for traction for the 2-speed 2WD layout; (**b**) Front-to-total wheel torque distribution map for traction for the single-speed 4WD layout.

4.3. Single-Speed 4WD Layout (L3)

For the single-speed 4WD vehicle, the optimization variable is the front-to-total wheel torque distribution. For the operating condition defined by the indices h, j, and l, the resulting power loss function is:

$$\begin{aligned} P_{loss,h,j,l} &= P^{f}_{loss,EM+D}\left(T^{f}_{EM,h,j,l}, \dot{\theta}^{f}_{EM,h}\right) + P^{r}_{loss,EM+D}\left(T^{r}_{EM,h,j,l}, \dot{\theta}^{r}_{EM,h}\right) + P^{f}_{loss,tr}\left(T^{f}_{tr,h,j,l}, \dot{\theta}^{f}_{EM,h}\right) \\ &+ P^{r}_{loss,tr}\left(T^{r}_{tr,h,j,l}, \dot{\theta}^{r}_{EM,h}\right) + P^{f}_{roll}\left(V_h, a_{h,j}\right) + P^{r}_{roll}\left(V_h, a_{h,j}\right) \end{aligned} \quad (27)$$

The optimal front-to-total torque distribution map in traction conditions is in Figure 12b, which shows: (i) an optimal even front-to-total wheel torque distribution region, i.e., with the same torque on the two axles, at low and high torque demands; and (ii) an optimal single axle region for medium values of torque demand. The rear axle was selected for single axle operation for the specific EV, given that the longitudinal load transfer in traction increases the longitudinal force capability of the rear axle. These results are consistent with the available literature, demonstrating that, under specific assumptions, for identical powertrains the energy-efficient torque distribution implies either single axle operation or even wheel torque distribution [38].

4.4. 2-Speed 4WD Layout (L4)

In the 2-speed 4WD vehicle, both relevant variables, i.e., the transmission state and front-to-total wheel torque distribution, are optimized. The optimal maps are presented in Figure 13. The results have been post-processed and filtered to avoid irregularities, since symmetric states on the front and rear drivetrains can provide the same optimal result while leading to frequent state switching.

The increased number of available gearbox states results in a more varied optimal state map. The front-to-total torque distribution map shows similar characteristics to the one of the single-speed 4WD vehicle (L3). Interestingly, a significant part of the operating region is characterized by single-axle operation, or even distribution operation. Different levels of front-to-total wheel torque distribution occur for the states with different gear ratios on the two axles.

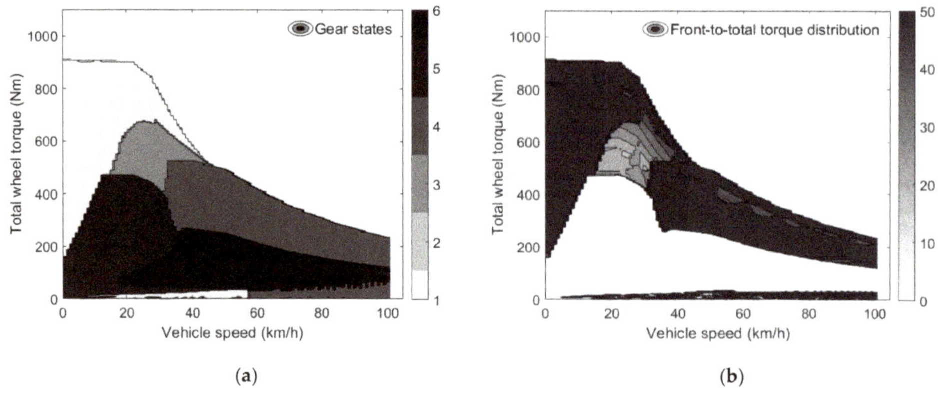

Figure 13. Optimization results: (**a**) State selection map for traction for the 2-speed 4WD layout; (**b**) Front-to-total wheel torque distribution map for traction for the 2-speed 4WD layout.

5. Acceleration, Gradeability, and Consumption Results

5.1. Acceleration and Gradeability Performance

The vehicle acceleration and top speed results for the considered EV configurations are presented in Table 4. The top speed is 78.7 km/h for the single-speed EVs, and 103.2 km/h for the 2-speed configurations; in all cases the top EV speed is limited by the maximum angular speed of the electric machines, rather than by the available motor power. Because of the higher gear ratio in first gear, the EV configurations with 2-speed transmissions show a significant improvement in the longitudinal acceleration performance below 30 km/h; on the other hand, they have higher 0–50 km/h and 0–70 km/h acceleration times, which are caused by the gearshifts and associated wheel torque gap, especially evident for the L2 case, see also Figure 9. The 0–30 km/h acceleration time is lower for L4 with respect to L2 because of the better traction capability of the 4WD layout, which does not require the intervention of the traction controller in high tire-road friction conditions. Another factor that affects the results, in particular in terms of 0–70 km/h acceleration time, is the mass of the different EV variants.

For each EV layout, Figure 14 reports the maximum achievable road gradient, as a function of: (a) the tire-road friction coefficient, from standstill conditions; and (b) vehicle speed, for high tire-road friction conditions. Although the performance of all cases is deemed adequate, the results show: (i) the improved traction capability in low friction conditions provided by the 4WD layouts, which can enhance active safety on critical terrains; and (ii) the enhanced gradeability in high tire-road friction conditions, offered by the layouts equipped with 2-speed transmission systems.

Table 4. Vehicle performance for the different layouts.

Vehicle	Top Speed (km/h)	Acc. Time 0–30 km/h (s)	Acc. Time 0–50 km/h (s)	Acc. Time 0–70 km/h (s)
L1: single-speed 2WD	78.7	3.1	5.4	9.1
L2: 2-speed 2WD	103.2	2.7	6.0	9.6
L3: single-speed 4WD	78.7	3.1	5.5	9.3
L4: 2-speed 4WD	103.2	2.4	5.7	9.4

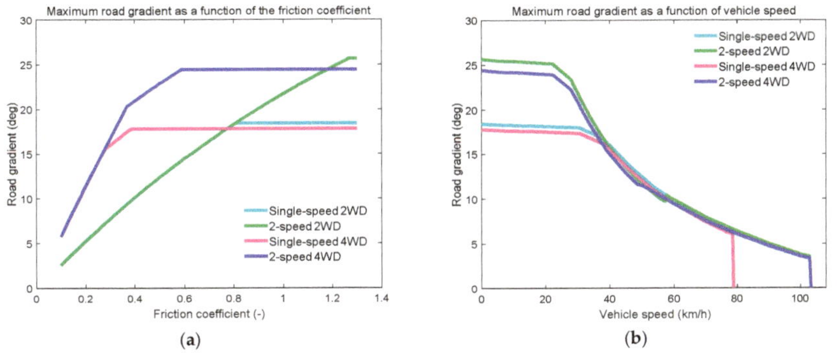

Figure 14. Gradeability performance characteristics: (**a**) Maximum achievable road gradient from standstill as a function of the available tire-road friction coefficient; (**b**) Maximum achievable road gradient as a function of vehicle speed, in high tire-road friction conditions.

5.2. Energy Consumption

This Section presents two sets of energy consumption values for L2-4, corresponding to: (i) the normal mode, which, for the 2-speed layouts, performs the gearshifts according to a typical gear selection map provided by the transmission manufacturer, using the current EV speed and axle torque demand. The upshift is carried out at relatively low motor speed for low torque demands, and at the speed limit of the first gear during operation at the maximum torque demand. In the 4WD configuration, to provide torque-fill functionality, the gearshift maps are characterized by an offset with respect to the EV speed, to ensure that only one gearshift is carried at a time at the vehicle level. In normal mode, the 4WD layouts operate with a 50% front-to-total wheel torque distribution; and (ii) the optimal mode, using the optimal gear states and front-to-total wheel torque distributions, according to the maps obtained in Section 4. For the single-speed 2WD EV (L1), there is a single operating state, so no optimal mode is available.

Table 5 presents the energy consumption results at constant speed, and their percentage difference with respect to the single-speed 2WD EV; positive differences, implying higher consumption, are in red, while negative differences are in green. The important conclusion is that, in most constant speed conditions, the additional mass of the complex powertrain configurations, which increases the wheel torque level to keep constant speed, is not compensated by the more efficient operation of the EV. This is especially evident in normal mode, in which the consumption increases with the complexity of the EV layout at all speeds, e.g., from 3.3% to 7.2% at 30 km/h, and from 0.9% to 2.8% at 70 km/h. The optimized energy management system consistently reduces the energy consumption, especially for the L4 EV, which is characterized by the highest level of operational flexibility. In L4, the energy consumption reduction associated with the optimal energy management, with respect to the normal mode for the same EV layout, is in excess of 4% at all considered speeds. In any case, at 30 km/h and 50 km/h, the most efficient configuration is the single-speed 2WD layout (L1), also in comparison with the optimal mode of the more complex EV configurations. On the other hand, interestingly, at 70 km/h in the optimal mode, both 4WD layouts consume less than L1, i.e., by 2.6% for L3, and 5.6% for L4.

The energy consumption performance of the EV configurations in normal and optimal modes was assessed through the nonlinear simulation model in Section 2.2, along: (i) the urban part of the New European Drive Cycle (NEDC), i.e., for four repetitions of the so-called ECE-15 cycle, and (ii) the Japanese 10–15 mode test cycle (J10–15). For example, Figure 15 reports the time profiles of EV speed, front and rear motor torque as well as gear state, during a section of the J10–15, for the two modes of the 2-speed 4WD EV configuration. The torque-fill capability of L4 during gearshifts at low-to-medium torque demands is evident from: (a) the regularity of the speed profile, with seamless EV acceleration also during the gearshifts, and (b) the ripples in the electric motor torque profiles of the axle not involved in the gearshift, which compensate for the torque gap during the gearshifts on the other axle. The inclusion of the gearshift dynamics in the simulation model allows to realistically account for the motor drive power losses also during the state transition phases.

Table 5. Energy consumption results at constant speed, and their percentage difference with respect to those of the single-speed 2WD EV (L1); positive differences, implying higher consumption, are in red, while negative differences are in green.

Consumption	30 km/h				50 km/h				70 km/h			
(Wh/km)	Normal		Optimal		Normal		Optimal		Normal		Optimal	
L1: single-speed 2WD	36.20		–		50.53		–		72.61		–	
L2: 2-speed 2WD	37.41	3.3%	37.18	2.7%	51.56	2.0%	50.92	0.8%	73.23	0.9%	73.23	0.9%
L3: single-speed 4WD	37.16	2.7%	37.16	2.7%	51.56	2.0%	51.47	1.9%	73.71	1.5%	70.70	−2.6%
L4: 2-speed 4WD	38.80	7.2%	37.31	3.1%	53.03	4.9%	51.02	1.0%	74.64	2.8%	68.53	−5.6%

Table 6 includes the complete set of energy consumption results. Along the selected driving cycles, L2-4 show significant energy consumption reductions, compared to the single-speed 2WD vehicle (L1); in particular, the benefit ranges from 5.7% to 8.2% in normal mode, and from 8.0% to 9.8% in optimal mode, during the NEDC; and from 3.1% to 5.9% in normal mode, and from 5.7% to 9.0% in optimal mode, during the J10–15. The analysis of the powertrain efficiency profiles in the time domain highlighted that, for the specific EV parameters, the 2-speed and 4WD configurations allow a major enhancement of the powertrain efficiency, including motor drive and mechanical transmission, for the torque levels that are typical of the acceleration/deceleration phases of the cycles. Very interestingly, in both driving schedules, the best energy consumption performance is achieved by the single-speed 4WD EV (L3), which, with respect to L1, provides the most effective trade-off between the efficiency enhancement allowed by the increased operational flexibility of the electric machines, and the penalty associated with the mass increase.

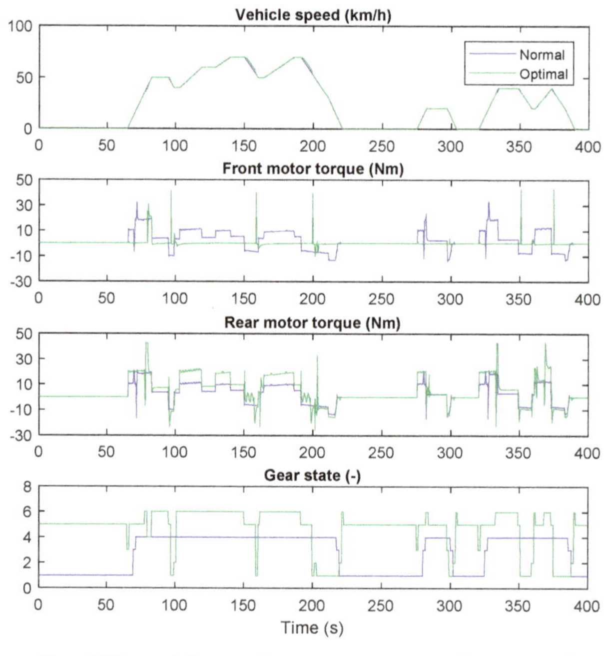

Figure 15. Time profiles of EV speed, front and rear motor torque, and gear state, for the 2-speed 4WD EV layout (L4) during a section of the J10–15 cycle.

Table 6. Energy consumption results along the selected driving cycles, and their percentage difference with respect to those of the single-speed 2WD EV (L1); negative differences, implying lower consumption, are in green.

Consumption	New European Drive Cycle (NEDC)—Urban Section (4 x ECE-15)				Japanese 10–15 Mode (J10–15)			
(Wh)	Normal		Optimal		Normal		Optimal	
L1: single-speed 2WD	240.1		–		418.2		–	
L2: 2-speed 2WD	220.4	−8.2%	219.1	−8.7%	393.4	−5.9%	394.3	−5.7%
L3: single-speed 4WD	223.2	−7.0%	216.5	−9.8%	402.5	−3.8%	380.4	−9.0%
L4: 2-speed 4WD	226.4	−5.7%	221.0	−8.0%	405.2	−3.1%	381.5	−8.8%

6. Conclusions

The implementation of 4-wheel-drive electric vehicle configurations with a single central motor per axle, as well as the adoption of 2-speed transmission systems, could lead to acceleration, gradeability and energy consumption benefits, in comparison with the conventional single-speed 2-wheel-drive electric vehicle configuration. However, the existing literature lacks analyses objectively comparing 2WD and 4WD EV layouts. To cover the gap, for a case study light passenger car prototype for urban mobility, this study compared four powertrain layouts. An optimization routine for the energy-efficient selection of the gear state and front-to-total wheel torque distribution was implemented to ensure fairness of the results. Based on the analysis for the specific EV, the main conclusions are as follows:

- The fast dynamics of typical traction motors, in conjunction with the low mechanical damping of electric drivetrains, can provoke significant resonances and torsional vibrations of the driveline. The electrical dynamics and mass moment of inertia of the electric machine/s, together with the transmission gear ratio and torsion stiffness of the half-shafts, are the key parameters affecting the magnitude and frequency of the resulting vibrations.
- 2-speed transmission systems and 4WD layouts add complexity and criticalities to the drivability response of the EV, and thus to the tuning of the anti-jerk controller. Without anti-jerk control, typical electric powertrains with central motor excite significant oscillations of the longitudinal acceleration of the EV during tip-in tests. The vibrations are effectively compensated by the proposed anti-jerk control scheme in all considered EV configurations, with a more reactive response of the 4WD layouts.
- At high torque demand, the gearshift dynamics significantly penalize the drivability and longitudinal acceleration performance of the EV layouts with the considered 2-speed transmission systems, as in these conditions the axle not involved in the gearshift cannot compensate for the wheel torque gap on the other axle. Despite this inconvenience, 2-speed transmissions expand the high efficiency operating region of the powertrain and increase top speed and gradeability.
- In terms of energy consumption, for the case study EV, during constant speed operation, i.e., at low torque demands, the prevailing effect is the mass increase associated with the advanced powertrain configurations. Hence, the best energy consumption performance is provided by the conventional single-speed 2WD EV layout. On the contrary, the increased flexibility in the operating conditions of the electric powertrains, provided by the more advanced EV architectures, reduces the energy consumption along driving cycles, with the single-speed 4WD vehicle achieving the lowest consumption.
- The single-speed 4WD layout does not require transmission controllers nor gearshift actuators, and, for the case study EV, can be considered an effective configuration, achieving a good compromise between predictable drivability, acceleration performance, gradeability for a variety of friction conditions, and energy consumption.

Future developments will extend the analysis of this paper to a wider range of case study EVs, with different electric motor technologies and performance characteristics of the powertrain components, to derive general conclusions and recommendations on the convenience of the analyzed layouts.

Author Contributions: Conceptualization and methodology, S.D.P., P.C., C.C., and A.S.; formal analysis, S.D.P., P.C., C.C., F.B., G.M., and A.S.; resources, P.P. and A.S.; writing—original draft preparation, S.D.P., P.C., C.C., F.B. and A.S.; writing—review and editing, A.S.; supervision, F.B., G.M. and A.S.; project administration and funding acquisition, P.P. and A.S. All authors have read and agreed to the published version of the manuscript.

Funding: The research leading to these results has received funding from the European Union's Seventh Framework Programme FP7/2007-2013 (PLUS-MOBY) under grant agreement n° 605502.

Conflicts of Interest: The authors declare no conflict of interest.

Nomenclature

a	Longitudinal vehicle acceleration.
$\widetilde{a}$	Longitudinal vehicle acceleration difference with respect to the steady-state value (see Section 3).
C	Longitudinal slip stiffness.
F	Force.
g	Gravitational acceleration.
h	Index corresponding to the vehicle speed value within the discrete grid (also used as a subscript).
H	Transfer function from wheel torque demand to angular vehicle acceleration.
i	Transmission gear ratio.
$IACA$	Integral (normalized with time) of the absolute value of the control action.
j	Index corresponding to the wheel torque value within the discrete grid (also used as a subscript).
J	Mass moment of inertia.
k	Torsional stiffness.
K	Anti-jerk controller gain.
l	Index indicating the front-to-total wheel torque distribution within the discrete grid (also used as a subscript).
L	Length.
m	Vehicle mass.
n	Index indicating the vehicle state number (also used as a subscript).
OS	Overshoot.
p	Torque distribution factor.
P	Power.
R	Radius.
RMS	Root mean square.
s, $j\omega$	Laplace operator and its imaginary part.
t	Time.
T	Torque.
V	Vehicle speed.
VDV	Fourth-power vibration dose value.
x	Gearbox actuator position.
α	Angle defining the longitudinal road gradient.
β	Torsional damping coefficient.
θ, $\dot{\theta}$, $\ddot{\theta}$	Angular position, speed, acceleration.
η	Efficiency.
σ	Slip ratio.
τ	Time constant or pure time delay.
ϕ	Phase angle difference between front and rear motor inputs.

The following subscripts and superscripts indices are used in this article:

act	Gearbox actuator.
$aero$	Aerodynamic drag.
app	Apparent.
AJ	Anti-jerk controller.
$AJ-off$	Deactivated anti-jerk controller.
$AJ-on$	Activated anti-jerk controller.
br	Brake.
$corr$	Corrective.
d	Delay.
dem	Demand.
des	Desired (or reference) value.
$diff$	Differential.
EM	Electric motor.
$EM+D$	Electric motor and drive.
eq	Equivalent.
f	Front.
$g1, g2, dis$	Gear 1, gear 2, disengaged gear.

HS	Half-shaft.
L	Left.
$loss$	Loss (referred to a power).
$peak$	Peak of the response.
PS, SS	Primary shaft, secondary shaft.
r	Rear.
rel	Tire relaxation.
req	Requested.
R	Right.
$resp$	Response time.
$roll$	Rolling resistance.
ss	Steady-state.
sel	Selected gear ratio (first, second or neutral).
$sett$	Settling time of the response.
tr	Transmission.
v	Vehicle.
w	Wheel.
$w\&f$	Windage and friction.
$*$	Optimal value.
0	Initial condition or linearization point.
1	Initial point of relevant interval.
2	Final point of relevant interval.

References

1. Wu, G.; Zhang, X.; Dong, Z. Powertrain architectures of electrified vehicles: Review, classification and comparison. *J. Frankl. Inst.* **2015**, *352*, 425–448. [CrossRef]
2. De Novellis, L.; Sorniotti, A.; Gruber, P.; Orus, J.; Rodriguez Fortun, J.M.; Theunissen, J.; De Smet, J. Direct yaw moment control actuated through electric drivetrains and friction brakes: Theoretical design and experimental assessment. *Mechatronics* **2015**, *26*, 1–15. [CrossRef]
3. Murata, S. Innovation by in-wheel-motor drive unit. *Veh. Syst. Dyn.* **2012**, *50*, 807–830. [CrossRef]
4. Shao, L.; Hartavi Karci, A.E.; Tavernini, D.; Sorniotti, A.; Cheng, M. Design approaches and control strategies for energy-efficient electric machines for electric vehicles—A review. *IEEE Access* **2020**, in press. [CrossRef]
5. Mutoh, N.; Kazama, T.; Takita, K. Driving characteristics of an electric vehicle system with independently driven front and rear wheels. *IEEE Trans. Ind. Electron.* **2006**, *53*, 803–813. [CrossRef]
6. Sinha, P.; Agrawal, V. Evaluation of electric-vehicle architecture alternatives. In Proceedings of the IEEE Vehicle Power and Propulsion Conference (VPPC), Chicago, IL, USA, 6–9 September 2011.
7. ISO. *Road Vehicles—Functional Safety*; ISO/FDIS Standard No. 26262:2018; ISO: Geneva, Switzerland.
8. Available online: https://www.rimac-automobili.com/en/hypercars/c_two/ (accessed on 15 June 2020).
9. Direct-Drive in-Wheel Motors. Available online: https://in-wheel.com/en/solutions/direct-drive-in-wheel-motors/ (accessed on 15 June 2020).
10. De Filippis, G.; Lenzo, B.; Sorniotti, A.; Gruber, P.; De Nijs, W. Energy-efficient torque-vectoring control of electric vehicles with multiple drivetrains. *IEEE Trans. Veh. Technol.* **2018**, *67*, 4702–4715. [CrossRef]
11. Tavernini, D.; Metzler, M.; Gruber, P.; Sorniotti, A. Explicit nonlinear model predictive control for electric vehicle traction control. *IEEE Trans. Control Syst. Technol.* **2019**, *27*, 1438–1451. [CrossRef]
12. De Novellis, L.; Sorniotti, A.; Gruber, P. Design and comparison of the handling performance of different electric vehicle layouts. *Proc. Inst. Mech. Eng. Part D J. Automob. Eng.* **2014**, *228*, 218–232. [CrossRef]
13. Ruan, J.; Walker, P.; Zhang, N. A comparative study energy consumption and costs of battery electric vehicle transmissions. *Appl. Energy* **2016**, *165*, 119–134. [CrossRef]
14. Sorniotti, A.; Loro Pilone, G.; Viotto, F.; Bertolotto, S.; Everitt, M.; Barnes, R.; Morrish, I. A novel seamless 2-speed transmission system for electric vehicles: Principles and simulation results. *SAE Int. J. Engines* **2011**, *4*, 2671–2685. [CrossRef]
15. Sorniotti, A.; Subramanyan, S.; Turner, A.; Cavallino, C.; Viotto, F.; Bertolotto, S. Selection of the optimal gearbox layout for an electric vehicle. *SAE Int. J. Engines* **2011**, *4*, 1267–1280. [CrossRef]
16. Spanoudakis, P.; Tsourveloudis, N.; Koumartzakis, G.; Krahtoudis, A.; Karpouzis, T.; Tsinaris, I. Evaluation of a 2-speed transmission on electric vehicle's energy consumption. In Proceedings of the IEEE International Electric Vehicle Conference (IEVC), Florence, Italy, 17–19 December 2014.
17. Gao, B.; Liang, Q.; Xiang, Y.; Guo, L.; Chen, H. Gear ratio optimization and shift control of 2-speed I-AMT in electric vehicle. *Mech. Syst. Signal Process.* **2015**, *50–51*, 615–631. [CrossRef]

18. Sorniotti, A.; Holdstock, T.; Loro Pilone, G.; Viotto, F.; Bertolotto, S.; Everitt, M.; Barnes, R.; Stubbs, B.; Westby, M. Analysis and simulation of the gearshift methodology for a novel two-speed transmission system for electric powertrains with a central motor. *Proc. Inst. Mech. Eng. Part D J. Automob. Eng.* **2012**, *226*, 915–929. [CrossRef]
19. Available online: https://www.porsche.com/international/aboutporsche/e-performance/magazine/e-performance-advantages/ (accessed on 9 May 2020).
20. De Pinto, S.; Sorniotti, A.; Gruber, P.; Camocardi, P.; Perlo, P.; Viotto, F. Gear shift control with torque-fill for a 4-wheel-drive fully electric vehicle. In Proceedings of the International Conference on Sustainable Mobility Applications, Renewables and Technology (SMART), Kuwait City, Kuwait, 23–25 November 2015.
21. De Pinto, S.; Camocardi, P.; Sorniotti, A.; Gruber, P.; Perlo, P.; Viotto, F. Torque-fill control and energy management for a four-wheel-drive electric vehicle layout with two-speed transmissions. *IEEE Trans. Ind. Appl.* **2016**, *53*, 447–458. [CrossRef]
22. Baraszu, R.; Cikanek, S. Torque fill-in for an automated shift manual transmission in a parallel hybrid electric vehicle. In Proceedings of the American Control Conference, Anchorage, AK, USA, 8–10 May 2002.
23. Kwon, K.; Seo, M.; Min, S. Efficient multi-objective optimization of gear ratios and motor torque distribution for electric vehicles with two-motor and two-speed powertrain system. *Appl. Energy* **2020**, *259*, 114190. [CrossRef]
24. Fiori, C.; Ahn, K.; Rakha, H.A. Power-based electric vehicle energy consumption model: Model development and validation. *Appl. Energy* **2016**, *168*, 257–268. [CrossRef]
25. Bottiglione, F.; De Pinto, S.; Mantriota, G.; Sorniotti, A. Energy consumption of a battery electric vehicle with infinitely variable transmission. *Energies* **2014**, *7*, 8317–8337. [CrossRef]
26. Holdstock, T.; Sorniotti, A.; Everitt, M.; Fracchia, M.; Bologna, S.; Bertolotto, S. Energy consumption analysis of a novel four-speed dual motor drivetrain for electric vehicles. In Proceedings of the IEEE Vehicle Power and Propulsion Conference (VPPC), Seoul, Korea, 9–12 October 2012; pp. 295–300.
27. Ruan, J.; Song, Q. A novel dual-motor two-speed direct drive battery electric vehicle drivetrain. *IEEE Access* **2019**, *7*, 54330–54342. [CrossRef]
28. Nguyen, C.T.; Walker, P.D.; Zhang, N.; Ruan, J. Efficiency improvement of a novel dual motor powertrain for plug-in hybrid electric buses. *Proc. Inst. Mech. Eng. Part D J. Automob. Eng.* **2020**, *234*, 1869–1882. [CrossRef]
29. De Carlo, M.; Mantriota, G. Electric vehicles with two motors combined via planetary gear train. *Mech. Mach. Theory* **2020**, *148*, 103789. [CrossRef]
30. Wu, J.; Liang, J.; Ruan, J.; Zhang, N.; Walker, P.D. Efficiency comparison of electric vehicles powertrains with dual motor and single motor input. *Mech. Mach. Theory* **2018**, *128*, 569–585. [CrossRef]
31. Scamarcio, A.; Gruber, P.; De Pinto, S.; Sorniotti., A. Anti-jerk controllers for automotive applications: A review. *Annu. Rev. Control* **2020**, in press.
32. Towards a Fast Uptake of mEdium/Low-Voltage Electric Powertrains (TELL). 2019. Available online: https://horizon2020-tell.eu/ (accessed on 9 May 2020).
33. Pacejka, H.B. *Tyre and Vehicle Dynamics*; Butterworth-Heinemann: Oxford, UK, 2005.
34. Amann, N.; Böcker, J.; Prenner, F. Active damping of drive train oscillations for an electrically driven vehicle. *IEEE/ASME Trans. Mechatron.* **2004**, *9*, 697–700. [CrossRef]
35. Rodriguez, J.; Meneses, R.; Orus, J. Active vibration control for electric vehicle compliant drivetrains. In Proceedings of the 39th Annual Conference of the IEEE Industrial Electronics Society (IECON), Vienna, Austria, 10–13 November 2013.
36. ISO. *Mechanical vibration and shock—Evaluation of human exposure to whole-body vibration—Part 1: General Requirements*; BS ISO 2631-1:1997; ISO: Geneva, Switzerland, 1997.
37. ISO. *Human Response to Vibration—Measuring Instrumentation— Part 1: General Purpose Vibration Meters*; BS EN ISO 8041-1:2017; ISO: Geneva, Switzerland, 2017.
38. Pennycott, A.; De Novellis, L.; Sabbatini, A.; Gruber, P.; Sorniotti, A. Reducing the motor power losses of a four-wheel drive, fully electric vehicle via wheel torque allocation. *Proc. Inst. Mech. Eng. Part D J. Automob. Eng.* **2014**, *228*, 830–839. [CrossRef]

energies

MDPI

Article

Co-Design of CVT-Based Electric Vehicles

Caiyang Wei [1,*], Theo Hofman [1] and Esin Ilhan Caarls [1,2]

[1] Department of Mechanical Engineering, Eindhoven University of Technology, P.O. Box 513, 5600 MB Eindhoven, The Netherlands; t.hofman@tue.nl (T.H.); Esin.IlhanCaarls@nl.bosch.com (E.I.C.)
[2] Bosch Transmission Technology, Postbus 500, 5000 AM Tilburg, The Netherlands
* Correspondence: c.wei.1@tue.nl; Tel.: +31-40-247-8325

Abstract: For an electric vehicle (EV) with a continuously variable transmission (CVT), a novel convex programming (CP)-based co-design method is proposed to minimize the total-cost-of-ownership (TCO). The integration of the electric machine (EM) and the CVT is the primary focus. The optimized system with co-design reduces the TCO by around 5.9% compared to a non-optimized CVT-based EV (based on off-the-shelf components) and by around 2% compared to the EV equipped with a single-speed transmission (SST). By taking advantage of the control and design freedom provided by the CVT, the optimal CVT, EM and battery sizes are found to reduce the system cost. It simultaneously finds the optimal CVT speed ratio and air-flow rate of the cooling system reducing the energy consumption. The strength of co-design is highlighted by comparing to a sequential design, and insights into the design of a low-power EV that is energy-efficient and cost-effective for urban driving are provided. A highly integrated EM-CVT system, which is efficient, low-cost and lightweight, can be expected for future EV applications.

Keywords: electric vehicle; co-design; continuously variable transmission; component sizing; optimal control; convex programming; energy efficiency

Citation: Wei, C.; Hofman, T.; Ilhan Caarls, E. Co-Design of CVT-Based Electric Vehicles. *Energies* **2021**, *14*, 1825. https://doi.org/10.3390/en14071825

Academic Editor: João Pedro Trovao

Received: 11 February 2021
Accepted: 19 March 2021
Published: 25 March 2021

1. Introduction

Growing concerns of environmental contamination and depletion of natural resources have led to the resurgence of electric vehicles (EVs). Admittedly, the emerging EV market is mainly dominated by single-speed transmissions (SSTs). Nevertheless, to optimize key performance indicators (KPIs), for example, energy consumption, system cost and performance, research on multi-speed transmissions is gaining popularity, such as two-speed transmissions and continuously variable transmissions (CVTs) [1–4].

To date, design of EVs (including CVT-based) is largely based on off-the-shelf components due to cost, and powertrain components are typically not optimized [5]. The component size, for example, the electric machine (EM) size in kW and battery size in kWh, is associated with the component cost. Finding optimal component sizes would contribute to cost reduction. The component size also influences the vehicle performance, such as top speed and acceleration time. Moreover, EV energy consumption is largely influenced by the driveline efficiency. In this respect, the integration of the EM and the CVT plays a key role [6]. The CVT could provide opportunities of optimizing the EM, thanks to the continuous ratio adjustment functionality. The wider power availability of the EM, in turn, could offer opportunities of optimizing the CVT. In literature, however, the EM and the CVT are often treated independently. Standard CVT controllers developed for conventional vehicles are used, where the speed ratio of the CVT (control) is selected to reduce the EM power losses. The influence of the CVT efficiency, CVT and EM size (design) are not always considered simultaneously [4]. In order to maximize system efficiency and minimize system cost, the coupling between the EM and the CVT from design and control perspectives has yet to be investigated. In order to address these issues, design and control frameworks are required.

This type of design problem can be tackled by different approaches. An overview of design and control optimization methodologies is given in [7,8]. As presented in [7], there are generally four approaches to solve the design (plant) and control (controller) problem. With respect to solution optimality, simultaneous and nested coordination schemes outperform iterative and sequential ones, although depending on the coupling and how sensitive the solution is to uncertainties in the design parameters. For instance, the optimal component sizes can be found iteratively, by using rule-based control [9]. However, these methods do not guarantee a globally optimal solution. To overcome this limitation, dynamic programming (DP), as an optimization-based strategy, can be used to find the optimal control inputs [10]. DP can also be utilized in combination with an evolutionary algorithm to find component sizes, such as particle swarm optimization (PSO). In the search space, PSO improves a candidate solution iteratively by moving particles towards the best known positions [11,12]. It can solve complex objective functions and problems with large design space. PSO is used in [13] to find optimal component sizes of an electrified powertrain. However, it requires tuning effort and has long computation time. Furthermore, an optimal solution is not guaranteed. An alternative is to use convex programming (CP) as a simultaneous approach, which is also a co-design optimization method. It finds optimal design parameters and control trajectories simultaneously without the need of checking optimality [14,15]. It ensures a unique optimum with a convex objective function and a convex feasible region. It is also computationally efficient. Due to the low computation time, CP enables optimization of problems with many dynamic states, such as thermal states, which may not be tractable by using DP for example [16].

Moreover, employing a CVT for an EV is a novel concept. Applying advanced optimization methods to optimize KPIs of CVT-based EVs have hardly been found [6]. Additionally, current research mainly concentrates on the energy domain to increase the energy efficiency, taking into account the mechanical and electrical energy flows [17]. The thermal domain, however, for instance, evaluation of cooling power consumption and temperature profile, has yet to be explored. It is also an integral part of an EV, which would affect the total energy consumption [18].

Considering the integration of the EM and CVT, as shown in Figure 1, the CVT speed ratio over time (γ_v) for a use case changes the EM and CVT operating points. Therefore, it influences the EM ($P_{m,loss}$) and CVT ($P_{c,loss}$) power losses. Furthermore, the desired CVT speed ratio affects the CVT size (s_γ), for example, the ratio coverage of the CVT, and the EM size (s_T) to meet the power demand for example. The CVT and EM sizes are related to their costs. The CVT size that determines the ratio range, in turn, affects the CVT speed ratio and the EM size. The EM size also has an effect on the CVT size and its speed ratio. In addition, the component size affects its own efficiency. Moreover, the battery size and losses are affected indirectly. Hence, the hypothesis is that there exists an optimal combination of the CVT speed ratio over time, battery size, EM size and the CVT size for an application, which results in the lowest total-cost-of-ownership (TCO). TCO consists of energy consumption and system cost. The system cost comprises the CVT, EM (including inverter) and battery prices. Given the interactions between these variables, this optimal solution for a use case can best be found by means of a simultaneous approach in an efficient manner.

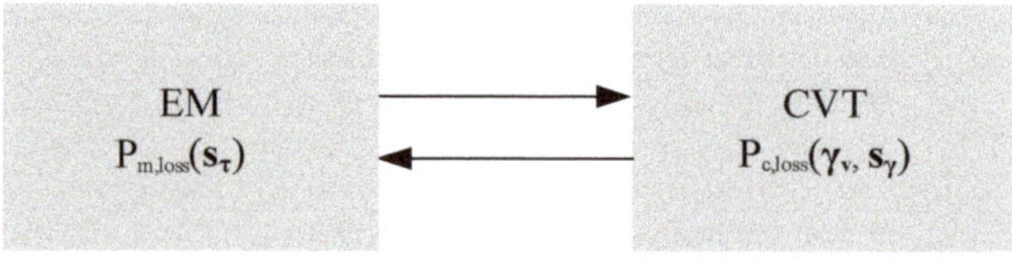

Figure 1. Schematic representation of the coupling between the continuously variable transmission (CVT) and electric machine (EM), where γ_v represents the CVT speed ratio, s_γ the CVT size, s_T the EM size, $P_{c,loss}$ the CVT power losses, and $P_{m,loss}$ the EM power losses.

Motivated by the above discussion, this study originally proposes a CP-based co-design optimization strategy for a CVT-based EV to reduce the TCO. It identifies the optimal CVT speed ratio over time and the desired air-flow rate of a thermal management system (TMS) to maintain the EM temperature below its thermal limit. Furthermore, it finds the optimal sizes of the CVT, EM and battery. The approach is as follows:

S_1: An SST-based EV model including energy dynamics and thermodynamics with reference to a series production vehicle is firstly created. It is developed based on static efficiency maps represented by lookup tables, which is validated against measurement data from real-world driving. It replicates the physical behavior of the vehicle in reality.

S_2: A CVT-based EV model is then developed based on S_1, where only the SST is replaced by a CVT (an off-the-shelf component, which is not optimized). Other components, for example, the battery and EM, are the same. The CVT model is created based on experimental data from a test rig.

S_3: Component models from S_2 are convexified to fit the measurement data from real-world driving and experimental data from the test rig. S_2 is subsequently optimized with the co-design optimization strategy.

All the systems (S_1, S_2 and S_3) have the same maximum EM power and similar vehicle performance (i.e., 0–100 km/h acceleration time below 11 s, top speed above 165 km/h, gradability above 30%). Therefore, the goal of this study is to show the advantages of the co-design approach in optimizing a CVT-based EV (e.g., sizing) and to compare TCO between S_1, S_2 and S_3. System cost comprising the expenses of the battery, EM and CVT are solely given for the implementation of the co-design approach. Moreover, the strengths of the co-design method are highlighted, by comparing with a sequential approach, where the EM size is fixed to meet performance requirements. Additionally, for urban driving that does not require high performance, insights into the design of a low-power EV are provided.

2. Problem Definition

The configuration of the considered EV is demonstrated in Figure 2. The integration of the EM and the CVT is the primary focus, which are highlighted in bold. The major components of the EV are the battery, DC (direct current)-DC converter, DC-AC (alternating current) inverter, EM, CVT, electric oil pump (ELOP), and vehicle. If not specified, the DC-AC inverter and EM are combined together in this study. The final drive (FD) that takes a constant value and a fixed efficiency is lumped into the variator (VA), which together is regarded as CVT. In order to change the CVT speed ratio, hydraulic actuation power is required from the ELOP. The ELOP power is supplied by the DC-DC converter onboard, which is assumed to be always charged. The EM is directly connected to the input shaft of the CVT without a pre-reduction gear. The battery provides the power requested by the EM. The EV model describes the longitudinal dynamics. It is backward-facing, that is, the drive cycle is given, with a discrete time-step of one second using time index k. The vehicle inertia is considered. The main model parameters are listed in Table A1 (Appendix D), including vehicle parameters and validated thermal parameters.

The main design criterion to find the optimal control and design variables is the minimization of the TCO, which consists of the consumed electricity cost J_e and system cost J_s, given by

$$\min_{\boldsymbol{s},\boldsymbol{x}(k),\boldsymbol{u}(k)} \quad J_e(\boldsymbol{s},\boldsymbol{x}(k),\boldsymbol{u}(k) \mid w(k)) + J_s(\boldsymbol{s} \mid w(k)), \tag{1}$$

$$\text{s.t.} \quad \boldsymbol{x}(k+1) = \boldsymbol{x}(k) + \boldsymbol{f}(\boldsymbol{s},\boldsymbol{x}(k),\boldsymbol{u}(k),w(k))\,\Delta t, \tag{2}$$

$$\boldsymbol{h}(\boldsymbol{s},\boldsymbol{x}(k),\boldsymbol{u}(k)) = 0, \tag{3}$$

$$\boldsymbol{g}(\boldsymbol{s},\boldsymbol{x}(k),\boldsymbol{u}(k)) \leq 0, \tag{4}$$

where Δt is the time step. The design variables $\boldsymbol{s}$ consist of the ratio coverage of the CVT s_γ, the scaling factor s_τ for scaling the maximum EM torque and the scaling factor s_b for scaling the battery cells, that is,

$$\boldsymbol{s} = [s_\gamma, s_\tau, s_b]^T. \tag{5}$$

The state variables $\boldsymbol{x}(k)$ are the state-of-energy of the battery and temperature states of the TMS, which are described in Section 3.7, given by

$$\boldsymbol{x}(k) = [E_b(k), \theta_m(k), \theta_c(k), \theta_o(k), \theta_i(k)]^T. \tag{6}$$

The control variables $\boldsymbol{u}(k)$ are the speed ratio of the CVT (γ_v) and the air-flow rate of the TMS (ϕ_a) to keep the EM temperature below its prescribed thermal limit, that is,

$$\boldsymbol{u}(k) = [\gamma_v(k), \phi_a(k)]^T. \tag{7}$$

The power balance of the vehicle is represented by (3), and (4) represents the feasible design space, where the design, state and control variables are bounded. Equation (4) also represents the component limits. The disturbance vector $w(k)$ contains vehicle speed (v_v) and acceleration (a_v), which are prescribed by the drive cycle, given by

$$w(k) = [v_v(k), a_v(k)]^T. \tag{8}$$

The consumed electricity cost J_e over the drive cycle represented by $w(k)$ starting at $k = 1$ and ending at $k = N$ is obtained by

$$J_e(\boldsymbol{s}, \boldsymbol{x}(k), \boldsymbol{u}(k)) = \sum_{k=1}^{N} \rho_e \, P_b(\boldsymbol{s}, \boldsymbol{x}(k), \boldsymbol{u}(k)) \, \Delta t, \tag{9}$$

where ρ_e is the price of electricity (€/kWh). The term $P_b(\boldsymbol{s}, \boldsymbol{x}, \boldsymbol{u})$ represents the battery output power. The system cost J_s over the drive cycle is calculated by

$$J_s(\boldsymbol{s}) = \frac{S_d}{S_v} \left(C_c(s_\gamma) + C_m(s_\tau) + C_b(s_b) \right), \tag{10}$$

where S_d is the length of the drive cycle (km) and S_v the traveled distance of the vehicle in its lifetime. The variable $C_c(s_\gamma)$ represents the CVT cost, $C_m(s_\tau)$ the EM cost (including inverter) and $C_b(s_b)$ the battery cost.

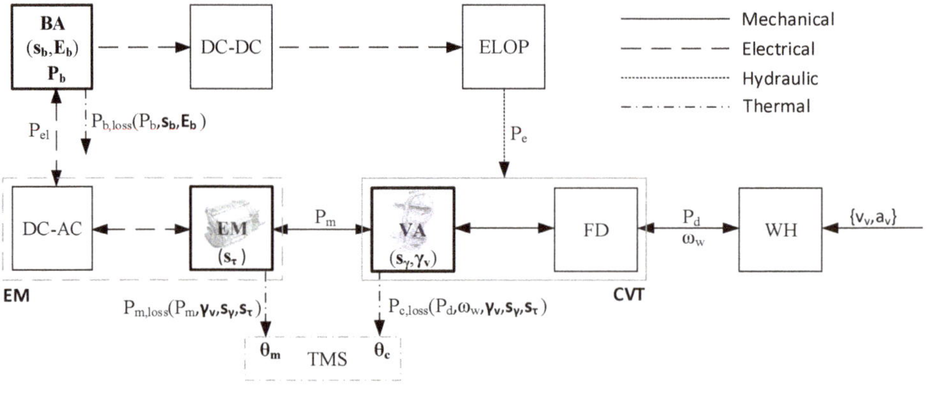

Figure 2. A CVT-based electric vehicle, where BA represents the battery, DC-AC the DC (direct current) to AC (alternating current) inverter, DC-DC the DC to DC converter, ELOP the electric oil pump, FD the final drive, VA the variator, and WH the wheel. Design, control, and state variables are highlighted in bold.

Based on the configuration shown in Figure 2, three systems, namely S_1, S_2 and S_3, are developed and compared in terms of TCO as in (1). Three systems have the same battery. Although battery aging is not the focus of this study, the methodology presented also applies to scenarios where battery needs to be changed. S_1 is developed with reference to the series production vehicle, which is described in Appendix A. S_2 is created based on S_1, where only the SST is replaced by the CVT, which is presented in Appendix B. In S_1 and S_2, the component models are represented by experimentally-based lookup tables (efficiency maps), which describe the power generation or power dissipation of each component. The design variables are fixed. In S_2, the CVT speed ratio is predetermined by a low-level CVT controller, which is developed based on [19]. The goal of the CVT controller is to reduce the EM power losses depending on the power demand, which is common in literature, such as [20]. The air-flow rate of the TMS is tuned to maintain the EM temperature below its thermal limit. In S_3, convex models are developed based on measurements. The design and control variables are to be determined by the co-design optimization strategy.

This section formulates the co-design optimization problem of a CVT-based EV. To solve this co-design optimization problem, the required convex EV model based on measurements, especially the mathematical coupling between the CVT and EM from design and control perspectives, is presented in more detail in the next section.

3. System Modeling

This section presents the convex EV model and cost models needed for solving the co-design problem defined in (1). An introduction to CP is first given in Appendix C, which serves as a guideline for developing convex models. A data-driven approach used to derive the convex models is then provided in Section 3.1. Subsequently, given the drive cycle in Section 3.2, the vehicle longitudinal dynamics are described in Section 3.3, which is an input to the convex models in S_3. Given the input, however, the torque input to the CVT is not available. Because the CVT speed ratio is a control variable, which is not known in advance and will be determined by the co-design optimization strategy. Dealing with torque information will lead to non-convexity (Appendix C). Therefore, to preserve convexity for the co-design optimization problem, all the relations are converted to power level. By utilizing the data-driven approach as mentioned above, three convex models are consequently developed, namely the CVT power loss model (Section 3.4), EM power loss model (Section 3.5) and the EM power limitation model (Section 3.6). Equality constraints are also relaxed with inequalities where applicable, and the equality holds at the optimum. Notice that, this study mainly focuses on the CVT, EM and battery sizing, and the ELOP sizing is not required at this level. Therefore, the ELOP is only considered in the calculation of energy consumption, and no convex representation is required. The ELOP power losses are computed offline, meaning the ELOP power losses are obtained based on the optimal CVT speed ratio found by the optimization algorithm and then added to the energy consumption. Furthermore, to remove the heat (power losses) generated by the EM and CVT, the TMS and its associated thermal model are presented in Section 3.7. Eventually, the required power is supplied by the battery, which is described in Section 3.8. The convex battery model is developed based on physics [21]. Additionally, in Section 3.9, convex mass and cost models related to CVT, EM and battery sizes are developed for the implementation of the co-design approach.

3.1. Derivation of Convex Models

As mentioned in Appendix C, the component models, for example, the power loss models, are required to be convex for the CP algorithm. The general idea to derive the convex models based on a data-driven approach can be seen in Figure 3, where measurements are used as inputs for the modeling. The measurement data (top) about the CVT and EM, for example, power losses, torque, speed, and ratio over time, are first served as inputs. Based on these measurements, a set of convex expressions capturing input and output relationships with sufficient accuracy is created. With different combinations of those expressions, for example, CVT and EM power loss models, are then developed. These parameterized models are subsequently fitted to capture the loss behaviors of components with sufficient accuracy, which have a negligible impact on the result. Fourth, the convex models are utilized by the co-design optimization method to find the optimal solution of the system.

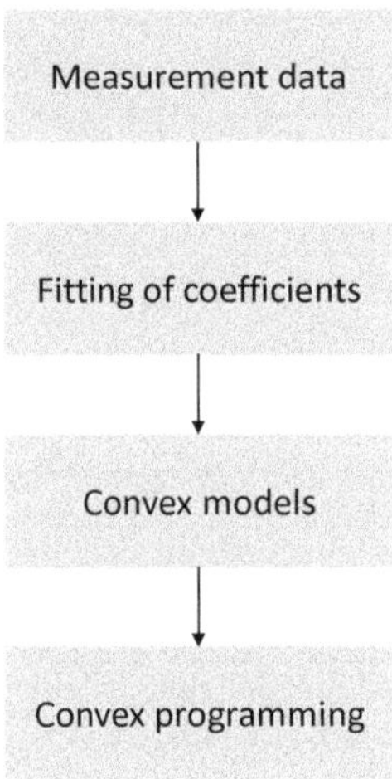

Figure 3. A data-driven approach to derive convex models. In order to apply convex programming, constraints and cost function are also required to be convex.

3.2. Drive Cycle

In this work, a realistic and representative real-world drive cycle, Worldwide Harmonized Light Vehicles Test Cycles (WLTC), is chosen. It is currently widely adopted in the automotive sector to certify energy consumption. It includes low, medium, high, and extra high speed scenarios, which can represent, for example, urban, rural, and highway driving conditions. In order to show the thermal effect, two repeated WLTC is used. The drive cycle contains the vehicle speed $v_{\mathrm{v}}(k)$ and acceleration $a_{\mathrm{v}}(k)$. Note that, although the quantity (e.g., energy consumption) might vary, the methodology presented also applies to other drive cycles.

3.3. Longitudinal Dynamics

Note that, in this case the total vehicle mass is a variable because of the scaling factors for the CVT, EM and battery, which will be presented in the next sections. Considering all the forces acting on the vehicle, the power demand for the known WLTC can be obtained by

$$P_{\mathrm{d}}(k) = \left(\frac{1}{2}\,\rho_{\mathrm{a}}\,c_{\mathrm{d}}\,A_{\mathrm{f}}\,v_{\mathrm{v}}^{2}(k) + c_{\mathrm{r}}\,m_{\mathrm{v}}\,g\,\mathrm{sign}(v_{\mathrm{v}}(k)) + \left(m_{\mathrm{v}} + 4\,\frac{J_{\mathrm{w}}}{r_{\mathrm{w}}^{2}}\right)a_{\mathrm{v}}(k)\right)v_{\mathrm{v}}(k), \tag{11}$$

where the total vehicle mass m_{v} is given by

$$m_{\mathrm{v}} = m_{\mathrm{cw}} + m_{\mathrm{c}} + m_{\mathrm{m}} + m_{\mathrm{b}} + m_{\mathrm{d}}, \tag{12}$$

where m_{cw} is the curb weight excluding the CVT (m_{c}), EM (m_{m}) and battery (m_{b}) mass (Table 1). m_{d} is the driver mass.

Table 1. Comparison of component parameters between S_1, S_2 and S_3.

Parameter	Unit	S_1	S_2	S_3
Transmission ratio	-	9.02	[4.47, 16.8]	[3.5, 11.68]
EM scaling factor	-	1	1	0.79
Battery cells	-	264	264	253
Maximum EM torque	Nm	290 ($\overline{\tau}_m$)	290	228
EM base speed	rpm	3293 ($\omega_{m,b}$)	3293	4188
Maximum EM power	kW	100 ($\overline{P}_m$)	100	100
Curb weight	kg	1252 (m_{cw})	1252	1252
Transmission mass	kg	26 (m_s)	56 ($\overline{m}_c$)	52
EM mass	kg	74 ($\overline{m}_m$)	74	58
Battery mass	kg	318 ($\overline{m}_b$)	318	303
Driver mass	kg	90 (m_d)	90	90

3.4. Convex CVT Model

Convex modeling of the CVT is shown in Figure 4, taking into account the effect of design (s_γ,s_τ) and control (γ_v) on the CVT power losses ($P_{c,loss}$).

Based on the measurement data, the CVT torque losses $\tau_{c,loss}$ can be expressed as a function of its input torque τ_c, input speed ω_p, and ratio over time γ_v. As shown in Figure 2, the relationships in S_3 are converted to the power domain. At given input speeds to the CVT, the CVT torque losses $\tau_{c,loss}$ are converted to their corresponding power losses $P_{c,loss}$, that is,

$$P_{c,loss}(k) = \tau_{c,loss}(k)\, \omega_p(k). \tag{13}$$

On the basis of these inputs (τ_c, ω_p, and γ_v) to the CVT, the outputs (P_d and w_w) of the CVT can be determined as

$$P_d(k) = \tau_c(k)\, \omega_p(k) - P_{c,loss}(k), \tag{14}$$

$$w_w(k) = \frac{\omega_p(k)}{\gamma_v(k)}. \tag{15}$$

The input-output relationship of the CVT implies that the CVT power losses can also be formulated on system level as (Figure 4a):

$$P_{c,loss}(k) = P_{c,loss}(P_d(k), w_w(k), \gamma_v(k)). \tag{16}$$

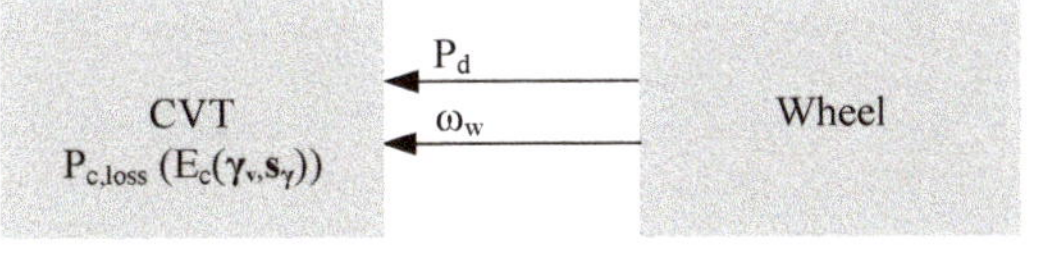

(a) Effect of design (s_γ) and control (γ_v) on the CVT power dissipation

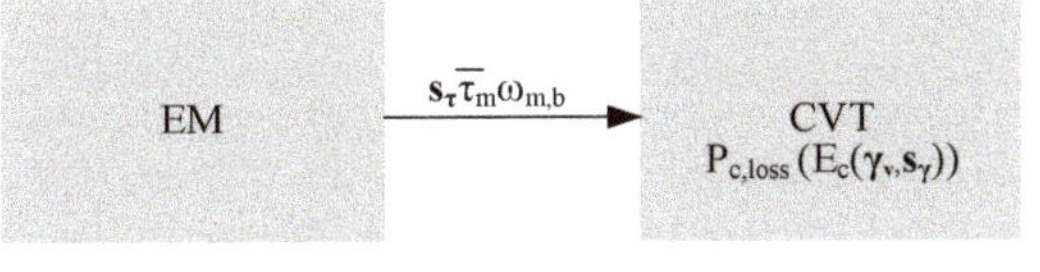

(b) Effect of EM design (s_τ) on the CVT power dissipation

Figure 4. Convex modeling of the CVT, where $\overline{\tau}_m$ represents the maximum EM torque and $\omega_{m,b}$ the EM base speed.

Clearly, the CVT power dissipation is influenced by the control variable γ_v. However, while P_d and w_w are known here, the CVT speed ratio γ_v will be decided by the optimization algorithm. Dealing with this speed ratio alone would lead to non-convexity (Appendix C). Hence, considering the co-design problem and in order to preserve convexity, the information of γ_v is embedded in another variable E_c, by applying a change of variables [22], given by

$$E_c(k) = \gamma_v^2(k)\ \omega_w^2(k). \tag{17}$$

Notice that there is no information loss and the optimization strategy will assess every possible combination of γ_v and ω_w. There are also two other reasons of selecting E_c. Firstly, the CVT is a rotating mechanical component, and this term is closely related to its kinetic energy. Secondly, in practice, the CVT speed ratio is often determined based on the EM speed $\gamma_v\ \omega_w$ and wheel speed ω_w, since there is always torque loss in between, but there is no speed loss. Therefore, the CVT power dissipation is affected by $E_c(\gamma_v)$ (Figure 4a), as it contains the information of speed ratio. Furthermore, the variable $E_c(\gamma_v)$ is influenced by the ratio coverage of the CVT s_γ, which is the range that the CVT can actually shift, as illustrated in Figure 5. In this figure, $\underline{\gamma}_v$ and $\overline{\gamma}_v$ are the overdrive ratio and underdrive ratio of the CVT, respectively. The ratio coverage of the CVT is defined by

$$s_\gamma = \overline{\gamma}_v / \underline{\gamma}_v. \tag{18}$$

The CVT sizing is carried out on the basis of the design variable s_γ, because it is one of the most influencing factors that affect cost, efficiency, drivability, and packaging of CVT and powertrain [6]. Note that there are other factors that affect the variator efficiency and hence the CVT loss, such as the center distance of the pulleys and variator asymmetry. Since this study focuses on system-level representation and there is no pre-reduction between the EM and CVT, these factors are not considered. The underdrive ratio $\overline{\gamma}_v$ and the overdrive ratio $\underline{\gamma}_v$ are known. In order to avoid non-convexity for the co-design problem (Appendix C), the overdrive ratio $\underline{\gamma}_v$ is fixed in this study. Therefore, constraints on γ_v, E_c, and s_γ are

$$\gamma_v(k) \in [\underline{\gamma}_v, s_\gamma\ \underline{\gamma}_v], \tag{19}$$

$$E_c(k) \in [\underline{\gamma}_v^2\ \omega_w^2(k), s_\gamma^2\ \underline{\gamma}_v^2\ \omega_w^2(k)], \tag{20}$$

$$s_\gamma^2 \in [\underline{s}_\gamma^2, \overline{s}_\gamma^2]. \tag{21}$$

The CVT power dissipation in (13) is measured based on an original CVT product (sold on the current market) with a full ratio range of [0.38, 2.63]. Therefore, it has a large ratio coverage, torque capacity ($\overline{\tau}_c$) and power capacity ($\overline{P}_c$). In case of a CVT with reduced ratio coverage (s_γ), smaller ELOP and on-demand actuation, resulting in a compact and small CVT (superscript "s") with smaller power capacity (P_c^s), a higher CVT efficiency is expected [6,23]. This higher efficiency potential means a lower power dissipation. Compared with the original CVT power losses $P_{c,loss}$ in (16), this lower power dissipation can be modeled by a multiplier μ_c, which is a function of the EM scaling factor, giving

$$P_{c,loss}^s(k) = \mu_c(s_\tau)\ P_{c,loss}(k). \tag{22}$$

This multiplier is sensitive to many parameters and technological advances. For example, as reported in [24], the maximum efficiency of the current variator could be above 98%. Notice that, the final drive does not benefit from this loss reduction. Additionally, the current CVT and EM are designed separately, which leads to a mismatch between the specifications. In this work, as shown in Figure 2, the EM is connected to the CVT without a pre-reduction gear. The required CVT torque capacity is determined by the output of the EM $s_\tau\ \overline{\tau}_m$, as demonstrated in Figure 4b. Here, s_τ, the scaling factor for the EM, is a design variable, which will be explained in Section 3.5. Thus, the multiplier $\mu_c(s_\tau)$ can be obtained by [23]

$$\begin{aligned}\mu_c(s_\tau) =&0.7 + 0.3\frac{\overline{P}_c^s}{\overline{P}_c}\\ =&0.7 + 0.3\frac{s_\tau\,\overline{\tau}_m\,\omega_{m,b}}{\overline{\tau}_c\,\omega_{m,b}}\\ =&0.7 + 0.3\frac{s_\tau\,\overline{\tau}_m}{\overline{\tau}_c},\end{aligned} \tag{23}$$

where $\overline{\tau}_m$ is the maximum EM torque when $s_\tau = 1$ and $\omega_{m,b}$ is the EM base speed. Similar to [23], this multiplier can be interpreted as a proportional improvement of the more efficient variator, which typically accounts for thirty percent of the total power losses in previous applications.

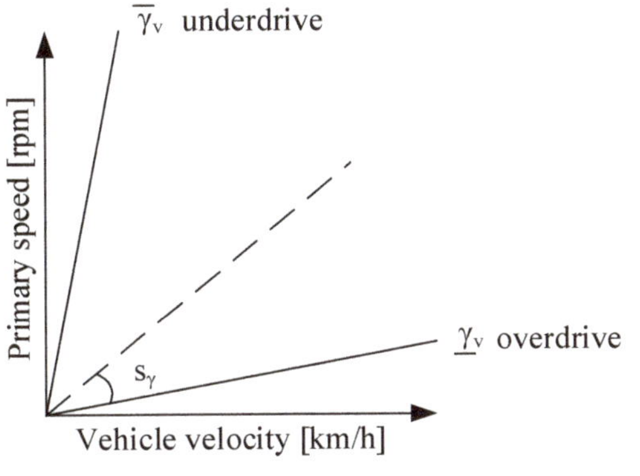

Figure 5. Ratio coverage of the CVT.

Consequently, the factors that influence the CVT power losses are P_d, w_w, $E_c(\gamma_v, s_\gamma)$, and s_τ (Figure 4a,b). Taking into account convexity and possible combinations of expressions, various models are developed to represent the CVT power losses. The models are fitted to capture the loss behavior of the CVT. Based on the evaluated fitting accuracy, the convex CVT model is identified as follows:

$$P^s_{c,loss}(k) = c_{c,0}\left(\frac{P_d(k)}{w_w(k)}\right)^2 + c_{c,1}\,E_c(k) + c_{c,2}\,|P_d(k)| + c_{c,3}\,s_\tau\,w_w(k) + c_{c,4}\,s_\tau + c_{c,5}, \tag{24}$$

where $c_{c,0}$, $c_{c,1}$, $c_{c,2}$, $c_{c,3}$, $c_{c,4}$ and $c_{c,5}$ are the corresponding coefficients. Notice that the power losses are always constrained to be equal to (i.e., vehicle velocity $v_v = 0$) or larger than zero. This model is convex (Appendix C) and has a fitting accuracy of around 98%. It should be noted that, information is exchanged between the terms. For example, the information of s_τ is not only explicitly expressed in s_τ and $s_\tau\,w_w$, but also implicitly embedded in other terms. The term $s_\tau\,w_w$ is important for CVT applications, as generally CVT does not operate at very high speeds, yet would lead to lower friction losses of the EM. Another design variable (s_γ) and control variable (γ_v) are reflected in E_c. While the information of speed and torque are conveyed already by $|P_d|$ and $s_\tau\,w_w$, it is further reinforced by torque squared $(\frac{P_d}{w_w})^2$ and speed squared E_c to capture essential CVT dynamics. An example of $\gamma_v = 0.7$ and $s_\tau = 1$ is shown in Figure 6, where the CVT power dissipation is a function of the power demand and $E_c(\gamma_v, s_\gamma)$. It can be seen that the developed convex CVT model and the original model based on the measurement data resemble well. The difference between them, which is represented by the absolute error, is small for the relevant range. It means that the developed convex model captures the loss behavior of the CVT with sufficient accuracy. Such correlations in data analysis can be done by clustering the data sets for certain physical attributes. In this case, the CVT power dissipation is considered as the attribute.

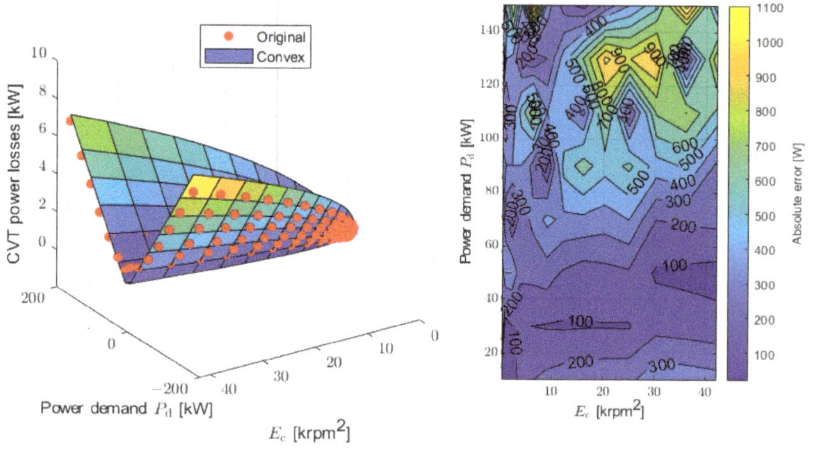

Figure 6. Convex CVT power loss model for $\gamma_v = 0.7$ and $s_\tau = 1$. The absolute error represents the difference in power loss between the convex model and the original model based on the measurement data.

3.5. Convex EM Model

Convex modeling of the EM power losses ($P_{\text{m,loss}}$) is illustrated in Figure 7, taking into consideration the interconnections between the EM and CVT from design and control perspectives.

To obtain the convex EM model, the steps in Figure 3 are followed, as it is done for the CVT in Section 3.4. As shown in Figure 2, the EM is directly linked to the input shaft of the CVT and provides the power requested by the CVT, referring to (22), that is,

$$P_{\text{m}}(k) = P_{\text{d}}(k) + P^{\text{s}}_{\text{c,loss}}(k). \tag{25}$$

Based on the measurement data, the EM power losses can be expressed as a function of its output torque τ_{m} and speed ω_{m}. The EM power dissipation is given by

$$P_{\text{m,loss}}(k) = P_{\text{m,loss}}(\tau_{\text{m}}(k), \omega_{\text{m}}(k)). \tag{26}$$

On the basis of the output torque and speed of the EM, its output power can be calculated by

$$P_{\text{m}}(k) = \tau_{\text{m}}(k)\, \omega_{\text{m}}(k). \tag{27}$$

Combing (26) and (27), as depicted in Figure 7, leads to

$$P_{\text{m,loss}}(k) = P_{\text{m,loss}}(\tau_{\text{m}}(k), \omega_{\text{m}}(k), P_{\text{m}}(k)). \tag{28}$$

Moreover, the EM power dissipation $P_{\text{m,loss}}$ is influenced by its size (in torque and speed). In this work, the maximum EM torque ($\overline{\tau}_{\text{m}}$) is scaled down by using the scaling factor s_τ, as shown in Figure 8. The base speed is increased ($\omega^s_{\text{m,b}}$) such that the maximum output power is maintained, that is,

$$\overline{P}_{\text{m}} = \overline{\tau}_{\text{m}}\, \omega_{\text{m,b}} = s_\tau\, \overline{\tau}_{\text{m}}\, \omega^s_{\text{m,b}}. \tag{29}$$

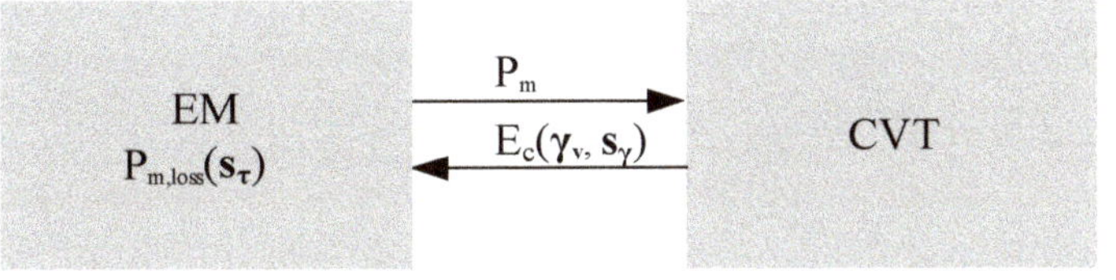

Figure 7. Convex modeling of the EM, where P_{m} represents the EM output power.

The EM sizing is performed based on the design variable s_{τ}, because it affects the EM efficiency, weight and cost. For example, reducing the maximum EM torque decreases the usage of active materials of the EM. Note that because of physical limitations of the CVT, the EM speeds above 6500 rpm are not used for the CVT application, which has a negligible impact on the result.

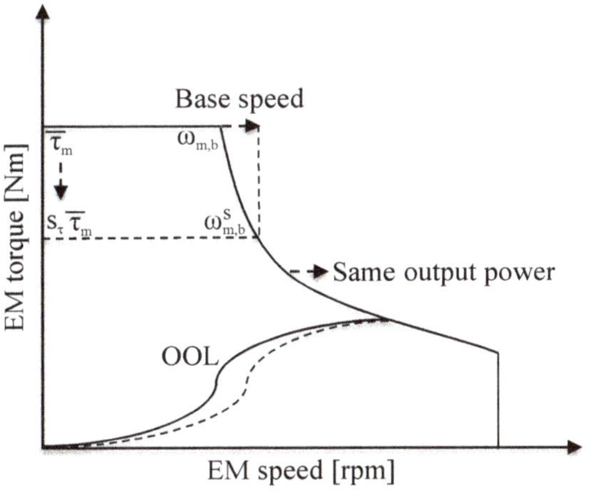

Figure 8. Scaling factor of the EM including the DC-AC inverter, where the solid lines represent the original EM, dashed lines the scaled EM while keeping the same output power, and optimal operating line (OOL) the optimal operating line minimizing the EM losses for every power demand.

The EM scaling factor s_{τ} influences the EM torque, optimal operating line (OOL) and power losses, which is bounded by

$$s_{\tau} \in [\underline{s}_{\tau}, \overline{s}_{\tau}]. \tag{30}$$

As illustrated in Figure 4b, the scaling of EM has a direct effect on the CVT design requirements. Furthermore, as explained before, the EM power dissipation $P_{\mathrm{m,loss}}$ is influenced by another design variable s_{γ} and control variable γ_{v}, which change its operating point and hence the power losses. The information of s_{γ} and γ_{v} are represented by E_{c}.

To sum up, the identified parameters that affect the EM power losses are P_{m}, s_{τ}, and $E_{\mathrm{c}}(\gamma_{\mathrm{v}}, s_{\gamma})$, which are shown in Figure 7. Models that contain these parameters and their combinations, which are also potentially convex, are built. In a similar fashion as with the convex CVT modeling (Section 3.4), the EM models are fitted to capture the loss behavior of the EM. On the basis of fitting accuracy, the convex EM model is found as follows:

$$P_{\mathrm{m,loss}}(k) = c_{\mathrm{m},0}\,\frac{P_{\mathrm{m}}^{2}(k)}{E_{\mathrm{c}}(k)} + c_{\mathrm{m},1}\,E_{\mathrm{c}}(k) + c_{\mathrm{m},2}\,|P_{\mathrm{m}}(k)| + c_{\mathrm{m},3}\,s_{\tau} + c_{\mathrm{m},4}, \tag{31}$$

where $c_{m,0}$, $c_{m,1}$, $c_{m,2}$, $c_{m,3}$ and $c_{m,4}$ are the corresponding coefficients. The information of speed and torque are conveyed already by $|P_m|$, and it is further reinforced by torque squared $\frac{P_m^2}{E_c}$ and speed squared E_c to capture essential EM dynamics. Based on Appendix C, this model is convex. It has a fitting accuracy of around 95%. An example of $s_\tau = 1$ is illustrated in Figure 9, where the EM power dissipation is a function of the EM power and $E_c(\gamma_v, s_\gamma)$. A good resemblance can be seen between the developed convex EM model and the original model on the basis of the measurement data. As represented by the absolute error, the difference between them is small. This implies that the developed convex model captures the loss behavior of the EM with sufficient accuracy. Notice that all the EM torque and speed combinations are taken into consideration in constructing the model, and some of them result in higher losses, leading to a larger discrepancy in the upper part of Figure 9. The EM, in practice, however, will not operate at those points. These points are outside the torque-speed envelope, which will be constrained by the EM power limitation model and will be described next. The relatively lower correlation accuracy in the lower part is due to the fact that both motoring and generating modes are considered. Taking into account limited operating points in this part for a long drive cycle in practice, the impact is small.

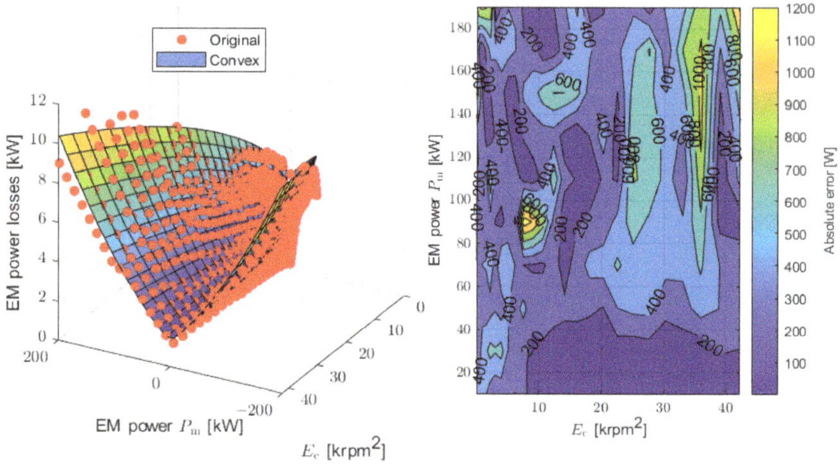

Figure 9. Convex EM power loss model for $s_\tau = 1$. The absolute error represents the difference in power loss between the convex model and the original model based on the measurement data.

3.6. Convex EM Power Limitation Model

As s_τ varies, the EM power limits change as well, referring to (29), as shown in Figure 10. Note that the EM power limits also consider the effect of the CVT ($E_c(\gamma_v, s_\gamma)$). Regarding the EM power limits, as can be seen from Figure 8, they mainly relate to two parts, that is, one before the base speed and the other after the base speed ($\omega_{m,b}$). Hence, the speed information is important. As speed information is mostly conveyed by E_c, it is identified as one of the key parameters. As a result, factors that influence the EM power limits could be, for example, s_τ and $E_c(\gamma_v, s_\gamma)$. Based on these parameters, possible models that preserve convexity are developed, which are fitted to represent the EM power limitation. Based on fitting accuracy, the convex EM power limitation model including motoring and generating modes is identified as follows:

$$\overline{P}_m = \min\{(c^+_{m,0}\, E_c(k) + c^+_{m,1}\, \sqrt{s_\tau\, E_c(k)} + c^+_{m,2}\, s_\tau + c^+_{m,3}), s_\tau\, \overline{P}_m\}, \quad (32)$$

$$\underline{P}_m = \max\{(c^-_{m,0}\, E_c(k) + c^-_{m,1}\, \sqrt{s_\tau\, E_c(k)} + c^-_{m,2}\, s_\tau + c^-_{m,3}), s_\tau\, \underline{P}_m\}, \quad (33)$$

where $c^+_{m,0}$, $c^+_{m,1}$, $c^+_{m,2}$, $c^+_{m,3}$, $c^-_{m,0}$, $c^-_{m,1}$, $c^-_{m,2}$ and $c^-_{m,3}$ are the corresponding coefficients. The plus sign (+) represents the motoring mode and the minus sign (-) the generating mode. This model is convex (Appendix C) and has a fitting accuracy of around 99%. The terms $\overline{P}_m$ and $\underline{P}_m$ represent the maximum and minimum power of the original EM, respectively, which can be obtained from Figure 8. An example of this model for $s_T = 1$ in motoring mode ($\overline{P}_m$) is depicted in Figure 11, which can be mirrored for generating mode ($\underline{P}_m$). It can be observed that the developed convex EM power limitation model and the original model based on the measurement data resemble well. It indicates that the developed convex model replicates the power limitation of the EM in reality. It should be noted that while the approximated convex model appears as straight lines in Figure 11 for $s_T = 1$, they may not be for other scaling factors, depending on (32) and (33). Notice that, this model is mainly used in Section 4.3.

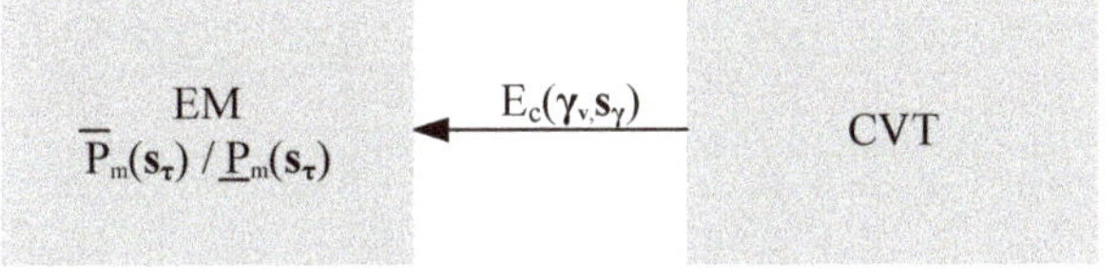

Figure 10. Convex modeling of the EM power limits.

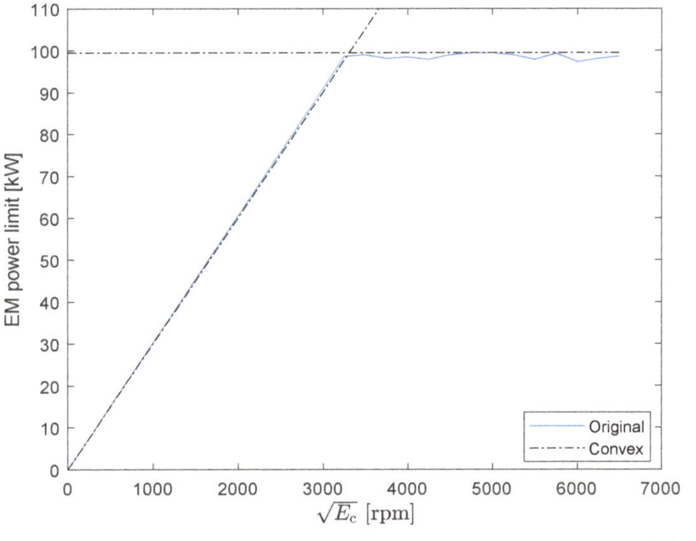

Figure 11. Convex EM power limitation model for s_T = 1 in motoring mode ($\overline{P}_m$). The dash-dot lines represent the approximated convex model before and after the base speed.

3.7. Thermal EM-CVT Model

The heat (power losses) generated by the CVT ($P^s_{c,loss}$ in (24)) and EM ($P_{m,loss}$ in (31)) is removed by a TMS, as demonstrated in Figure 12. The main difference between Figures 12 and A4 is that there is an extra small off-the-shelf heat exchanger in Figure 12. It enables heat exchange between the CVT cooling medium and the EM cooling medium. The EM and the CVT are physically attached. The heat from the CVT is removed directly by its cooling medium, which exchanges that with the EM cooling medium. Furthermore, the heat from the EM is taken away by its cooling medium driven by a pump, which is eventually removed by the radiator with a fan providing the required air-flow rate. The EM and CVT dissipate heat to the ambient air due to convection. The aim of the TMS is to find the desired air-flow rate (ϕ_a) that maintains the EM temperature (θ_m) below its prescribed thermal limit (65 °C) dictated by the manufacturer. A lumped-parameter approach is utilized to describe the thermal behavior of the EM and CVT. Based on first principles of thermodynamics, referring to Appendix A.4, the thermal EM-CVT model is given by

$$c_{\mathrm{m}}\, m_{\mathrm{m}}\, \dot{\theta}_{\mathrm{m}}(k) = P_{\mathrm{m,loss}}(k) - h_{\mathrm{m}}\, A_{\mathrm{m}}\, (\theta_{\mathrm{m}}(k) - \theta_{\mathrm{o}}(k)) - k_{\mathrm{m}}\, (\theta_{\mathrm{m}}(k) - \theta_{\mathrm{c}}(k)) - h_{\mathrm{a}}\, A_{\mathrm{a}}\, (\theta_{\mathrm{m}}(k) - \theta_{\mathrm{a}}), \tag{34}$$

$$c_{\mathrm{f}}\, m_{\mathrm{f}}\, \dot{\theta}_{\mathrm{o}}(k) = h_{\mathrm{m}}\, A_{\mathrm{m}}\, (\theta_{\mathrm{m}}(k) - \theta_{\mathrm{o}}(k)) - \phi_{\mathrm{f}}\, c_{\mathrm{f}}\, (\theta_{\mathrm{o}}(k) - \theta_{\mathrm{i}}(k)), \tag{35}$$

$$c_{\mathrm{f}}\, m_{\mathrm{f}}\, \dot{\theta}_{\mathrm{h}}(k) = k_{\mathrm{h}}\, (\theta_{\mathrm{c}}(k) - \theta_{\mathrm{h}}(k)) - \phi_{\mathrm{f}}\, c_{\mathrm{f}}\, (\theta_{\mathrm{h}}(k) - \theta_{\mathrm{o}}(k)), \tag{36}$$

$$c_{\mathrm{h}}\, c_{\mathrm{c}}\, m_{\mathrm{c}}\, \dot{\theta}_{\mathrm{c}}(k) = P^{\mathrm{s}}_{\mathrm{c,loss}}(k) + k_{\mathrm{e}}\, (\theta_{\mathrm{m}}(k) - \theta_{\mathrm{c}}(k)) - k_{\mathrm{h}}\, (\theta_{\mathrm{c}}(k) - \theta_{\mathrm{h}}(k)) - h_{\mathrm{c}}\, A_{\mathrm{c}}\, (\theta_{\mathrm{c}}(k) - \theta_{\mathrm{a}}), \tag{37}$$

$$c_{\mathrm{f}}\, m_{\mathrm{f}}\, \dot{\theta}_{\mathrm{i}}(k) = \phi_{\mathrm{c}}\, c_{\mathrm{f}}\, (\theta_{\mathrm{h}}(k) - \theta_{\mathrm{i}}(k)) - \epsilon\, \phi_{\mathrm{a}}(k)\, c_{\mathrm{a}}\, (\theta_{\mathrm{h}}(k) - \theta_{\mathrm{a}}). \tag{38}$$

The thermal variables are restricted by

$$\theta_{\mathrm{m}}(k) \in [\underline{\theta}_{\mathrm{m}}, \overline{\theta}_{\mathrm{m}}], \tag{39}$$

$$\theta_{\mathrm{o}}(k) \in [\underline{\theta}_{\mathrm{o}}, \overline{\theta}_{\mathrm{o}}], \tag{40}$$

$$\theta_{\mathrm{i}}(k) \in [\underline{\theta}_{\mathrm{i}}, \overline{\theta}_{\mathrm{i}}]. \tag{41}$$

This model is validated against measurement data in terms of temperature, which is presented in Appendix A.4. Note that since detailed pump and fan signals are not available in the measurement data, validation of cooling power consumption is not performed. Hence, the cooling power consumption is not added to the overall energy consumption. It does not influence the purpose of this study, focusing on evaluation of thermal performance (Section 4.4).

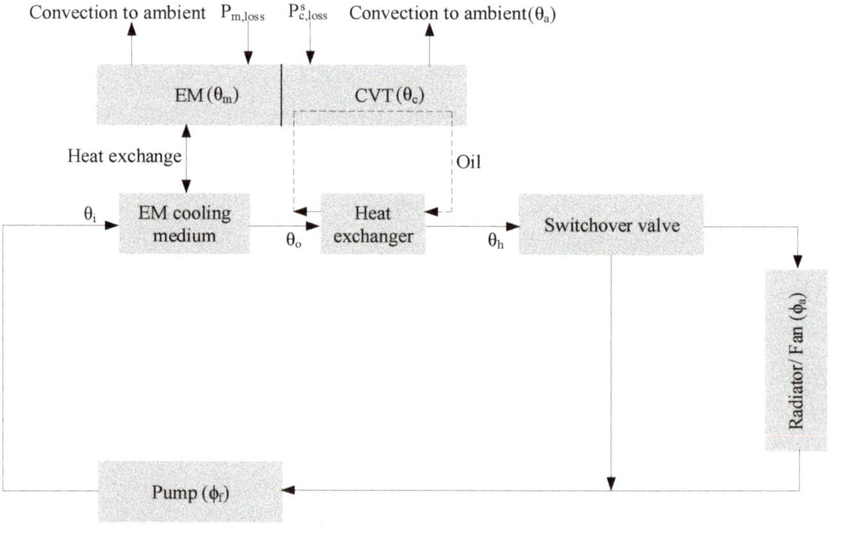

Figure 12. Thermal management configuration for the EM-CVT.

3.8. Convex Battery Model

The electric power of the EM is provided by the battery. The battery model is based on lithium-ion technology with identical cells. The required battery power is given by

$$\begin{aligned} P_{\mathrm{b}}(k) &= P_{\mathrm{el}}(k) + P_{\mathrm{b,loss}}(k) \\ &= P_{\mathrm{m}}(k) + P_{\mathrm{m,loss}}(k) + P_{\mathrm{b,loss}}(k), \end{aligned} \tag{42}$$

where $P_{\mathrm{b,loss}}$ represents the battery losses, which is calculated by

$$\begin{aligned} P_{\mathrm{b,loss}}(k) =& s_{\mathrm{b}}\ N_0\ I_{\mathrm{c}}^2(k)\ R_{\mathrm{c}} \\ =& \frac{P_{\mathrm{b}}^2(k) R_{\mathrm{c}}}{s_{\mathrm{b}}\ N_0\ V_{\mathrm{c}}^2(k)}, \end{aligned} \tag{43}$$

where I_{c} is the cell current and R_{c} its resistance. The battery sizing is performed based on the scaling factor s_{b} for scaling the battery cells, which linearly influences the battery energy, weight and cost. N_0 is the original number of battery cells when $s_{\mathrm{b}} = 1$. To preserve convexity for the co-design problem (Appendix C), the open circuit voltage of a battery cell (V_{c}) is approximated as a linear function of the state-of-charge of the battery (ξ), given by

$$V_{\mathrm{c}}(k) = \frac{Q_{\mathrm{c}}}{F_{\mathrm{c}}}\ \xi(k) + V_0, \tag{44}$$

where Q_{c} is the cell capacity and F_{c} the capacitance. The battery energy can then be calculated by

$$\begin{aligned} E_{\mathrm{b}}(k) =& s_{\mathrm{b}}\ N_0 \int_0^{\xi} V_{\mathrm{c}}(k)\ Q_{\mathrm{c}} \mathrm{d}\xi \\ =& \frac{F_{\mathrm{c}}}{2}\ s_{\mathrm{b}}\ N_0\ (V_{\mathrm{c}}^2(k) - V_0^2). \end{aligned} \tag{45}$$

Taking the derivative of E_{b} yields

$$\dot{E}_{\mathrm{b}}(k) = -P_{\mathrm{b}}(k). \tag{46}$$

In order to model the battery on a pack level instead of a cell level, which does not require information of series-parallel connection, the cell voltage V_{c} is replaced by a new variable U_{b}, by applying a change of variables, given by

$$U_{\mathrm{b}}(k) = s_{\mathrm{b}}\ N_0\ V_{\mathrm{c}}^2(k) = \frac{2}{F_{\mathrm{c}}}\ E_{\mathrm{b}}(k) + s_{\mathrm{b}}\ N_0\ V_0^2. \tag{47}$$

The battery power losses can then be expressed as

$$\begin{aligned} P_{\mathrm{b,loss}}(k) =& \frac{P_{\mathrm{b}}^2(k)\ R_{\mathrm{c}}}{U_{\mathrm{b}}(k)} \\ =& \frac{P_{\mathrm{b}}^2(k)\ R_{\mathrm{c}}\ F_{\mathrm{c}}}{2\ E_{\mathrm{b}}(k) + s_{\mathrm{b}}\ N_0\ F_{\mathrm{c}}\ V_0^2}. \end{aligned} \tag{48}$$

The battery energy and power are constrained by

$$E_{\mathrm{b}}(k) \in [\frac{F_{\mathrm{b}}}{2}\ s_{\mathrm{b}}\ N_0\ (\underline{V}_{\mathrm{c}}^2 - V_0^2), \frac{F_{\mathrm{b}}}{2}\ s_{\mathrm{b}}\ N_0\ (\overline{V}_{\mathrm{c}}^2 - V_0^2)], \tag{49}$$

$$P_{\mathrm{b}}(k) \in [\underline{I}_{\mathrm{c}}\ V_{\mathrm{b}}(k), \overline{I}_{\mathrm{c}}\ V_{\mathrm{b}}(k)], \tag{50}$$

where V_{b} is given by

$$\begin{aligned} V_{\mathrm{b}}(k) =& s_{\mathrm{b}}\ N_0\ V_{\mathrm{c}}(k) \\ =& \sqrt{s_{\mathrm{b}}\ N_0\ U_{\mathrm{b}}(k)}. \end{aligned} \tag{51}$$

3.9. Convex Mass and Cost Models

Apart from powertrain and thermal models, convex mass and cost models are required for the implementation of the co-design method. This section describes the convex mass and cost models associated with the CVT, EM and battery sizes. The CVT mass includes the mass of the variator and final drive. The EM mass includes the mass of the EM and inverter.

For a generic CVT, its weight can be approximated as a function of its torque capacity $\overline{\tau}_c$ based on existing CVTs [25]. The CVT weight m_c is modeled by

$$m_c = 0.28\,\overline{\tau}_c + 23.21. \tag{52}$$

The original production CVT with a full ratio range of [0.38, 2.63], implying a ratio coverage of around 7, has a torque capacity of 250 Nm. On the basis of this CVT, assume the ratio affects the torque capacity proportionally. For each torque capacity (ratio coverage), the corresponding weight can be computed based on (52). Specifically for the CVT in this study, a one-on-one mapping between the weight m_c and ratio coverage s_γ can thus be expressed as

$$m_c = 1.19\,s_\gamma^2 + 39.12. \tag{53}$$

Therefore, given the specific cost of CVT [23] (a_c in Table A1), the CVT cost C_c (in €) is given by

$$C_c = a_c\,(1.19\,s_\gamma^2 + 39.12). \tag{54}$$

The prediction (54) is valid on the basis of this CVT, since the ratio coverage is scaled down.

The EM weight is estimated as a function of its scaling factor [26], that is,

$$m_m = s_\tau\,\overline{m}_m, \tag{55}$$

where $\overline{m}_m$ is the original EM mass when $s_\tau = 1$, which is provided in Table 1.

The EM cost C_m (in €) is estimated as a function of its scaling factor [27,28], giving

$$C_m = b_m\,s_\tau. \tag{56}$$

The battery weight is described as a function of its scaling factor, given by

$$m_b = s_b\,\overline{m}_b, \tag{57}$$

where $\overline{m}_b$ is the original battery mass when $s_b = 1$, which is provided in Table 1. The battery cost C_b (in €) is approximated by [28,29]

$$C_b = s_b\,c_b\,\overline{E}_b. \tag{58}$$

Note that, the price of CVT could be lowered, as some parts of the CVT for conventional applications are not needed in EVs, such as torque converter and DNR (drive, neutral and reverse). However, since part of this information is taken into account in (53) and the exact weight is not known before optimization, it is not further addressed. Moreover, currently, there is no consensus on the component price. The numbers used in this study are only indicative figures (e.g., b_m, $c_b\,\overline{E}_b$), which have no direct relation to possible market prices. They do not affect the comparison, as the same scale is employed in all the systems (S_1, S_2 and S_3).

Additionally, notice that the component specifications appear large in this case. The components are scaled down, which means that, for example, $s_\tau > 1$ and $\gamma_v > 2.63$, are not necessary. The models are always valid within their feasible ranges based on the measurement data. The right combination of the speed ratio of the CVT over time γ_v, air-flow rate of the cooling system ϕ_a, ratio coverage s_γ, the scaling factor for the EM s_τ and the scaling factor for the battery s_b will be determined simultaneously by the optimization algorithm, which is discussed in the next section.

This section presents a convex EV model and its associated cost models for the co-design optimization problem. A data-driven approach is used to derive the convex CVT and EM models. In particular, the coupling between the EM and CVT from design and control perspectives is described in detail. Compared to the original model based on measurements, the convex CVT power loss model is developed with a correlation accuracy of 98%. The convex EM power loss and power limitation models are developed with correlation accuracies of 95% and 99%, respectively. An experimentally validated thermal EM-CVT model is subsequently presented. A convex battery model is also developed. Finally, convex mass and cost models that are size-dependent for the CVT, EM and battery are created. The overall system has a large number of states, and there is a strong coupling between the CVT and EM. By applying CP, not only will it find an optimal solution but also it will find the solution in a computationally efficient manner. It allows for extensive parameter variation studies and evaluation of diverse design aspects on system and component level.

4. Optimization Results and Discussion

On the basis of the convex models developed in the last section, the objective of the co-design optimization strategy is to minimize the TCO. It aims at generating an optimal control trajectory of the speed ratio of the CVT (γ_v) and finding the desired air-flow rate of the cooling system (ϕ_a). Furthermore, it aims to identify the optimal sizes of the CVT (s_γ), EM (s_τ) and battery (s_b). Based on (1), the overall co-design optimization problem in S_3 can be written as follows:

$$\min \sum_{k=1}^{N} \rho_e \, P_b(s_\gamma, s_\tau, s_b, E_b(k), \theta_m(k), \theta_c(k), \theta_o(k), \theta_i(k), \phi_a(k), \gamma_v(k) \mid v_v(k), a_v(k)) \, \Delta t \\ + \frac{S_d}{S_v} \left(C_c(s_\gamma) + C_m(s_\tau) + C_b(s_b) \right), \tag{59}$$

$$\text{s.t.} \quad (11) - (58). \tag{60}$$

The final state of the battery energy is not constrained, considering the battery capacity and the power demand of the drive cycle. The overall optimization problem is convex, including convex cost function, models and constraints. Basic convex functions, for example, linear, quadratic, quadratic-over-linear and opposite of geometric mean functions, and operations that preserve convexity, for example, nonnegative weighted sums and pointwise maximum, are used to verify model convexity. For example, the battery mass model (57) is linear and the CVT cost model (54) is quadratic. A quadratic-over-linear term is used in the EM loss model (31). The overall co-design optimization problem is solved by using SDPT3 [30]. Specifically, the problem can be recognized as a semidefinite program. It is then translated automatically by a tool CVX into a form required by SDPT3 [30]. This permits the problem to be written in a readable form, for example, using expressions/symbols to hold operations over variables. It solves the dual problem for improved efficiency [22].

4.1. Control and Design Freedom

The co-design optimization method tends to obtain a globally optimal solution by simultaneously optimizing the design and control variables to minimize the TCO. Figure 13 shows the EM operating points on the WLTC for the three systems. It can be seen from Figure 13 that the EM operation in S_1 is relatively fixed, which has no control freedom because of the fixed gear ratio (9.02 in Table 1). The EM has to operate according to the driving conditions, which can hardly be efficient in consideration of real-world dynamic behavior. In contrast, the EM in S_2 has relatively more freedom to adjust operating points to reduce the power dissipation. This is realized by changing the speed ratio of the CVT, depending on the loading conditions. Yet, implementing the standard CVT controllers developed for conventional vehicles would reduce only the EM power losses (Section 2), regardless of the CVT efficiency. While the EM operation in S_2 is efficient, which follows its OOL (Figure 13), the system (combined EM and CVT) is not able to operate efficiently over a dynamic cycle (WLTC). The combined EM and transmission losses are thus higher in S_2 than that in S_1, as shown in Figure 14. The co-design optimization strategy, however, takes full advantage of control freedom (continuous ratio adjustment depending on the driving conditions) provided by the CVT. Specifically, apart from the EM power dissipation, the CVT power losses are also monitored in S_3. The CVT speed ratio over time γ_v (19) is selected to improve the overall system efficiency. Ratio variation that results in a higher loss is penalized. Smooth ratio change improves efficiency, drivability and reduced ELOP power losses. This effect can be seen in Figure 14, where the CVT power dissipation in S_3 is significantly lower than that in S_2 and the system efficiency is higher. The EM efficiency is also high in S_3, as demonstrated in Figure 13. The battery power dissipation is similar in three systems.

Additionally, the control freedom offered by the CVT creates design flexibility, which is not explored by S_2, as the component sizes are fixed. The design space is larger in S_3, with different combinations of CVT (21) and EM (30) sizes, which bring opportunities of optimizing the system from design perspective. The optimal component sizes, namely the right combination of the scaling factor for the battery (s_b^o), scaling factor for the EM (s_τ^o) and the ratio coverage of the CVT (s_γ^o), are eventually determined by the co-design optimization method. The co-design optimization strategy takes into account the coupling between the EM and CVT, cost function and the drive cycle.

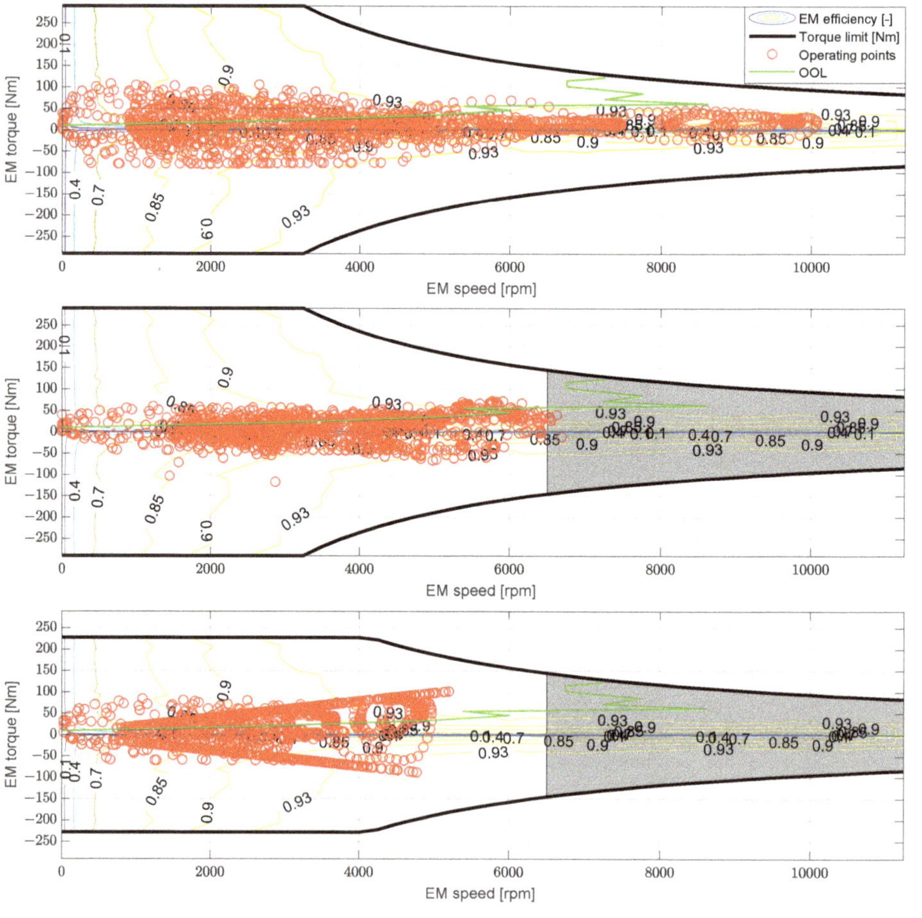

Figure 13. EM operating points on the Worldwide Harmonized Light Vehicles Test Cycles (WLTC). (**Top** to **bottom**) S_1, S_2, and S_3.

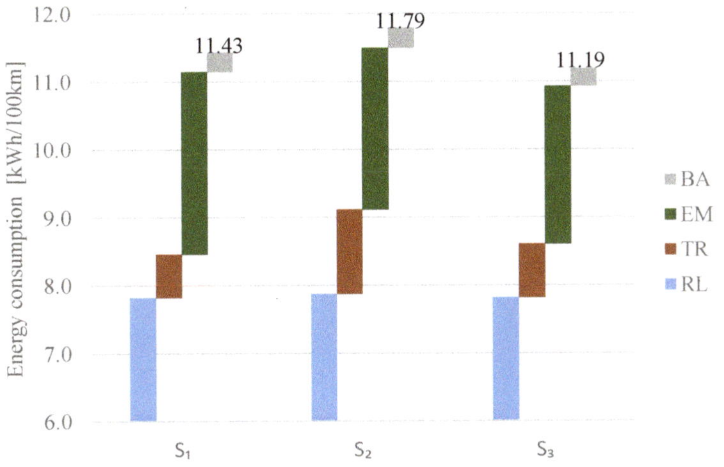

Figure 14. Comparison of energy consumption between S_1 (with a total vehicle mass of 1670 kg), S_2 (with a total vehicle mass of 1700 kg) and S_3 (with a total vehicle mass of 1666 kg) on the WLTC, where RL represents road load. The transmission (TR) losses include the ELOP power dissipation, which is zero in S_1. The EM power losses include the effect of the DC-AC inverter.

As shown in Table 1, the maximum EM torque is reduced from $s_T = 1$ in S_1 and S_2 to $s_T = 0.79$ in S_3. Reduced EM torque and increased base speed decrease the EM losses (Figure 13). The optimal battery size is found and the number of battery cells is reduced because of improved efficiency (Table 1). The EM in S_3 has the same maximum power as the other systems. Because the same EM from the SST-based EV on the market (S_1) is utilized in S_2, which is not optimized for the CVT application. Even if the EM is made smaller in S_2, the overall system efficiency is not necessarily always be higher, see Figure 8 about the OOL, as the CVT efficiency is not considered.

This reduced EM size is achieved due to a variable ratio coverage. The right ratio coverage of the CVT can also thus be determined in combination with the EM size in terms of torque and power. It is reduced from [0.7, 2.63] with a ratio coverage of 3.76 before optimization to [0.7, 2.34] with a ratio coverage of 3.34 after optimization (Table 1). In S_2, the CVT is oversized considering the WLTC. First, the CVT used in S_2 is based on an off-the-shelf component, which is not optimized. Second, the CVT is controlled only to reduce the EM power dissipation, regardless of the CVT size. The system mass including transmission, EM and battery (Table 1) is thus reduced in S_3 due to the reduced component sizes, resulting in a decrease in power demand (RL in Figure 14). The curb weight and driver mass are the same in the three systems.

Owing to high CVT power losses in S_2, the energy consumption is reduced by around 3.1% in S_1. Compared with S_2, because of the reduction in the system power losses (Figure 14), the energy consumption is decreased by around 5.1% in S_3. The decrease in the CVT power dissipation contributes more to the energy saving. Given the component sizes, the component costs are calculated based on (54), (56) and (58). As shown in Figure 15, compared with S_1, the system cost is increased by around 5.1% in S_2, as the CVT cost is higher than the SST price. Due to the reduction in the component sizes (Table 1), compared to S_1, the system cost is reduced by around 1.8% in S_3. Reduced EM torque decreases the EM mass and cost. The decrease in the battery and EM prices contributes more to the cost saving.

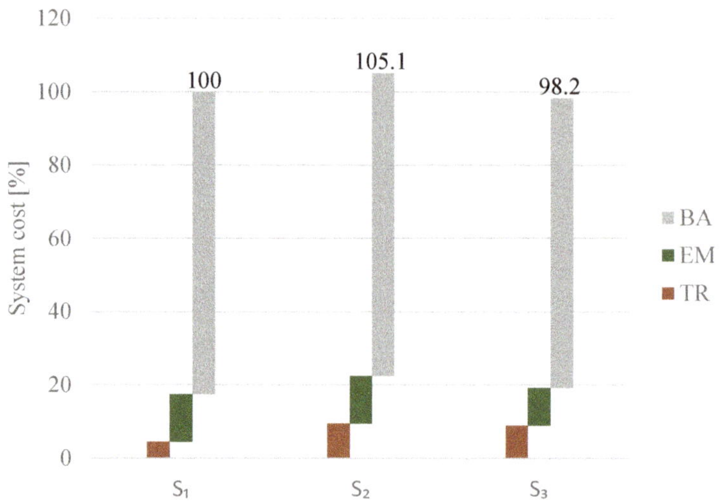

Figure 15. Comparison of normalized system cost between S_1, S_2 and S_3, where the EM cost includes the DC-AC inverter.

Overall, as shown in Figure 16, S_3 has the lowest TCO because of reduced energy consumption and system cost, which is around 2% lower than S_1. The optimization results demonstrate that the optimized EM-CVT system can be compact, lightweight, energy-efficient and cost-effective. It is very different from its traditional image. Additionally, it can be suggested that automotive suppliers could greatly benefit from highly integrated components and systems to maximize their system efficiency and minimize cost targets. It can also be observed that reducing the maximum EM torque and increasing the base speed while having a higher slope of OOL compared with that of SST are beneficial for CVT-based EVs (Figure 13). Most importantly, it shows the importance of cycle-driven (Section 3.2) and co-design (Section 2) in identifying the optimal control trajectories and component sizes. Specifically, it finds the trade-off between CVT power losses (24), CVT size (21), EM power losses (31), and EM size (30), based on the combined EM-CVT characteristics (Figure 7) and cost function (1).

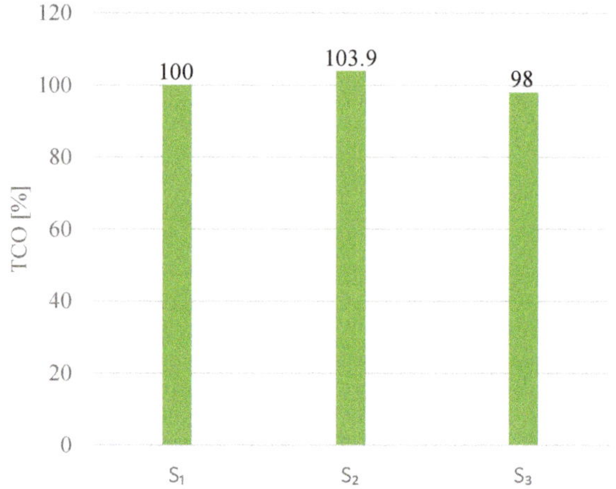

Figure 16. Comparison of normalized TCO between S_1, S_2 and S_3.

4.2. Sequential Design versus Simultaneous Design

The strength of co-design (simultaneous design in this case) lies in the fact that it finds the optimal design and control variables simultaneously. It minimizes the cost function for a strongly coupled problem, which is the case for the integration of the EM and CVT (Figure 1). The CVT speed ratio (control) influences the CVT and EM sizes (design/plant), and vice versa (e.g., (24) and (31)). The cost function, namely the minimization of the TCO (1), consists of the energy consumption (9) and system cost (10), and both are affected by the design and control variables. In order to demonstrate the effectiveness of the co-design approach in S_3, it is compared to a sequential design (SD) method, where the plant and controller are determined sequentially (Section 1). Note that for purposes of comparison, not all the plant parameters are predetermined. SD is defined as follows:

- SD: Based on S_3, assuming that the EM size is fixed (i.e., $s_\tau = 1$) in order to achieve the required performance (Section 1), the goal is to find the CVT speed ratio over time, CVT and battery size reducing the TCO.

Referring to (1), the corresponding cost function is given by

$$\min \sum_{k=1}^{N} \rho_e \, P_b(s_\gamma, s_b, x(k), u(k) \mid s_\tau) \, \Delta t + \frac{S_d}{S_v} \left(C_c(s_\gamma) + C_m + C_b(s_b) \right), \tag{61}$$

where C_m is a constant because of the fixed EM size. Other constraints remain the same. The comparison between SD and S_3 is shown in Figure 17.

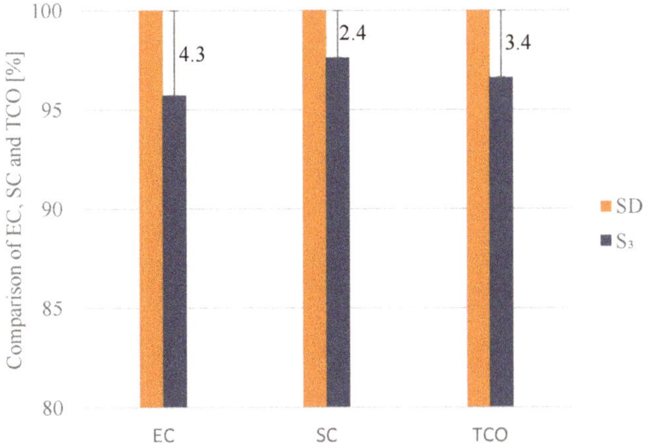

Figure 17. Comparison of normalized EC, SC and TCO between SD (with a total vehicle mass of 1677 kg) and S_3, where EC represents the energy consumption and SC the system cost.

It can be seen from Figure 17 that the energy consumption (EC) and system cost (SC) are reduced in S_3 compared with SD. Overall, the TCO is decreased by around 3.4% in S_3. The primary reason is that the control and design freedom provided by the CVT cannot be fully exploited in SD because of the fixed EM size. Admittedly, the CVT is optimized in SD, especially the CVT cost because of a reduced ratio coverage of around 2, which is significantly lower than that of 3.34 in S_3. Nevertheless, the EM and battery optimization are not taken into account, leading to a higher system cost. The control trajectory identified may not be efficient for the overall system. In SD, the problem is decoupled and the interconnections between the CVT (24), EM (31) and battery (42) are not considered, resulting in a higher TCO. For an inherently coupled problem (i.e., the integration of the EM and CVT), SD creates a separation and cannot guarantee an optimal solution. This issue is tackled by the co-design approach (S_3), where the right combination of the control policy and component sizes reducing the TCO is identified. It considers the coupling between the components from design and control perspectives.

4.3. Towards a Low-Power Application

As illustrated in Figure 13, the usage of the available EM torque is low considering WLTC. It implies that powertrain components are typically oversized (considering WLTC) to meet certain vehicle performance requirements. The selection of electric powertrain components (e.g., EM) for future EV applications is still an ongoing question. Current selection criteria are largely based on performance requirements, such as top speed, acceleration time and gradability (Section 1). It usually leads to oversized components, considering drive cycles used for efficiency indicators. Therefore, even with the co-design approach, the downsizing potential is limited (Section 4.1). High performance, however, may not be required in urban driving. In order to see the downsizing potential, a low-power (LP) design is utilized as an example. Note that for purposes of comparison, the EM design is not restricted by (29). LP is defined as follows:

- LP: Based on S_3, assuming that there are no performance requirements (Section 1), the aim is to find the optimal design and control variables reducing the TCO while satisfying drive cycle requirements.

The objective function in this case is similar to (1). The comparison between LP and S_3 is shown in Figure 18. It can be observed from Figure 18 that compared with S_3, the energy consumption (EC) and system cost (SC) are decreased in LP. Overall, in LP, the TCO is reduced by around 4%. The main reason is that LP takes full advantage of the control and design freedom provided by the CVT without performance constraints. Compared to S_3, all the parameters as presented in Table 1, Figures 14 and 15 are reduced in LP. For example, the maximum EM power is reduced from 100 kW in S_3 to 54 kW in LP with a CVT ratio coverage of around 2.06. The battery size is decreased in LP as well. The minimum required component size is found in LP. For a specific use case (WLTC) and a given EM-CVT system (combined EM-CVT characteristics), there is a lower bound for the system (i.e., the CVT, EM and battery) so as to complete the driving mission. The result also demonstrates the importance of co-design optimization in determining the system size for a representative use case, considering the coupling between the EM and CVT (combined EM-CVT characteristics). Although reduction of peak power has a negative effect on vehicle performance (e.g., the 0–50 km/h and 0–100 km/h acceleration times of LP are around 6 s and 18 s, respectively), LP is energy-efficient and cost-effective. It can be used for urban driving that does not require high performance.

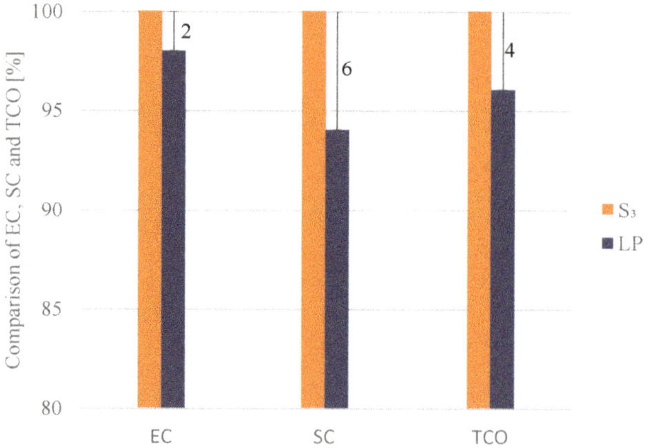

Figure 18. Comparison of normalized EC, SC and TCO between LP (with a total vehicle mass of 1635 kg) and S_3, where EC represents the energy consumption and SC the system cost.

4.4. Thermal Performance

The EM is a power source. It is also a heat source, and so does the transmission. They generate heat during operation, which needs to be taken away efficiently by a cooling system. For purposes of comparison, S_1, S_2 and S_3 have the same control target, maintaining the EM temperature below its thermal limit 65 °C dictated by the manufacturer at the end of the drive cycle. All the other conditions are the same, including the coolant flow rate. For simplicity, the problem is translated to finding a constant air-flow rate for each system over the WLTC. Assume the air-flow rate is proportional to the cooling power consumption. They are compared in terms of cooling power consumption and temperature profile.

It is found that $\phi_a(S_2) = 0.89\ \phi_a(S_1)$, which means that compared to S_1, the cooling power consumption is reduced in S_2, as shown in Figure 19. Recall that the system losses are higher in S_2 (Figure 14). The EM power dissipation in S_2 is less than that in S_1 but the transmission power dissipation is much higher. In this case, the EM is more dominant in determining the level of the cooling power consumption. Another reason is that the thermal mass (EM and CVT) is higher in S_2 (Table 1). Moreover, because of topology difference, the extra small off-the-shelf heat exchanger in S_2 enables heat exchange between the EM and CVT. It changes the overall thermal behavior of the cooling system. It is also calculated that $\phi_a(S_3) = 0.78\ \phi_a(S_2)$, which implies that compared to S_2, the cooling power consumption is decreased in S_3. They have the same topology. The primary reason is that the power losses are less in S_3, which requires less cooling power to remove the heat, although the system mass is reduced in S_3. The corresponding temperature profiles are illustrated in Figure 20.

Additionally, an important finding from Figure 20 is that the EM temperature and the CVT temperature are very similar in S_3. The primary reason is that the small heat exchanger provides extra heat exchange between the EM and the CVT. It is a crucial step towards a thermally integrated EM and CVT, for example, using a combined cooling loop with a dedicated cooling medium, taking into account such as corrosion, viscosity and conductivity [31–33]. It would make the system even more compact and efficient.

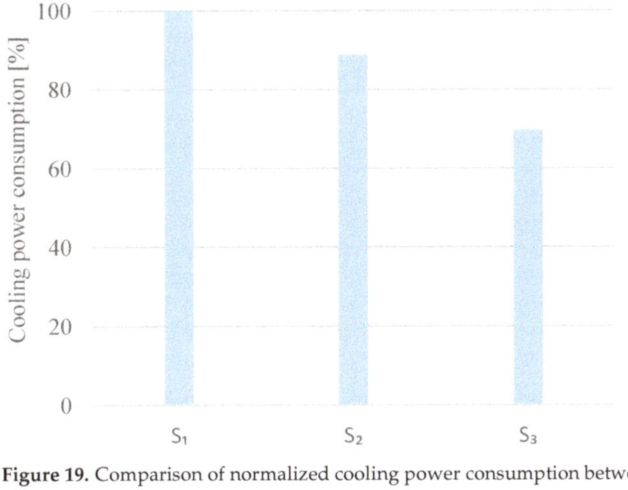

Figure 19. Comparison of normalized cooling power consumption between S_1, S_2 and S_3.

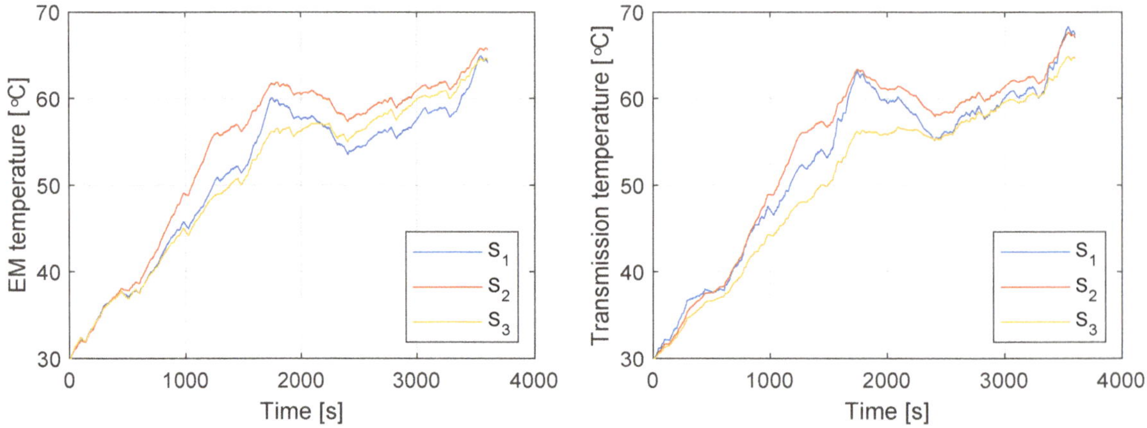

Figure 20. Comparison of temperature profile between S_1, S_2 and S_3.

Note that as the systems, overall, are efficient, this amount of reduction will not significantly influence the total energy consumption, although depending on the use case. Furthermore, the exact value of cooling power is not provided, because detailed pump and fan signals are not available in the controller area network (CAN) data, as mentioned in Section 3.7. However, based on physics and model validation (Appendix A.4), the result is representative. Previously, it is shown that there is a strong coupling between the EM and CVT, and the combined system can be lightweight, efficient and low-cost. Now, it is demonstrated that they can also be integrated thermally, and a further reduction in weight, energy usage and cost can be expected. In the future, a highly integrated EM-CVT system can be anticipated for EV applications.

5. Conclusions

A co-design optimization method based on CP is proposed for a CVT-based EV to minimize the TCO, focusing on the integration of the EM and CVT. The co-design optimization method finds the optimal CVT speed ratio, air-flow rate of the cooling system, CVT size, EM size and battery size simultaneously. It takes full advantage of the control and design freedom provided by the CVT. The optimized system with the co-design approach decreases the TCO by around 2% compared with an SST-based EV with reference to a series production vehicle and by around 5.9% compared with a non-optimized CVT-based EV (based on off-the-shelf components). The advantages of the co-design approach are also highlighted by comparing to a sequential method. Moreover, for urban driving, insights into the design of a low-power EV are given based on the co-design approach, which finds the minimum required component size. It can be concluded that although the current EV market is dominated by SSTs, multi-speed transmissions, for example, CVTs, are competitive alternatives for EVs in terms of TCO, due to continuous ratio adjustment depending on driving conditions and the resulting design freedom. For EV applications, a highly and thermally integrated EM-CVT system, which is low-cost, efficient and lightweight, can be anticipated.

Author Contributions: C.W. proposed the concepts and methodologies. He wrote and edited the original draft; T.H. and E.I.C. were responsible for supervision, internal review and editing of the draft and concepts. All authors have read and agreed to the published version of the manuscript.

Funding: This research received no external funding.

Institutional Review Board Statement: Not applicable.

Informed Consent Statement: Not applicable.

Data Availability Statement: No new data were created or analyzed in this study. Data sharing is not applicable to this article.

Acknowledgments: This research was carried out under project number T18001 in the framework of the Research Program of the Materials innovation institute (M2i) (www.m2i.nl accessed on 11 February 2021) supported by the Dutch government.

Conflicts of Interest: The authors declare no conflict of interest.

Abbreviations

The following abbreviations are used in this manuscript:

AC	Alternating Current
BA	Battery
CAN	Controller Area Network
CP	Convex Programming
CVT	Continuously Variable Transmission
DC	Direct Current
DNR	Drive, Neutral and Reverse
DP	Dynamic Programming
EC	Energy Consumption
ELOP	Electric Oil Pump
EM	Electric Machine
EV	Electric Vehicle
FD	Final Drive
ICDC	Intercity Drive Cycle
KPI	Key Performance Indicator
LP	Low Power
OOL	Optimal Operating Line
PSO	Particle Swarm Optimization
RL	Road Load
SC	System Cost
SD	Sequential Design
SST	Single Speed Transmission
TCO	Total Cost of Ownership
TMS	Thermal Management System
TR	Transmission
VA	Variator
VCU	Vehicle Control Unit
WH	Wheel
WLTC	Worldwide Harmonized Light Vehicles Test Cycle

Appendix A. SST-Based EV Model

The SST-based EV model (S_1) is validated by measurement data obtained from the series production vehicle driven on a real-world drive cycle, namely intercity drive cycle (ICDC). The measurement data are extracted from the CAN signals of the vehicle control unit (VCU) in the series production vehicle driven on the ICDC. The sampling time is 0.01 s. The vehicle specification is provided in Table A1.

Appendix A.1. Longitudinal Dynamics

Given the drive cycle represented by vehicle velocity v_v and acceleration a_v and taking into account aerodynamic drag force, rolling resistance and inertia force, the demanded wheel torque τ_w and speed w_w can be calculated by

$$\tau_\text{w}(k) = \left(\frac{1}{2}\,\rho_\text{a}\,c_\text{d}\,A_\text{f}\,v_\text{v}^2(k) + c_\text{r}\,m_\text{v}\,g\,\text{sign}(v_\text{v}(k)) + \left(m_\text{v} + 4\,\frac{J_\text{w}}{r_\text{w}^2}\right)a_\text{v}(k)\right) r_\text{w}, \tag{A1}$$

$$\omega_{\mathrm{w}}(k) = \frac{v_{\mathrm{v}}(k)}{r_{\mathrm{w}}}, \tag{A2}$$

where the total vehicle mass m_{v} is given by

$$m_{\mathrm{v}} = m_{\mathrm{cw}} + m_{\mathrm{s}} + \overline{m}_{\mathrm{m}} + \overline{m}_{\mathrm{b}} + m_{\mathrm{d}}, \tag{A3}$$

where m_{s} is the SST mass (Table 1).

Table A1. Main parameters of EV model.

Parameter	Value	Unit	Description
ρ_{a}	1.18	kg/m^3	Density of air
c_{d}	0.27	-	Aerodynamic drag coefficient
A_{f}	2.21	m^2	Frontal area
c_{r}	0.00724	-	Rolling resistance coefficient
J_{w}	1	kgm^2	Wheel inertia
r_{w}	0.312	m	Wheel radius
η_{g}	0.98	-	Fixed gear efficiency
η_{f}	0.985	-	Final drive efficiency
$\overline{\gamma}_{\mathrm{v}}$	2.63	-	Underdrive ratio
$\underline{\gamma}_{\mathrm{v}}$	0.7	-	Overdrive ratio
c_{m}	430 [19]	J/kgK	Specific heat capacity of EM
c_{c}	630 [23]	J/kgK	Specific heat capacity of CVT
c_{s}	630 [23]	J/kgK	Specific heat capacity of SST
c_{f}	4090	J/kgK	Specific heat capacity of EM cooling medium
c_{a}	1000	J/kgK	Specific heat capacity of air
c_{h}	0.62 [23]	-	CVT heating coefficient
h_{m}	2000	W/m^{2}K	Heat transfer coefficient between EM and its cooling medium
h_{a}	10	W/m^{2}K	Heat transfer coefficient between EM and ambient air
h_{c}	10	W/m^{2}K	Heat transfer coefficient between CVT and ambient air
k_{m}	111	W/K	Heat transfer coefficient between EM and CVT
k_{h}	125	W/K	Heat transfer coefficient between EM cooling medium and CVT oil
A_{m}	0.2	m^2	Heat exchange area between EM and its cooling medium
A_{a}	0.32	m^2	Heat exchange area between EM and ambient air
A_{c}	0.17	m^2	Heat exchange area between CVT and ambient air
m_{f}	1.5	kg	Cooling medium mass
ϕ_{f}	0.35	kg/s	Coolant flow rate
ϵ	0.6 [34]	-	Radiator effectiveness
$\overline{\theta}_{\mathrm{m}}$	65	°C	Maximum EM temperature
$\overline{\theta}_{\mathrm{o}}$	65	°C	Maximum cooling medium temperature at EM outlet
$\overline{\theta}_{\mathrm{i}}$	65	°C	Maximum cooling medium temperature at EM inlet
S_{d}	46.532	km	Two repeated WLTC length
S_{v}	300,000	km	Traveled distance of vehicle in its lifetime
p_{e}	0.23	€/kWh	Price of electricity
a_{c}	13 [23]	€/kg	Specific cost of CVT
b_{m}	1000	€	Specific cost of EM
c_{b}	250	€/kWh	Specific cost of battery
$\overline{E}_{\mathrm{b}}$	25.4	kWh	Battery energy
N_0	264	-	Battery cells

Appendix A.2. Single-Speed Transmission

The SST including the final drive provides a fixed speed ratio γ_s between the EM and the wheel. To meet the torque and speed at the wheels given by (A1) and (A2), the required torque and speed of the SST are obtained by

$$\tau_s(k) = \begin{cases} \frac{\tau_w(k)}{\eta_s\ \gamma_s}, & \text{if } \tau_w(k) > 0, \\ \frac{\eta_s\ \tau_w(k)}{\gamma_s}, & \text{if } \tau_w(k) \le 0, \end{cases} \tag{A4}$$

$$\omega_s(k) = \gamma_s\ \omega_w(k), \tag{A5}$$

where η_s is the combined efficiency of the fixed gear η_g and the final drive η_f (Table A1).

Appendix A.3. Electric Machine

The EM is a permanent magnet synchronous machine, featuring motoring and generating modes. Its specification is provided in Table 1. The torque and speed of the EM are computed by

$$\tau_m(k) = \tau_s(k), \tag{A6}$$

$$\omega_m(k) = \omega_s(k). \tag{A7}$$

Given the same inputs (vehicle speed and acceleration) from the ICDC, the outputs of the EM from the developed model (simulation) are compared with the corresponding CAN signals (measurement). These signals are the EM torque and speed, and the comparisons are shown in Figures A1 and A2, respectively. In Figures A1 and A2, the right subplot zooms in on the left subplot, showing the details of the comparison. It should be noted that the measurement data are presented as they are in this study. In consideration of noise, driving environment, data recording, and effect of CAN signals, the simulation and measurement resemble well.

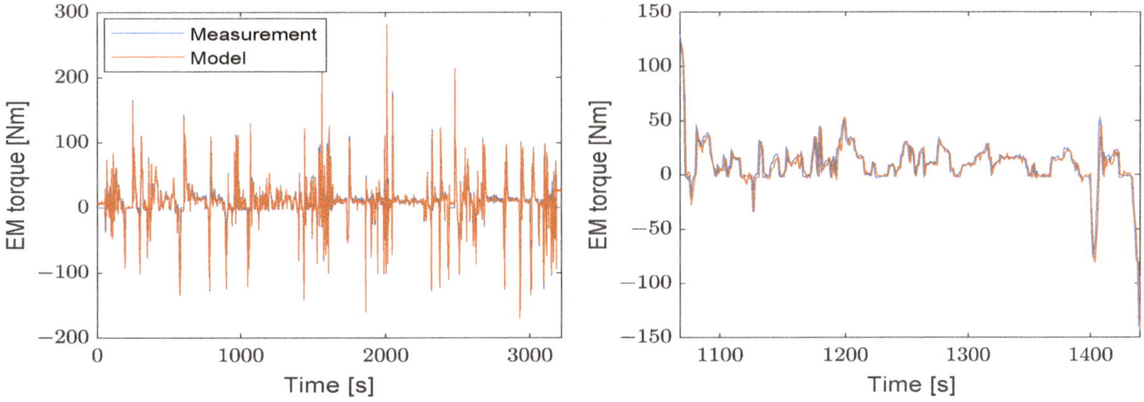

Figure A1. Validation of the EM torque for the ICDC.

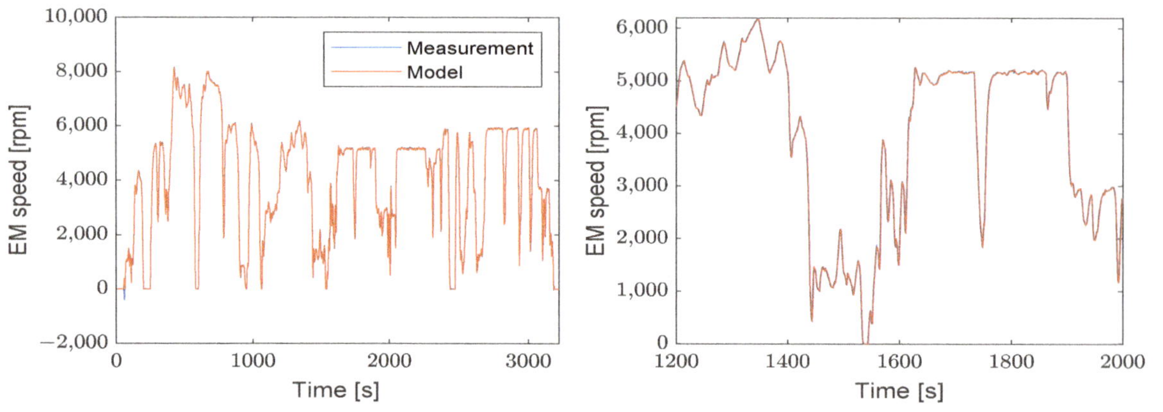

Figure A2. Validation of the EM speed for the ICDC.

The mechanical power of the EM is then given by

$$P_{\mathrm{m}}(k) = \tau_{\mathrm{m}}(k)\,\omega_{\mathrm{m}}(k). \tag{A8}$$

An efficiency map is used to calculate the power losses of the EM, as shown in Figure A3. This efficiency map includes the effect of the DC-AC inverter. The EM power dissipation is described as a function of its torque and speed, that is,

$$P_{\mathrm{m,loss}}(k) = P_{\mathrm{m,loss}}(\tau_{\mathrm{m}}(k), \omega_{\mathrm{m}}(k)). \tag{A9}$$

Therefore, the electrical power supplied to/by the EM is expressed as

$$P_{\mathrm{m,el}}(k) = P_{\mathrm{m}}(k) + P_{\mathrm{m,loss}}(k). \tag{A10}$$

The EM torque and speed are bounded by

$$\tau_{\mathrm{m}}(k) \in [\underline{\tau}_{\mathrm{m}}(\omega_{\mathrm{m}}(k)), \overline{\tau}_{\mathrm{m}}(\omega_{\mathrm{m}}(k))], \tag{A11}$$

$$\omega_{\mathrm{m}}(k) \in [\underline{\omega}_{\mathrm{m}}, \overline{\omega}_{\mathrm{m}}]. \tag{A12}$$

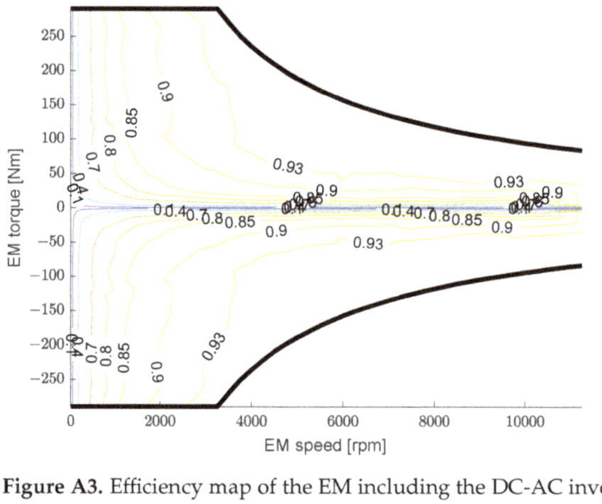

Figure A3. Efficiency map of the EM including the DC-AC inverter.

Appendix A.4. Thermal EM-SST Model

The SST ($P_{s,loss}$) and EM ($P_{m,loss}$) power losses appear as heat, which is removed by a TMS, as demonstrated in Figure A4. The EM and the SST are physically attached. The SST is cooled based on oil splashing, where its heat is taken away by the EM cooling medium indirectly and the ambient air directly due to convection. The EM dissipates heat to the ambient air. Moreover, the cooling medium driven by a pump removes heat from the EM. When the EM temperature is higher than a predefined threshold, the EM is cooled down with a radiator. The goal of the TMS is to maintain the EM temperature below its prescribed thermal limit. A lumped-parameter method is used to capture the thermal behavior of the EM and SST. On the basis of first principles, the thermal EM-SST model is described by

$$c_m\, m_m\, \dot{\theta}_m(k) = P_{m,loss}(k) - h_m\, A_m\, (\theta_m(k) - \theta_o(k)) - k_m\, (\theta_m(k) - \theta_s(k)) - h_a\, A_a\, (\theta_m(k) - \theta_a), \tag{A13}$$

$$c_f\, m_f\, \dot{\theta}_o(k) = h_m\, A_m\, (\theta_m(k) - \theta_o(k)) - \phi_f\, c_f\, (\theta_o(k) - \theta_i(k)), \tag{A14}$$

$$c_h\, c_s\, m_s\, \dot{\theta}_s(k) = P_{s,loss}(k) + k_m\, (\theta_m(k) - \theta_s(k)) - h_c\, A_c\, (\theta_s(k) - \theta_a), \tag{A15}$$

$$c_f\, m_f\, \dot{\theta}_i(k) = \phi_f\, c_f\, (\theta_o(k) - \theta_i(k)) - \epsilon\, \phi_a(k)\, c_a\, (\theta_o(k) - \theta_a). \tag{A16}$$

The thermal EM-SST model is validated by measurement data. Since the SST temperature is not available in the CAN data, the EM temperature is used for validation. The measurement data are extracted from the CAN signal regarding the EM temperature of the VCU in the series production vehicle driven on the ICDC. The estimated thermal parameters are provided in Table A1. As shown in Figure A5, a good resemblance can be seen between the model and measurement.

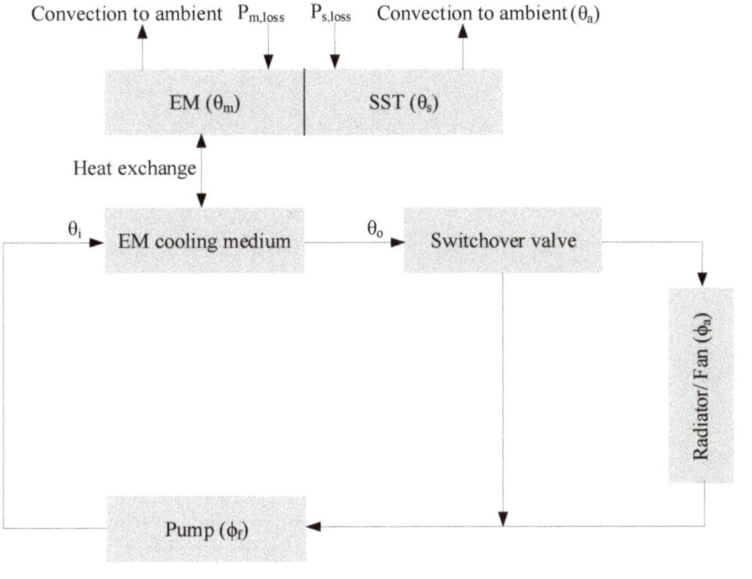

Figure A4. Thermal management architecture for the EM-SST.

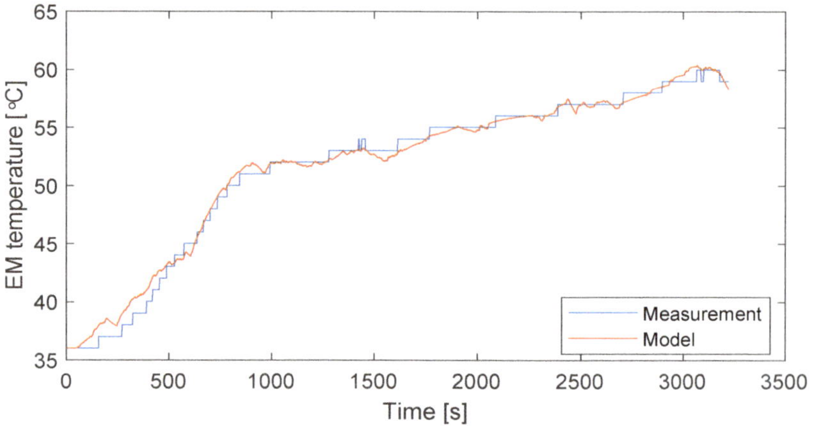

Figure A5. Validation of the EM temperature for the ICDC.

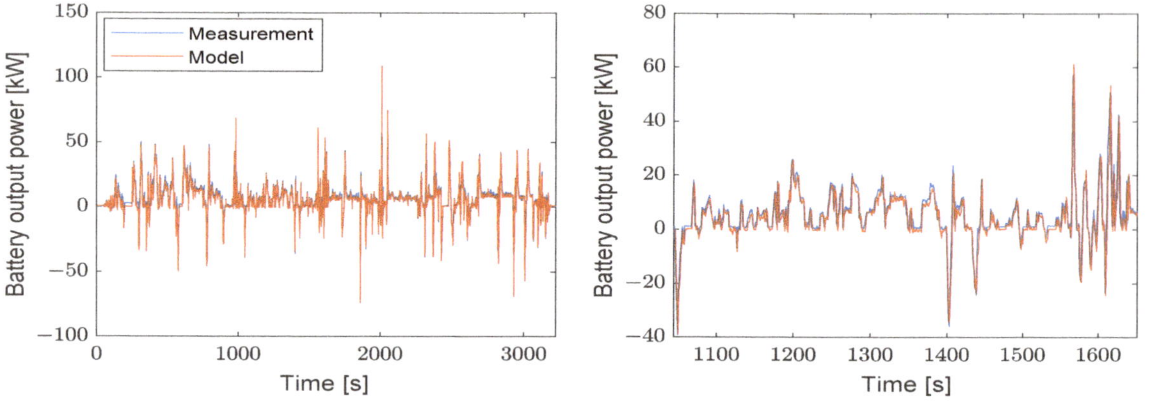

Figure A6. Validation of the battery output power for the ICDC.

Appendix A.5. Battery

The battery provides the power required by the EM, that is,

$$P_{\mathrm{b}}(k) = P_{\mathrm{m,el}}(k), \tag{A17}$$

$$\dot{E}_{\mathrm{b}}(k) = -P_{\mathrm{b}}(k). \tag{A18}$$

For the same inputs from the ICDC, the battery output power from the model is compared to the corresponding CAN signal regarding the battery output power in terms of voltage and current (measurement). The comparison can be seen in Figure A6, where the right subplot zooms in on the left subplot. It can be observed that the model and measurement resemble well.

In summary, the SST-based EV model represented by S_1 is validated against the measurement data, which replicates the physical behavior of the series production vehicle in reality.

Appendix B. CVT-Based EV Model

As presented in Section 1, a CVT-based EV model (S_2) is created based on S_1 (Appendix A). The main difference is that the SST model as described in Appendix A.2 is replaced by a

CVT model. The SST mass in (A1) is replaced by $\overline{m}_c$ (Table 1). The thermal model is the same as that in Section 3.7. All the other component models remain the same as S_1.

In this context, the variator of the pushbelt CVT has two pulleys, a primary pulley (subscript "p") and a secondary pulley (subscript "s"), which are connected by a pushbelt. The CVT provides a continuous variable speed ratio γ_v between the primary pulley and the secondary pulley. It permits the EM speed to be independent of the wheel speed to optimize its operating point. Given the required torque τ_w (A1) and speed ω_w (A2) at the wheels, the torque and speed of the primary pulley are obtained by

$$\tau_p(k) = \begin{cases} \frac{\tau_w(k)}{\eta_f\,\gamma_v(k)}, & \text{if } \tau_w(k) > 0, \\ \frac{\eta_f\,\tau_w(k)}{\gamma_v(k)}, & \text{if } \tau_w(k) \leq 0, \end{cases} \tag{A19}$$

$$\omega_p(k) = \gamma_v(k)\,\omega_w(k), \tag{A20}$$

The total torque input to the CVT is thus given by

$$\tau_c(k) = \tau_p(k) + \tau_{c,loss}(k), \tag{A21}$$

where $\tau_{c,loss}$ represents the torque loss in the CVT, which is described by a lookup table, that is,

$$\tau_{c,loss}(k) = \tau_{c,loss}(\tau_p(k), \omega_p(k), \gamma_v(k)). \tag{A22}$$

The power losses of the CVT can thus be calculated by

$$P_{c,loss}(k) = \tau_{c,loss}(\tau_p(k), \omega_p(k), \gamma_v(k))\,\omega_p(k). \tag{A23}$$

The primary torque and speed ratio are constrained by

$$\tau_p(k) \in [\underline{\tau}_p, \overline{\tau}_p], \tag{A24}$$

$$\gamma_v(k) \in [\underline{\gamma}_v, \overline{\gamma}_v], \tag{A25}$$

where $\underline{\gamma}_v$ is the overdrive ratio and $\overline{\gamma}_v$ the underdrive ratio. Bounds on the primary speed will be implicitly taken into consideration in the constraints on the EM speed. Additionally, to maintain the CVT speed ratio γ_v, the corresponding hydraulic actuation power can be calculated.

Appendix C. Convex Programming

CP demands all the models to be convex. A brief introduction to CP is given here, and interested readers are referred to [22] for a rigorous treatment. A convex optimization problem can be formulated as follows:

$$\begin{aligned} \min \quad & f_0(x), \\ \text{s.t.} \quad & f_i(x) \leq 0, \quad i = 1, ..., m, \\ & h_j(x) = A_j^T(x) - B_j = 0, \quad j = 1, ..., n, \end{aligned} \tag{A26}$$

where $f_i(x)$ are convex functions and $h_j(x)$ are affine functions. The feasible set of this optimization problem is convex with m convex sublevel sets and n hyperplanes. A convex function can be described as

$$f(\beta\, x_1 + (1-\beta)\, x_2) \leq \beta\, f(x_1) + (1-\beta)\, f(x_2), \tag{A27}$$

where $\beta \in [0, 1]$, and it means that the line segment between any two points lies above the graph. Models that are originally non-convex can be reformulated based on approximations, relaxations, and change of variables. Model convexity can be verified, by using basic convex functions, for example, linear functions, quadratic functions, quadratic-over-linear functions and opposite of geometric mean functions, and operations that preserve convexity, such as nonnegative weighted sums and pointwise maximum.

Appendix D. Main Parameters of EV Model

The main EV model parameters are provided in the following table, including vehicle parameters with reference to the series production vehicle and validated thermal parameters.

References

1. Kwon, K.; Seo, M.; Min, S. Efficient Multi-Objective Optimization of Gear Ratios and Motor Torque Distribution for Electric Vehicles with Two-Motor and Two-Speed Powertrain System. *Appl. Energy* **2020**, *259*, 114190. [CrossRef]
2. Karki, A.; Phuyal, S.; Tuladhar, D.; Basnet, S.; Shrestha, B. P. Status of Pure Electric Vehicle Power Train Technology and Future Prospects. *Appl. Syst. Innov.* **2020**, *3*, 35. [CrossRef]
3. Han, J.; Shin, J.; Kim, J.; Oh, S. Design 2-Speed Transmission for Compact Electric Vehicle Using Dual Brake System. *Appl. Sci.* **2019**, *9*, 1793. [CrossRef]
4. Hupkes, I. Variable Drive EV: Comfort Solution for Full Electric Vehicles. In Proceedings of the 3rd International VDI Conference: CVT in Automotive Applications, Baden-Baden, Germany, 19–20 March 2019.
5. Anselma, P. G.; Belingardi, G. Comparing Battery Electric Vehicle Powertrains through Rapid Component Sizing. *Int. J. Electr. Hybrid Veh.* **2019**, *11*, 36–58. [CrossRef]
6. Van der Sluis, F.; Römers, L.; van Spijk, G.; Hupkes, I. *CVT, Promising Solutions for Electrification*; SAE Technical Paper 2019-01-0359; SAE International: Warrendale, PA, USA, 2019.
7. Wei, C.; Hofman, T.; Ilhan Caarls, E.; van Iperen, R. A Review of the Integrated Design and Control of Electrified Vehicles. *Energies* **2020**, *13*, 5454. [CrossRef]
8. Silvas, E.; Hofman, T.; Murgovski, N.; Etman, L.F.P.; Steinbuch, M. Review of Optimization Strategies for System-Level Design in Hybrid Electric Vehicles. *IEEE Trans. Veh. Technol.* **2017**, *66*, 57–70. [CrossRef]
9. Singh, K.V.; Bansal, H.O.; Singh, D. Feed-forward Modeling and Real-time Implementation of an Intelligent Fuzzy Logic-based Energy Management Strategy in a Series–parallel Hybrid Electric Vehicle to Improve Fuel Economy. *Electr. Eng.* **2020**, *102*, 967–987. [CrossRef]
10. Lee, H.; Song, C.; Kim, N.; Cha, S.W. Comparative Analysis of Energy Management Strategies for HEV: Dynamic Programming and Reinforcement Learning. *IEEE Access* **2020**, *8*, 67112–67123. [CrossRef]
11. Qi, Z.; Shi, Q.; Zhang, H. Tuning of Digital PID Controllers Using Particle Swarm Optimization Algorithm for a CAN-Based DC Motor Subject to Stochastic Delays. *IEEE Trans. Ind. Electron.* **2020**, *67*, 5637–5646. [CrossRef]
12. Zhang, Y.; Wang, S.; Ji, G. A Comprehensive Survey on Particle Swarm Optimization Algorithm and Its Applications. *Math. Probl. Eng.* **2015**, *2015*, 931256. [CrossRef]
13. Pourabdollah, M.; Silvas, E.; Murgovski, N.; Steinbuch, M.; Egardt, B. Optimal Sizing of a Series PHEV: Comparison between Convex Optimization and Particle Swarm Optimization. *IFAC-PapersOnLine* **2015**, *48*, 16–22. [CrossRef]
14. East, S.; Cannon, M. Energy Management in Plug-In Hybrid Electric Vehicles: Convex Optimization Algorithms for Model Predictive Control. *IEEE Trans. Control Syst. Technol.* **2020**, *28*, 2191–2203. [CrossRef]
15. Pourabdollah, M.; Murgovski, N.; Grauers, A.; Egardt, B. Optimal Sizing of a Parallel Phev Powertrain. *IEEE Trans. Veh. Technol.* **2013**, *62*, 2469–2480. [CrossRef]
16. Wei, C.; Hofman, T.; Ilhan Caarls, E.; van Iperen, R. Evolution and Classification of Energy and Thermal Management Systems in Electrified Powertrains. In Proceedings of the 2019 IEEE Vehicle Power and Propulsion Conference, Hanoi, Vietnam, 14–17 October 2019.
17. Luin, B.; Petelin, S.; Al-Mansour, F. Microsimulation of Electric Vehicle Energy Consumption. *Energy* **2019**, *174*, 24–32. [CrossRef]
18. Wei, C.; Hofman, T.; Ilhan Caarls, E.; van Iperen, R. Integrated Energy and Thermal Management for Electrified Powertrains. *Energies* **2019**, *12*, 2058. [CrossRef]
19. Markel, T.; Brooker, A.; Hendricks, T.; Johnson, V.; Kelly, K.; Kramer, B.; O'Keefe, M.; Sprik, S.; Wipke, K. ADVISOR: A Systems Analysis Tool for Advanced Vehicle Modeling. *J. Power Sources* **2002**, *110*, 255–266. [CrossRef]
20. Ruan, J.; Walker, P.; Zhang, N. A Comparative Study Energy Consumption and Costs of Battery Electric Vehicle Transmissions. *Appl. Energy* **2016**, *165*, 119–134. [CrossRef]
21. Murgovski, N.; Johannesson, L.; Sjöberg, J. Convex Modeling of Energy Buffers in Power Control Applications. *IFAC Proc. Vol.* **2012**, *45*, 92–99. [CrossRef]
22. Boyd, S.; Vandenberghe, L. *Convex Optimization*; Cambridge University Press: Cambridge, UK, 2009.
23. van Berkel, K. Control of a Mechanical Hybrid Powertrain. Ph.D. Thesis, Eindhoven University of Technology, Eindhoven, The Netherlands, 2013.
24. van der Sluis, F. The Pushbelt designed for Change. In Proceedings of the 3rd International VDI Conference: CVT in Automotive Applications, Baden-Baden, Germany, 19–20 March 2019.

25. van der Sluis, F.; Yildiz, S.; Brandsma, A.; Veltmans, P.; Kunze, M. The CVT Pushbelt reinvented for Future Compact and Efficient Powertrains. In Proceedings of the 2017 JSAE Annual Congress, Yokohama, Japan, 24–26 May 2017.
26. Pourabdollah, M. Optimization of Plug-in Hybrid Electric Vehicles. Ph.D. Thesis, Chalmers University of Technology, Gothenburg, Sweden, 2015.
27. *Electrical and Electronics Technical Team Roadmap*; Technical Report; U.S. DRIVE: Washington, DC, USA, 2017.
28. Lutsey, N.; Nicholas, M. *Update on Electric Vehicle Costs in the United States through 2030*; Technical Report; International Council on Clean Transportation: Washington, DC, USA, 2019.
29. Mahmoudzadeh Andwari, A.; Pesiridis, A.; Rajoo, S.; Martinez-Botas, R.; Esfahanian, V. A Review of Battery Electric Vehicle Technology and Readiness Levels. *Renew. Sustain. Energy Rev.* **2017**, *78*, 414–430. [CrossRef]
30. Software for Disciplined Convex Programming. Available online: http://cvxr.com/ (accessed on 11 February 2021).
31. Tang, T.; Devlin, M.; Mathur, N.; Henly, T.; Saathoff, L. Lubricants for (Hybrid) Electric Transmissions. *SAE Int. J. Fuels Lubr.* **2013**, *6*, 289–294. [CrossRef]
32. McFadden, C.; Hughes, K.; Raser, L.; Newcomb, T. Electrical Conductivity of New and Used Automatic Transmission Fluids. *SAE Int. J. Fuels Lubr.* **2016**, *9*, 519–526. [CrossRef]
33. Bennion, K.; Moreno, G. Convective Heat Transfer Coefficients of Automatic Transmission Fluid Jets with Implications for Electric Machine Thermal Management. In Proceedings of the ASME 2015 International Technical Conference and Exhibition on Packaging and Integration of Electronic and Photonic Microsystems InterPACK2015, San Francisco, CA, USA, 6–9 July 2015.
34. Salah, M. Nonlinear Control Strategies for Advanced Vehicle Thermal Management Systems. Ph.D. Thesis, Clemson University, Clemson, SC, USA, 2007.

energies

MDPI

Article

Power Management Strategy of a Parallel Hybrid Three-Wheeler for Fuel and Emission Reduction

Waruna Maddumage [1,*], Malika Perera [1], Rahula Attalage [2] and Patrick Kelly [3]

[1] Faculty of Engineering, Sri Lanka Institute of Information Technology, Malabe 10115, Sri Lanka; malika.p@sliit.lk
[2] Faculty of Graduate Studies and Research, Sri Lanka Institute of Information Technology, Malabe 10115, Sri Lanka; rahula.a@sliit.lk
[3] Wolfson School of Mechanical, Electrical and Manufacturing Engineering, Loughborough University, Loughborough LE11 3TU, UK; p.kelly@lboro.ac.uk
* Correspondence: waruna.m@sliit.lk

Abstract: Millions of three-wheelers in large cities of Asia and Africa contribute to the already increasing urban air pollutants. An emerging method to reduce adverse effects of the growing three-wheeler fleet is hybrid-electric technology. The overall efficiency of a hybrid electric vehicle heavily depends on the power management strategy used in controlling the main powertrain components of the vehicle. Recent studies highlight the need for a comprehensive report on developing an easy-to-implement and efficient control strategy for hybrid electric three-wheelers. Thus, in the present study, a design methodology for a rule-based supervisory controller of a pre-transmission parallel hybrid three-wheeler based on an optimal control strategy (i.e., dynamic programming) is proposed. The optimal control problem for minimizing fuel, emissions (i.e., HC, CO and NOx) and gear shift frequency are solved using dynamic programming (DP). Numerical issues of DP are analyzed and trade-offs between optimizing objectives are presented. Since DP strategy cannot be implemented as a real-time controller, useful strategies are extracted to develop the proposed rule-based strategy. The developed rule-based strategy show performance within 10% of the DP results on WLTC and UDC-NEDC drive cycles and has the clear advantage of being near-optimal, easy-to-implement and computationally less demanding.

Keywords: hybrid electric vehicle; auto-rickshaw; energy management strategy; multi-objective optimization; rule-based control; dynamic programming; fuel economy; backward-facing model; forward-facing model

Citation: Maddumage, W.; Perera, M.; Attalage, R.; Kelly, P. Power Management Strategy of a Parallel Hybrid Three-Wheeler for Fuel and Emission Reduction. *Energies* **2021**, *14*, 1833. https://doi.org/10.3390/en14071833

Academic Editors: João Pedro Trovao, Riccardo Berta and Rui Esteves Araújo

Received: 21 November 2020
Accepted: 16 March 2021
Published: 25 March 2021

Publisher's Note: MDPI stays neutral with regard to jurisdictional claims in published maps and institutional affiliations.

1. Introduction

The ever-increasing cost of fuel, air pollution and consumer market trends have forced the automobile industry towards investing in greener vehicles. Currently, over 90% of the transport sector is reliant on oil and 49% of the total oil production is consumed by the transportation sector alone. Responsible for one-quarter of energy-related unwanted greenhouse gas emissions in 2009, the transport sector is the fastest-growing energy-consuming source in the world [1]. Among the transportation sector, three-wheelers show a significant growth rate. In India alone, during 2018–2019 fiscal year, domestic sales of three-wheelers increased by 10% and exports of manufactured three-wheelers increased by 49%; overall sales increased from 1.017 million units to 1.269 million units [2]. Even though considerable work is being done in four-wheel category to reduce emissions and fuel consumption, little work is been conducted on three-wheelers. Figure 1 shows a typical motorized three-wheeler by Bajaj Motor company.

Figure 1. BAJAJ RE 205 cc motorized three-wheeler [3].

Three-wheelers are found in most Asian, South American and African countries [4]. In large cities of Asia and Africa, millions of auto-rickshaw three-wheelers function as taxicabs [5]. Three-wheelers in developing countries offer a low volume, but high-frequency service and they are efficient feeders that can connect bigger public transport systems and provide last-mile connectivity to the doorstep. Three-wheelers can weave through dense urban traffic conditions and a solution for the growing parking problem in urban cities due to its small size. Even though three-wheelers are known to be highly pollutant vehicles due to the use of two-stroke engines and small-scale diesel engines, with stringent government regulations most of the three-wheeler fleet has moved to use four-stroke engines and fuels such as Petrol, CNG and LPG to reduce overall emissions. However, to improve the air quality of worlds most polluted cities such as Delhi, Mumbai and Jakarta, emission levels of the growing three-wheeler fleet should be further reduced and Electric Vehicle (EV) technology is an emerging solution to address this issue.

Hybrid EVs, plug-in EVs and battery EVs are common types of electric vehicles [1]. However, only the battery EVs can give potentially zero emissions. In recent years, the battery EV three-wheelers market has grown considerably. In India alone, during 2018–2019 fiscal year, the sale of battery EV three-wheelers was 630,000 units [2]. However, the emission reduction from battery EVs is offset by coal-burning power plants since these vehicles are charged using the national power grid. A case study made in West Bengal state, India show that when the emissions from the coal-fired thermal power plants are considered, the use of battery EV three-wheelers reduced the specific CO_2 emissions (gm/passenger-km) by only 11% compared to a diesel three-wheeler [6]. In addition, to the carbon intensity of electricity generation, battery manufacturing and disposal must be considered. An effective alternative option is the reuse of batteries and it is found that the carbon emission reduction potential of reusing batteries is similar to moving from an oil-based vehicle to an EV [7]. In the global context, high infrastructure cost, anxiety on the resale market value, increase in demand for power generation, strong oil-base vehicle market and unit cost are some of the resistive forces for EVs in developing countries [8]. For instance, developing countries like Pakistan cannot transition directly from oil-based vehicles to battery EVs due to factors like charging infrastructure, load shedding of electricity and high cost of non-renewable energy [9]. In addition, battery EV three-wheeler owners in developing countries face challenges such as poor drivability characteristics [6], electricity blackouts, limited driving range and long charging times. An interim solution that mitigates these challenges and reduces overall fuel consumption and emissions is hybrid-electric technology.

The benefits of a fuel-efficient and eco-friendly hybrid electric three-wheeler to reduce adverse effects from the growing motorized three-wheeler fleet is well understood. Thus, the design of hybrid electric three-wheelers has been extensively investigated in the literature under different hybrid configurations. For instance, Vezzini et al. investigated a series hybrid electric three-wheeler and developed a prototype in [10]. Developing on the series hybrid concept, Amjad et al. investigated plug-in series hybrid technology for three-wheelers in [11]. Another popular hybrid configuration studied for three-wheelers is the parallel powertrain configuration. For example, Hofman et al. in [5] studied parallel micro-hybrid architecture for three-wheelers. Roy and Indulal in [12] extended the study of parallel hybrid technology for three-wheelers by investigating a parallel configuration where motor and engine power is coupled through road. Use of plug-in technology for parallel hybrid electric three-wheelers is studied by Padmanaban et al. [13]. Comparative analysis of hybrid configurations of three-wheelers is carried out by a recent study from Maddumage et al. [14]. The study examines the effect of hybrid configuration on fuel and emissions by studying conventional, parallel, series and complex powertrain designs for hybrid electric three-wheelers. These studies indicate that the use of hybrid technology for three-wheelers is a viable solution to reduce overall fuel consumption and exhaust emissions of the vehicle.

Regardless of the intensive studies on developing hybrid-electric three-wheelers, reports on parallel hybrid electric three-wheelers have not thoroughly dealt with designing an efficient and easy-to-implement power management strategy. The power management strategy of a parallel hybrid mainly involves in the management of power flow, namely, the power split strategy between the motor and engine. Various methods in developing power management strategies for hybrid electric vehicles have been proposed in the literature. These methods can be classified into two main groups, i.e., optimal control-based strategies and heuristic control strategies. Optimal control-based strategies use global optimization and instantaneous optimization methods based on optimal control tools. Limitations of optimal control-based strategies are that their solutions rely on future unknown information [15] and require high computational power. An alternate approach is heuristic control strategies. Heuristic power management strategies do not employ explicit optimization; instead, power management is carried out through heuristic rules [15]. Heuristic control methods have monopolized the production vehicle market due to the low computational power demand, natural adaptability to online-applications, good reliability and satisfactory fuel consumption results [16]. Fuzzy logic [17] and deterministic rule-based control belong in this category. One of the main limitations of this type of control strategies is the sub-optimal results compared to optimal control-based strategies. Nevertheless, by using results of a global optimization algorithm like dynamic programming (DP) as a template to design empirical rules for the control strategy, it is possible to improve the results of the rule-based control [18].

The possibility of deriving useful real-time control strategies from the global optimal results of the DP algorithm has been widely investigated in literature [19–22]. For instance, Wang and Lukic [23], applied DP to find the real-time optimal split between engine and motor in a series-parallel powertrain. Similarly, Lin et al. [24], found that optimal control rules can be extracted from DP results and used to design a near-optimal rule-based control strategy for a parallel hybrid. Results show a 28% increase in fuel economy with the near-optimal rule-based strategy, compared to a conventional vehicle. In another study by Lin et al. [18], a simple approach (based on the power request ratio from engine and motor to transmission speed) for extracting empirical rules for the control strategy from DP results is investigated. Results from the study showed a 50–70% reduction in the performance gap between the DP results. The combination of DP results and empirical rules of the rule-based strategy for real-time charge sustaining control of hybrid vehicles is studied by Biasini et al. in [15]. This study used a penalty function to control power split ratio in relation to battery state of charge (SoC) level. Results show that the performance of the improved control strategy is within 3% of the DP results. These studies show, extracting empirical rules from the DP results (optimal controller) is a viable approach to improve overall performance of a rule-based controller.

In spite of the research advances made, some of these power management strategies have been found to yield selective performance, which is charge-depleting in highway driving conditions and charge-hoarding in urban driving conditions [16]. In general, three-wheelers are used in urban areas. Thus, the use of DP to design a rule-based controller for hybrid electric three-wheelers should be carried out carefully to avoid charge-hoarding. A recent study from Asghar et al. [25] has attempted to implement a sub-optimal rule-based power management strategy based on DP results for a parallel hybrid electric three-wheeler. Despite the extensive methodology followed in extracting rules from the DP results; report has not thoroughly dealt with transmission gear strategy, numerical issues of DP, exhaust emissions and powertrain component sizing techniques in designing the control strategy of the hybrid electric three-wheeler.

The aim of the present paper is to design a rule-based control strategy for a parallel hybrid electric three-wheeler based on DP results to reduce fuel consumption and exhaust emissions, i.e., HC, CO and NOx. These exhaust emissions have serious impact on the urban air quality and public health [26]. Although HC, CO and NOx emissions can be reduced with emission control technologies [27]; in general, motorized three-wheelers on the market are not equipped with such measures. Three-wheelers are low-cost price-sensitive vehicles with a high price elasticity of demand. Larger social benefits such as harmful emissions are not relevant to the average user [28]. This is one of the factors for not having a stringent emission control strategy in motorized three-wheelers yet. Therefore, in the present study, reduction of HC, CO and NOx emission via hybrid technology is investigated in addition to the fuel economy. Overall fuel consumption and emissions of the powertrain are affected by gear shift strategy and component sizes of the hybrid power train. Since parallel hybrid electric vehicles use a motor and an engine in harmony to propel the vehicle, it is harder to develop an efficient gear shift strategy based on traditional methodologies. Thus, in the present study transmission gear strategy is developed based on DP results. In addition, a sequential design methodology [29] (sizing and control is solved separately in a sequential manner) is used in developing the hybrid powertrain. The power management strategy is developed in the present study for the sized parallel hybrid powertrain from [14].

Contributions of the study is of interest due to the systematic procedure presented to establish an easy to implement real time control strategy for parallel hybrid three-wheelers; which can be used by a developer to implement an eco-friendly parallel hybrid three-wheeler. The proposed control strategy uses only 2D maps and simple rules. Unlike with control strategies such as time delay neural network (TDNN) [30] and model predictive control (MPC) [31], proposed control law requires low computational power to operate. Furthermore, online control approaches like adaptive equivalent consumption minimization strategy (ECMS) [32] require predictive equipment like GPS; which add cost and complexity to vehicle development. The proposed strategy can be implemented without such equipment or sensors. The developed strategy is easy to implement and it does not require specialize knowledge to install for a practical application.

Striking a good balance to optimize all the objectives is difficult in a high dimensional objective space with more than three objectives using rule-based control law. Usually, three or less multi-objective are considered in developing DP based heuristic control laws [18,19,21,23,25]. The present study considers a multi-objective optimization problem with more than three objectives; which are defined as many objective optimization problems [33]. In addition, the DP method use quantization and interpolation techniques and these methods introduce numerical errors. Therefore, present study investigates numerical issues of DP in implementing the optimal control strategy. Furthermore, rules extracted from DP results does not necessarily constitute a comprehensive control law that can be implemented in a real-life vehicle. Thus, present study considers additional rules such as gearshift strategy in full EV mode.

The contributions of the study can be summarized as follows:

- Investigation of numerical errors of the DP algorithm in implementing an optimal control strategy for hybrid vehicles.
- Comprehensive rule extraction from DP results to establish a near optimal rule-based strategy for multi-objective (many objectives).

- Establish additional rules to overcome shortcomings of the useful strategies extracted from DP results.
- A systematic methodology to develop an easy to implement, real time, near optimal power management strategy for parallel hybrid three-wheelers.

This paper is organized as follows: In Section 2, simulation models for the control strategy analysis is developed. First, the main powertrain components of a parallel hybrid are modelled. Using the modelled powertrain components two quasi-static hybrid electric vehicle models, i.e., a backward-facing model and forward-facing model, are developed to simulate the DP-based optimal controller and proposed rule-based controller, respectively. In Section 3, the DP-based optimal controller is implemented. DP for the optimal control problem of the parallel hybrid is formulated and numerical issues of the DP algorithm are analyzed. The controller is simulated to reduce five objectives (i.e., gear shift frequency, fuel, HC, CO and NOx) and the trade-offs between the objectives are presented. In Section 4, results from the DP-based controller are extracted and used to develop the proposed rule-based strategy. In addition, the performance of the proposed rule-based strategy is analyzed and compared with DP results. Finally, conclusions are presented in Section 5.

2. Hybrid Electric Three-Wheeler Models

DP results are used to identify the empirical rules for the rule-based control strategy proposed for the hybrid electric three-wheeler. DP algorithm requires knowledge of future driving characteristics. Thus, the DP controller is implemented in a backward-facing hybrid powertrain model. Based on the DP results, a sub-optimal rule-based strategy is developed. To perform the energy analysis and the fuel consumption evaluation of the control strategies, the simulation method is generally used and the quasi-static models are usually selected [34,35]. Since only the energy flow is focused, the accuracy of the quasi-static models is sufficient for the simulation time step as large as 1 s [36]. Thus, in the present study to evaluate the proposed sub-optimal rule-based strategy a forward-facing quasi-static model is developed (time step of the used drive cycles is 1 s). Note, the developed quasi-static models do not consider dynamic characteristics such as response time of the engine and motor. The operational characteristics of the powertrain components are taken as instantaneous. Even though quasi-static models are adequate to analyze and optimize fuel economy and performance it is not sufficient to evaluate drivability characteristics.

In this section, mathematical models for main powertrain components of a parallel hybrid electric vehicle are modelled and based on the powertrain models two hybrid electric vehicle models are developed, i.e., forward-facing model and backward-facing model. [18,22,25,36,37] used a similar methodology.

2.1. Parallel Hybrid Electric Powertrain Model

Control strategy investigation is conducted for a pre-transmission parallel hybrid powertrain with a degree of hybridization of 0.3, the powertrain configuration is shown in Figure 2. Type of powertrain configuration is chosen by considering the volume required and cost. The three-wheeler is a small low-power and low-cost vehicle used in developing countries. The design of the hybrid powertrain should be carried out within the volume constraints of the existing three-wheeler design. Thus, the volume and cost required are major concerns in selecting the hybrid configuration.

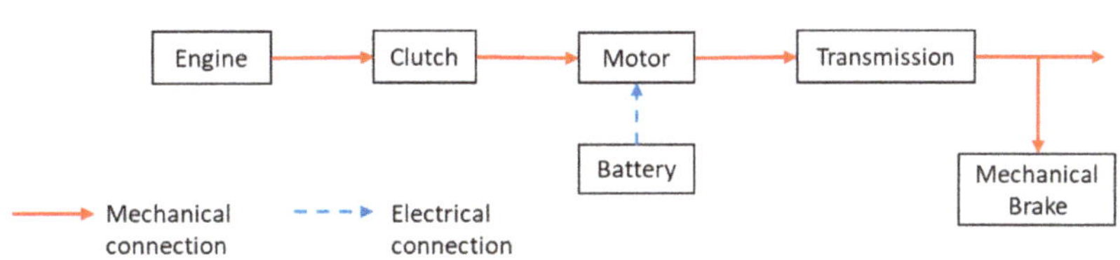

Figure 2. Schematic of the pre-transmission parallel hybrid powertrain.

Both hybrid and plug-in hybrid topologies have excellent fuel economy and environmental benefits [38]. In addition, to the electricity generation from the motor in a hybrid, the plug-in hybrids can externally charge the battery by connecting to the electrical grid. Although plug-in hybrids have a high fuel economy compared to a hybrid [39], plug-in hybrids are equipped with a large energy storage system [38]. In addition, the batteries account for a significant amount of the initial cost of a plug-in hybrid [39,40]. Thus, in the present study a non-plug-in hybrid topology is selected.

Out of the three traditional hybrid configurations (i.e., series, parallel and series-parallel), the parallel hybrid configuration has the least number of powertrain components. In general, the series-parallel configuration requires two electric motors, engine and a planetary gear set, which make the powertrain complex and costly [41]. The series hybrid also requires two motors and an engine. Even though for short trips engine can be downsized relatively easily, sizing the motors and battery is still a challenge, making the series hybrids expensive [41]. The parallel hybrid has the advantage of a downsized motor and drive compared to the series configuration [42]. Since the mechanical power is shared between the engine and motor, the parallel configuration does not require a large motor [43].

The powertrain consists of an internal combustion engine, an electric motor, a battery pack, a clutch and an automatically controlled manual transmission. As mentioned before, powertrain design follows a sequential design approach, component sizing and design of power management strategy is carried out successively [29]. Powertrain component sizes are found from previous work [14]. Basic parameters of the hybrid powertrain components and the vehicle are summarized in Table 1.

Table 1. Basic vehicle parameters.

Powertrain	Parameter	Value
Engine	Maximum power (kW)	6
Electric motor	Maximum power (kW)	3
Battery	No. of Modules	6
	Capacity (Ah)	6
	Nominal voltage (volt/module)	10.8
Vehicle	Frontal area (m^2)	1.86
	Tire radius (m)	0.2
	Coefficient of drag	0.44
	Rolling resistance coefficient	0.015
	Glider mass–without propulsion (kg)	280
	Curb weight (kg)	448

As shown in Figure 2, the output of the engine is connected to a motor via a clutch operated by the control unit. The automatically controllable clutch is used to connect or disconnect the engine torque with the electric motor torque. The clutch between the motor and internal combustion engine allows the powertrain to operate in full electric mode (motor only) without the engine. When the powertrain is operating on parallel mode (motor and engine operate in harmony) both electric motor torque and engine torque are coupled. Furthermore, the engine, motor and battery models are developed based on ADvanced VehIcle SimulatOR (ADVISOR) library maps/data. This approach already known in literature, is used to develop hybrid vehicle simulation models for the present study. Some examples where ADVISOR powertrain models are used for similar hybrid powertrain investigations are [18,20,22–24,44].

2.1.1. Longitudinal Vehicle Model

The vehicle model is defined with the characteristics of a motorized three-wheeler, with a glider mass of m_{gm} (without the powertrain components). The total mass of the vehicle is,

$$m_{veh} = m_{gm} + m_{ice} + m_{mot} + m_{batt} + m_{gb} + m_{cargo} \tag{1}$$

where m_{ice} mass of the engine, m_{mot} mass of the motor, m_{batt} mass of the battery, m_{gb} mass of the gearbox (transmission) and m_{cargo} mass of three passengers including the driver.

The movement characteristics of a vehicle along its moving direction are determined by all the forces acting in that direction. Forces acting on the vehicle are modelled as a representation of the force balance at the tire patch. In the longitudinal direction, the main external forces acting on the vehicle are aerodynamic drag force F_a, rolling resistance force F_r, climbing resistance F_c, acceleration resistance force F_i and traction force at the wheel F_t. The dynamic equation for the longitudinal hybrid electric vehicle model is expressed by,

$$F_t = F_a + F_r + F_c + F_i \tag{2}$$

$$F_t = \frac{1}{2}\cdot C_d\cdot A\rho\cdot {v_{veh}}^2 + m_{veh} g C_r\cdot \cos\alpha + m_{veh} g\cdot \sin\alpha + m_{veh}\cdot \frac{dv}{dt} \tag{3}$$

where C_d is the aerodynamic drag coefficient, A is the frontal area of the vehicle, ρ is the air density, v_{veh} is the linear velocity of the vehicle, g is the acceleration due to gravity, C_r is the rolling resistance coefficient, α is the road gradient and $\frac{dv}{dt}$ is the linear acceleration of the vehicle.

For the simulation, the vehicle speed is extracted from the drive cycle at each time interval. Traction force required to overcome driving resistance forces depend on the current characteristics of the vehicle and driver demand at the next moment.

The vehicle simulation is a discrete-time system; therefore, the current acceleration $a_{veh,k}$ is defined as a discrete function as follows,

$$a_{veh,k} = \frac{v_{veh,k+1} - v_{veh,k}}{t_{step}} \tag{4}$$

where $v_{veh,k}$ is the current vehicle velocity. $v_{veh,k+1}$ is the target velocity in the next moment and t_{step} is the time step of the simulation.

By considering Equations (1) and (3), traction torque required by the powertrain at the vehicle wheel during the current step in the discrete-time space T_{wh}, is expressed as,

$$T_{wh} = r_{wh}\left(\frac{1}{2}\cdot C_d A\rho {v_{veh,k}}^2 + m_{veh} g C_r\cdot \cos\alpha_k + m_{veh} g\cdot \sin\alpha_k + m_{veh}\cdot a_{veh,k}\right) \tag{5}$$

where α_k is the road gradient at the particular timestamp and r_{wh} is the vehicle's wheel radius. The angular velocity ω_{wh} and angular acceleration $\dot{\omega}_{wh}$ of the wheel is expressed as,

$$\omega_{wh} = \frac{v_{veh}}{r_{wh}} \tag{6}$$

$$\dot{\omega}_{wh} = \frac{a_{veh,k}}{r_{wh}} \tag{7}$$

The vehicle model assumes a constant rolling resistance coefficient and a constant tire radius. Effect of rotational inertia in the linear direction is taken as negligible. In addition, system vibrations and wheel slip are neglected.

2.1.2. Transmission Model

The transmission assembly is modelled as a four-gear automatically controlled manual transmission. The gearbox efficiency is considered equal for all the gears $\eta_{gb} = 0.95$. The powertrain torque on the clutch side of the gearbox T_{gb}, is modelled as,

$$T_{gb} = \frac{T_{wh}}{\eta_{gb}\cdot \gamma(i)}\ ,\ T_{wh} \geq 0 \tag{8}$$

$$T_{gb} = \frac{T_{wh}\cdot \eta_{gb}}{\gamma(i)}\ ,\ T_{wh} < 0 \tag{9}$$

where $\gamma(i)$ is the gear ratio for each gear including the final drive ratio. The angular velocity ω_c and angular acceleration $\dot{\omega}_c$ of the crankshaft is expressed as,

$$\omega_c = \gamma(i) \cdot \omega_{wh} \tag{10}$$

$$\dot{\omega}_c = \gamma(i) \cdot \dot{\omega}_{wh} \tag{11}$$

The gear switching strategy for the transmission is modelled using a discrete-time dynamic model. The transmission gear number at $k+1$ time step, $N_{gear}(k+1)$ is expressed as,

$$N_{gear}(k+1) = \begin{cases} 4, & N_{gear}(k+1) + shift(k) > 4, \\ 1, & N_{gear}(k+1) + shift(k) < 1, \\ N_{gear}(k) + shift(k), & otherwise \end{cases} \tag{12}$$

where $N_{gear}(k)$ is the transmission gear number at kth time step and $shift(k)$ is the gear shift command, defined as $shift(k) \in \{-1, 0, 1\}$.

- $shift(k) = -1$, gear downshift.
- $shift(k) = 1$, gear upshift.
- $shift(k) = 0$, maintain the current gear.

The gearbox model assumes no energy losses during gear shifting and inertia of the gearbox and final drive (differential) is neglected.

2.1.3. Clutch Model

The clutch between the motor and internal combustion engine allows the powertrain to operate in full electric mode without the resistive inertial torque of the engine. During the parallel operational mode, both electric motor resistive inertial torque T_{mot_0} and engine resistive inertial torque T_{ice_0}, are considered. The total torque demand from the electric motor and engine T_{dem}, is expressed as follows,

$$T_{dem} = T_{gb} + T_{mot\_0}\text{ , } clutch\ disengaged \tag{13}$$

$$T_{dem} = T_{gb} + T_{mot\_0} + T_{ice\_0}\text{ , } clutch\ engaged \tag{14}$$

$$T_{ice\_0} = I_{ice} \cdot \dot{\omega}_c \tag{15}$$

$$T_{mot\_0} = I_{mot} \cdot \dot{\omega}_c \tag{16}$$

where I_{ice} is the moment of inertia of the engine and I_{mot} is the moment of inertia of the motor. T_{dem} is split between the internal combustion engine and electric motor depending on the torque ratio determined by the power management strategy. The model assumes dynamics of the clutch are negligible.

2.1.4. Internal Combustion Engine Model

The internal combustion engine model is generated with a static fuel consumption map, developed using experimental modelling method based on the Willians approximation. By using Willians line scaling approach, based on known steady-state efficiency data of a reference machine, efficiency of a new machine in the same category can be estimated [35]. In the present work, the engine model uses a scaled fuel consumption map of a gasoline SI Geo Metro 1.0 L engine from the ADVISOR library. The fuel consumption map of the engine is expressed as a relationship between engine torque, speed and brake specific fuel consumption (BSFC) as shown in Figure 3a. BSFC is a measure of fuel efficiency: the rate of fuel consumption divided by the power produced. Like fuel consumption of the engine, exhaust gas emissions (i.e., HC, CO and NOx) are estimated using experimental static gas emission maps. As shown in Figure 3b–d the emissions maps are expressed as a relationship between engine torque, speed and brake specific exhaust gas emission.

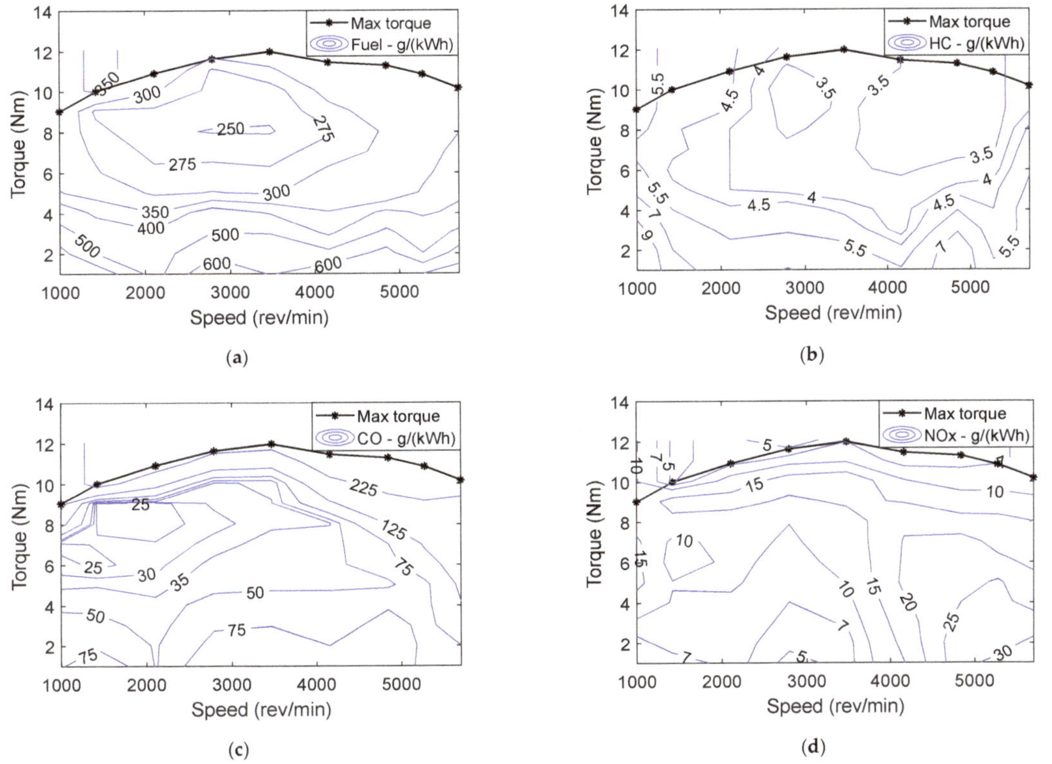

Figure 3. Engine fuel and emission maps: (**a**) Fuel efficiency map; (**b**) HC emission map; (**c**) CO emission map; (**d**) NOx emission map.

The engine model obtains the brake specific fuel consumption $O_{BSFC}(\omega_e, T_e)$ at the engine operating point (ω_e, T_e), where the engine outputs torque T_e and speed ω_e, using cubic spline data interpolation method,

$$O_{BSFC}(\omega_e, T_e) = f(\omega_e, T_e) \tag{17}$$

Fuel consumption rate $\dot{m_f}$ (g/s) is expressed as,

$$\dot{m_f} = \frac{1}{3600 \times 1000} \cdot \omega_e \cdot T_e \cdot O_{BSFC}(\omega_e, T_e) \tag{18}$$

The fuel power P_{fuel} given to the engine is calculated by,

$$P_{fuel} = \dot{m_f} \cdot Q_{lhv} \tag{19}$$

Engine model considers exhaust gas emissions: HC, CO and NOx. The brake specific exhaust gas emission rate $E_a(\omega_e, T_e)$, "a" represent the emission gas (i.e., HC, CO and NOx), at the engine operating point (ω_e, T_e) is obtained using the cubic spline data interpolation method,

$$E_a(\omega_e, T_e) = f(\omega_e, T_e) \tag{20}$$

Exhaust gas rate $m_{e\dot{x}h,a}$ (g/s), "a" represent the emission gas (i.e., HC, CO and NOx), is expressed as,

$$m_{e\dot{x}h,a} = \frac{1}{3600 \times 1000} \cdot \omega_e \cdot T_e \cdot E_a(\omega_e, T_e) \tag{21}$$

Engine model assumes the effect of change of temperature on the engine and frictional force of the engine are negligible. In addition, the dynamic characteristics of the engine are not considered.

2.1.5. Electric Motor Model

The motor model is also developed using a static efficiency map. As shown in Figure 4, the efficiency map is expressed as a relationship between engine torque, speed and motor efficiency.

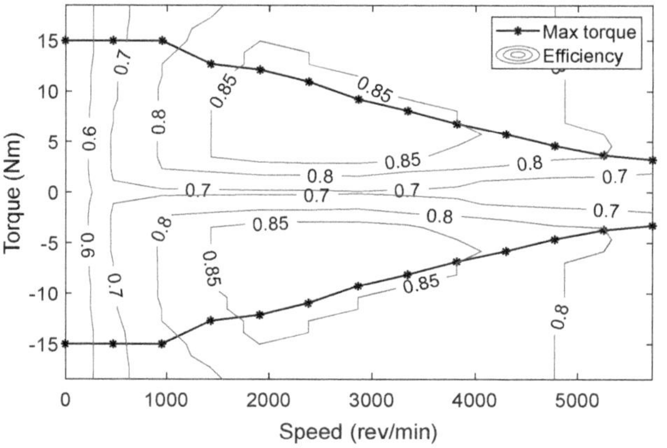

Figure 4. Motor efficiency map.

The electric motor efficiency $\eta_{mot}(\omega_{mot}, T_{mot})$, at the motor operating point (ω_{mot}, T_{mot}), where the motor outputs torque T_{mot} and speed ω_{mot}, is obtained using the cubic spline data interpolation method:

$$\eta_{mot}(\omega_{mot}, T_{mot}) = f(\omega_{mot}, T_{mot}) \tag{22}$$

The electric power needed from or supplied to the battery by the electric motor P_{elec_mot}, is expressed by

$$P_{elec\_mot} = \eta_{mot}(\omega_{mot}, T_{mot})\cdot\omega_{mot}\cdot T_{mot},\ T_{mot} < 0 \tag{23}$$

$$P_{elec\_mot} = \frac{\omega_{mot}\cdot T_{mot}}{\eta_{mot}(\omega_{mot}, T_{mot})},\ T_{mot} > 0 \tag{24}$$

The electric motor model does not consider the dynamics of the machine. In addition, the model assumes the frictional force of the electric motor is negligible.

2.1.6. Battery Model

The battery model is generated as multiple modules with a combination of parallel and series cells to represent the 6 Ah Saft Li-Ion battery from the ADVISOR library. The battery input or output power P_{batt}, is expressed by,

$$P_{batt} = P_{elec\_mot} + P_{aux} \tag{25}$$

where P_{aux} is a constant auxiliary power demand that represents the power consumption of all the secondary systems of the vehicle. The battery pack is modelled as an equivalent circuit comprising an open circuit voltage V_{oc} in series with an internal resistance R_{int} as illustrated in Figure 5. In addition, U is the terminal voltage of the battery. The model is based on experimental data of battery charging-discharging.

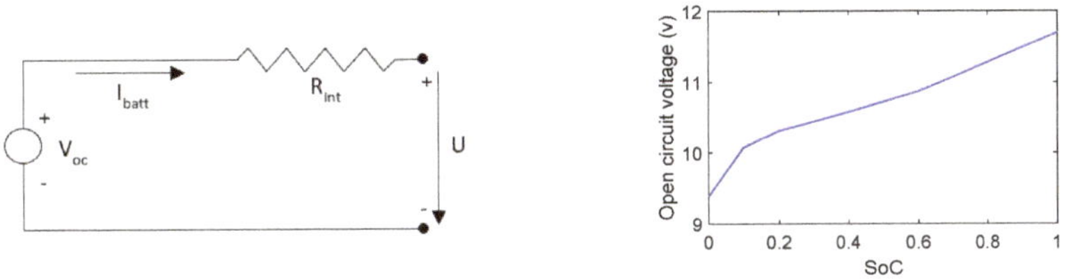

(a) (b)

Figure 5. Battery model and electrical performance: (**a**) The equivalent battery circuit model; (**b**) Battery module open circuit voltage with SoC at 41 °C.

In general, three-wheelers are widely used in countries such as India, Sri Lanka and Pakistan. Where temperature variation over the year is not significant. Thus, in the present study a constant temperature of 41 °C is considered to model the battery. The battery open-circuit voltage V_{oc}, which is a function of battery SoC and battery temperature T_{batt}, is shown in Figure 5b and expressed by the interpolation function,

$$V_{oc} = f(SoC, T_{batt}) \tag{26}$$

Battery internal resistance R_{int}, which include Ohmic resistance R_o and polarization resistance R_p, is expressed by the interpolation function,

$$R_{int} = f(SoC, T_{batt}) \tag{27}$$

Charging and discharging current of the battery, which is defined as positive during discharging and negative during charging, is expressed by,

$$I_{batt} = \frac{V_{oc} - \sqrt{{V_{oc}}^2 - 4R_{int} \cdot P_{batt}}}{2R_{int}} \tag{28}$$

The battery state of charge SoC, in the discrete system, is expressed as,

$$SoC_{k+1} = SoC_k - \left(\frac{V_{oc,k} - \sqrt{{V_{oc,k}}^2 - 4R_{int,k} \cdot P_{batt,k}}}{2R_{int,k}} \right) \cdot \frac{\eta_{batt}}{3600 \cdot Q_{batt}} \tag{29}$$

where SoC_{k+1} is the battery SoC at the (k + 1) time step and SoC_k, $V_{oc,k}$, $R_{int,k}$ and $P_{batt,k}$ are the battery SoC, open-circuit voltage, the internal resistance of the battery and battery power at the kth time step, respectively. In addition, Q_{batt} is the battery capacity and η_{batt} is the battery charging-discharging efficiency. Similar to [22] η_{batt} is defined as,

$$\eta_{batt} = \begin{cases} 1.0\,, & I_{batt} \geq 0 \\ 0.9\,, & I_{batt} < 0 \end{cases} \tag{30}$$

Note that an alternate approach to calculate battery efficiency is using battery circuit equivalent model data. Similar to [45] open-circuit voltage, internal resistance and output power can be used to better approximate the battery charging-discharging efficiency.

2.2. Quasi-Static Hybrid Electric Vehicle Models

Forward-facing model and backward-facing model [46] are developed with the above powertrain component models. The forward-facing model is developed in MATLAB/Simulink environment and backward-facing model is developed in MATLAB application. As mentioned above, the optimal controller (DP based) is simulated on the backward-facing model and the proposed rule-based controller is simulated on the forward-facing model. Figure 6 shows the basic schematics of the backward-facing model and forward-facing model.

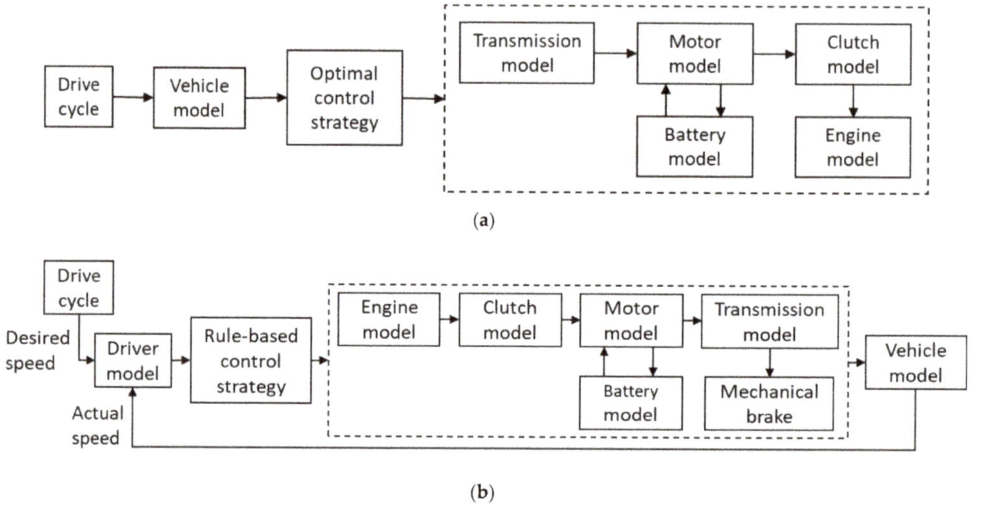

Figure 6. Schematics of Quasi-static models: (**a**) backward-facing model; (**b**) forward-facing model.

As shown in Figure 6b, the forward-facing model is implemented with a driver model. The driver model simulates the manipulation of the accelerator pedal and the brake pedal by the driver. The model is implemented as a PI controller, which compares the actual velocity (velocity of the model) with the desired velocity (velocity from the drive cycle) to generate acceleration and brake commands. Acceleration and brake commands are given to the supervisory controller which formulates the clutch command, gear shift command, engine torque command and the motor torque command.

The backward-facing model is implemented with two control commands, i.e., gear shift command and torque split command (torque split factor between motor and engine). These commands are generated by the optimal controller (DP-based controller). The torque split factor is defined as a continuous variable $u \in [-1\ 1]$.

- $u = 1$, only the motor provides necessary torque or full brake energy recuperation;
- $0 < u < 1$, both the motor and engine provide the necessary torque;
- $u = 0$, only the engine provides the necessary torque;
- $-1 < u < 0$, engine provides surplus torque and motor is in generator mode;
- $u = -1$, engine provides surplus torque and motor is in maximum generator mode (full recharge).

To gauge the comparability of the two models, the backward-facing model and forward-facing model is simulated with the same parameters. The cumulative error on fuel consumption, drive cycle and battery *SoC* are 1.7%, 0.75% and 0.01%, respectively.

3. Global Optimal Control Strategy Based on DP

DP is a powerful numerical method for solving optimal control problems. One of the main advantages of the DP technique is that global optimality of the found solution is guaranteed even for nonlinear dynamic systems with constraints [21]. The main limitations of the technique are the in-ability to solve causal problems and the exponential computational power requirement growth with the increase of state variables and input variables of the analyzing dynamic problem [47].

In the present study, the dynamic program algorithm is implemented based on the MATLAB DP function developed by Sundstrom and Guzzella in [48]. DP technique is used to design the control strategy of the hybrid electric three-wheeler. Using a drive cycle, DP based optimizing algorithm can find the optimal power management strategy. Thus, the results of the DP optimization can be used for developing less computationally demanding online control strategies. The drive cycle used in this study, is the worldwide harmonized light vehicles test cycle (WLTC) class 1 test cycle. In this section, firstly, the theoretical framework of the implemented DP controller is presented. Secondly, the formulation of the hybrid electric three-wheeler optimization problem is shown. Thirdly, the numerical issues of the implemented DP algorithm are analyzed. Finally, results of the multi-objective optimization are examined.

3.1. Implementation of DP

DP solves the complex optimal problem by breaking it down to a collection of sub-problems. To solve the DP problem, the main continuous problem is discretized, i.e., hybrid vehicle model. The discrete hybrid vehicle model is expressed as,

$$x_{k+1} = f_k(x_k, u_k),\ k = 0, 1, \ldots k, k+1, \ldots, N-1 \tag{31}$$

where x_k and u_k are the state variables (such as: vehicle speed, transmission gear position, battery state of charge) and control variables (such as: gear shift command, motor torque, motor speed, engine speed and engine torque) at the kth time stamp, respectively. x_{k+1} is the state variable at the $k+1$ time step. The goal of the DP algorithm is to minimize a cost function over a given test drive cycle by varying the control variables $u(k)$. The cost function of the optimization problem is expressed as,

$$J = g(x_N) + \varepsilon(x_N) + \sum_{k=0}^{N-1} h_k(x_k, u_k) \tag{32}$$

where $g_N(x_N) + \varepsilon_N(x_N)$ is the cost at the terminal timestamp of the drive cycle. $g_N(x_N)$ is the final cost and $\varepsilon_N(x_N)$ is the additional cost incurred due to the partially constrained final state, $x\left(t_f\right) \in \left(x_{f,min}, x_{f,max}\right)$. $x\left(t_f\right)$ is the state variable value at the terminal time step of the drive cycle. $x_{f,min}$ and $x_{f,max}$ is the final expected state variable range, minimum value of the final state variable and maximum value of the final state variable, respectively. $h_k(x_k, u_k)$ is the instantaneous cost incurred when the control policy represented by the control variable u_k, is applied at x_k, grid point. The grid points are the intersection of discretizing lines of the state space and time space [37]. N is the time length of the driving cycle.

DP algorithm solves the optimization problem by finding the optimal control signal map and optimal control sequence. Based on the principles of optimality, DP algorithm evaluates the optimal cost-to-go function $J^*{}_k(x_k)$ at every node in the discretized state time space by going backwards in time [49].

The final cost at the Nth time step is calculated by

$$J^*{}_N(x_N) = g(x_N) + \varepsilon(x_N) \tag{33}$$

For the kth ($k = N-1, N-2, \ldots, 0$) time step, cost-to-go function is calculated by

$$J^*{}_k(x_k) = min_{u_k}[h_k(x_k, u_k) + J^*{}_{k+1}(f_k(x_k, u_k))] \tag{34}$$

The optimal control is given by the argument that gives the minimum value for Equation (34) at every grid point of the discretized state time space. By solving Equations (33) and (34), the optimal control signal map is found. Which is used to find the optimal control signal by a forward simulation of the model in Equation (31), starting from a given x_0, to generate the optimal state trajectory [49]. Note that principles of the implemented DP algorithm are already reported in [22,48,49] and thus only a summary is presented here.

3.2. Problem Formulation

To implement DP technique to the hybrid electric three-wheeler model, state and control variables need to be found. State of the vehicle model can be expressed through state variables such as vehicle speed, transmission gear number and SoC. The vehicle speed is a known value at every stage of the drive cycle. Therefore, in the present study, the state of charge of the battery and transmission gear number are defined as the state variables.

SoC of the vehicle model over its drive cycle is continuous, for the DP, the continuous variable must be discretized. The continuous state variable SoC is discretized into finite number of points. Where the number of grid points of the discretized state variable, SoC, is expressed as

$$S_g = \frac{SoC_{max} - SoC_{min}}{\Delta SoC} \tag{35}$$

where SoC_{max} is the maximum SoC value and SoC_{min} is the minimum SoC value. ΔSoC is the increment value of the discretized SoC. The state variable, transmission gear number, $N_{gear}(k)$, is defined as a discrete function with four values, to represent the gear number at kth time step:

$$N_{gear}(k) \in [1, 2, 3, 4] \tag{36}$$

Several control variables can be found for the vehicle model, such as gear shift command, motor torque, motor speed, engine speed and engine torque. In the present study, gear shift command, motor torque and engine torque are considered as control variables. The two variables, motor torque and engine torque are expressed by a single control variable defined as the torque split factor. The control variable torque split factor, $u \in [-1\ 1]$, is the ratio between the motor torque and engine torque. Torque split factor is a continuous variable. For the DP algorithm, it is necessary to discretize the control variable into finite number of points. The discretization resolution for the control variable is expressed by, Δu, which is the increment value of the discretized control variable. The number of grid points of the discretized control variable, u, is expressed as

$$u_g = \frac{u_{max} - u_{min}}{\Delta u} \tag{37}$$

where u_{max} is the maximum torque split factor value and u_{min} is the minimum torque split factor value. The control variable, gear shift command, $shift(k)$, is defined as a discrete function to represent the shift command at kth time stamp. Variable is restricted to three values, -1, 0 and 1, to represent downshift, same gear and upshift, respectively. Table 2 summarizes the key variables of the DP optimization algorithm.

Table 2. Variables and grids of the dynamic programming (DP) problem.

	Variables	Grid
Stage (k)	Time	0:1:1612
State (x)	SoC	0.4:0.0027:0.7
	Gear number	1, 2, 3, 4
Control (u)	Torque split factor	−1:0.028:1
	Gear shift command	−1, 0, 1

In general, in optimizing hybrid electric vehicles, only the fuel consumption is considered. In the present work, both fuel consumption and exhaust emissions are considered. The multi-objective optimization approach results in Pareto optimal solutions instead of one single solution. To generate the Pareto front, a scalarization method is used. The multi-objective problem is transformed into a series of single objectives by the classic weighted sum approach [50]. Even though the cost of each objective, i.e., Fuel consumption, HC, CO and NOx emissions, have the same unit of measurement, i.e., grams per second, the range is different. Therefore, prior to the optimization process costs are normalized [51]. The normalized cost f_i^{norm}, where $i \in \{Fuel,\ HC, NOx, CO\}$, is expressed as,

$$f_i^{norm} = \frac{f_i(x_k, u_k) - \min(f_i)}{\max(f_i) - \min(f_i)} \tag{38}$$

where $f_i(x_k, u_k)$ represents the cost incurred at kth time stamp, namely instantaneous fuel consumption rate for fuel and instantaneous exhaust emission rate for emissions, i.e., HC, CO, NOx. $\max(f_i)$ and $\min(f_i)$ is the maximum and minimum cost values of each objective, respectively. Normalized cost f_i^{norm}, is a non-dimensional function with upper and lower bounds.

The gear shifting control is governed by the DP algorithm and if the gear shift control is not restricted, DP chooses the gear position to reduce fuel consumption and emissions at each time step resulting in a gear position map with frequent shifting. Therefore, a fifth objective is added to restrict the gear shifting frequency.

$$f_{gear} = |shift(x_k, u_k)| \tag{39}$$

The cost-to-go function at kth time step is defined as

$$J_k(x_k) = min_{u_k}\left[\sum w_i \cdot f_i^{norm} + w_{gear} \cdot f_{gear} + J_{k+1}(x_{k+1})\right] \tag{40}$$

where $w_i > 0$ for all $i \in \{Fuel,\ HC, NOx, CO\}$, $w_{gear} > 0$ and $\sum w_i + w_{gear} = 1$ [52]. The $J_{k+1}(x_{k+1})$, is the cost of the cost-to-go function at $k + 1$ time step.

The objective of the dynamic program is to minimize the cost-to-go function within defined constraints by varying the control variables. During the optimization, constraints relating to power ratings and characteristics of the powertrain components should be guaranteed to ensure the smooth operation of components such as the engine, motor and battery. Following are the imposed constraints,

$$\begin{gathered} SoC_{min} \le SoC \le SoC_{min} \\ \omega_{e\_min} \le \omega_e \le \omega_{e\_max} \\ T_{e\_min} \le T_e \le T_{e\_max} \\ \omega_{mot\_min} \le \omega_{mot} \le \omega_{mot\_max} \\ T_{mot\_min} \le T_{mot} \le T_{mot\_max} \end{gathered} \tag{41}$$

where $\omega_{e\_min}, \omega_{e\_max}, T_{e\_min}$ and T_{e_max} is the engine minimum speed, maximum speed, minimum torque and maximum torque, respectively. In addition, ω_{mot_min}, ω_{mot_max}, T_{mot_min} and T_{mot_max} is the motor minimum speed, maximum speed, minimum torque and maximum torque, respectively.

3.3. *Numerical Issues of DP*

DP is a numerical algorithm, which is generally solved using quantization and interpolation methods. To solve the optimization problem, continuous variables such as state, control and time are discretized. Discretization inherently introduces numerical errors, which degrade the accuracy of the DP results. Therefore, to avoid these numerical errors without increasing computational effort, careful implementation of the DP algorithm is important [47]. In this section, numerical issues of DP are investigated, i.e., the boundary line issue and resolution of the discretized variables.

3.3.1. Boundary Issue and Resolution of the Study

One of the numerical issues resulting from the interpolation is the boundary line error. Infeasible states and inputs are infinitely expensive. Therefore, the cost incurred at such a point should have an infinite cost. Using an infinite cost value make it impossible to use linear interpolation to calculate the boundary line of the infeasible region. Therefore, a finite cost, which is bigger than the maximum value of the cost-to-go function is used for the infeasible points to reduce the blurring effect of the boundary line.

The resolution of the total vehicle energy demand over a drive cycle increases with the resolution of the time space, thus increasing the resolution of the fuel consumption and amount of emissions released. However, the minimum sampling period of the time space is restricted by the sampling period of the drive cycle. Therefore, in the present study, the resolution of the time space is taken as the resolution of the drive cycle.

3.3.2. Resolution of State Variable

The only continuous state variable and control variable in the present study is the battery state of charge and torque split factor. According to the mathematical principles of DP, the discretization resolution of the state space and the discretization resolution of the control variables are independent [37]. Therefore, discretization resolution of the state variable SoC can be investigated independently of the control variable torque split factor and vice versa.

The resolution of the SoC is determined by the number of grid points S_g, the state variable is divided. Figure 7 represents the relationship between the number of grid points in the state space, normalized DP results and computational effort. The number of grid points indicates the resolution of the state space. Higher the number of grid points, higher the resolution of the state space. Normalized DP result change represents the change of the solution of the optimization problem, namely the output of the cost function. The relative computational effort represents the change of the computational time requirement as a multiple of the computational time required for the minimum grid number analyzed. Note that DP problem is solved in the present study with an Intel® Core™ i7-7700HQ CPU and a 16 GB RAM.

Figure 7 shows that normalized DP results decrease with the increase of the number of grid points in the state space. However, when the resolution of the state space is sufficiently high, DP results tend to be stable. In contradiction, with increased state-space resolution, relative computational effort increases by many folds; computational effort increased from 30 grid points: 37 s to 300 grid points: 135 s. Even though the computational effort increases with the accuracy of the DP result, the appropriate number of grid points for the state variable that ensures a satisfactory DP result with a minimum computational load can be obtained. Onwards from 110 grid points, DP results change is less than 0.1, while computational effort increases by fourfold from 110 grid points to 300 grid points. In the present study, the number of grid points for the state space is selected as 110 points, resulting in a ΔSoC of 0.0027, the increment value of the discretized SoC.

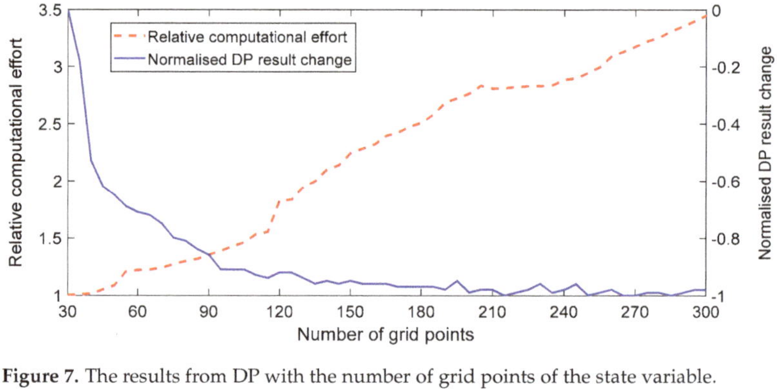

Figure 7. The results from DP with the number of grid points of the state variable.

3.3.3. Resolution of Control Variable

The number of grid points of the control variable u_g, indicate the resolution of the control variable. Higher the number of grid points of the control variable, higher the resolution. Figure 8 represents the relationship between the number of grid points of the control variable, DP results and computational effort. Similar to the analysis on the state space resolution, the number of grid points indicate the resolution of the control variable. Normalized relative DP result change represents the change of the solution of the optimization problem. The relative computational effort represents the change of the computational time requirement as a multiple of the computational time required for the minimum grid number analyzed.

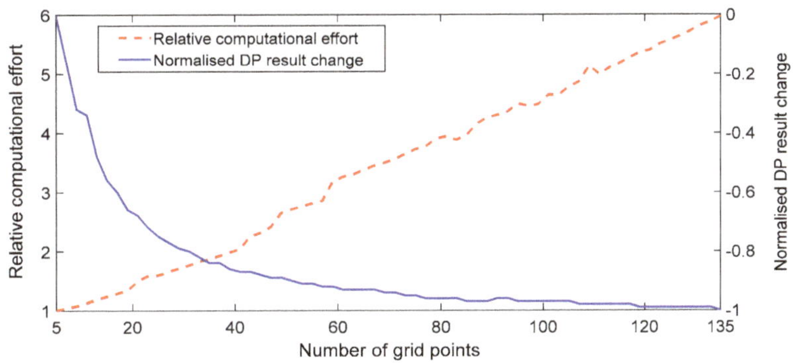

Figure 8. DP results with the number of grid points of the control variable.

Results show that the normalized DP result is improved with the resolution and tends to be stable after the resolution of the control variable is sufficiently high. In contradiction, the computational effort has increased six-fold from 5 grid points: 28 s to 135 grid points: 167 s. Even though the computational effort increases with the accuracy of the DP result, the appropriate number of grid points for the control variable that ensures a satisfactory DP result with a minimum computational load can be obtained. Between 71 and 135 grid points, DP result is improved by only 0.08, but computational effort doubled. In the present study, the number of grid points for the state space is selected as 71 points, resulting in a Δu of 0.028, the increment value of the discretized torque split factor. Note that the fuel consumption map of the engine should be considered when selecting the discretization resolution of the control variable [37]. The amount of fuel consumed and emissions released are related to the characteristics of the engine. The relationship between DP results and the number of grid points of the control variable is affected by the fuel consumption map and emission maps of the engine.

3.4. Results of DP Based Control Strategy

The implemented DP algorithm produces Pareto optimal solutions for the cost function. Prior to running the DP algorithm weighting factors of the objective function should be determined. The weight selection procedure assumes that the solution set of the present work has a convex Pareto front [53]. The points of the Pareto front can be found by varying the weighting factors of the objective function. In the present study, 896 Pareto points were found by varying the weighting factors for the five objectives, i.e., gear shift, fuel, HC, CO and NOx. The sizes of the weighting factors are decided by comparing the cost incurred by each objective at each Pareto point except for the gear shift.

In the DP algorithm, gear shift command is a control variable and the optimization procedure identify the optimal gear position to reduce fuel consumption and exhaust emissions. However, each gear change increases the cost incurred by the gear shift objective. Therefore, lower the weightage of the gear shift objective, higher the frequency of gear shifting. Since there is no cost for changing gear positions, DP algorithm changes gears to reduce fuel consumption and emissions with no restrictions, resulting in a gear trajectory map with frequent shifting of gears. Since such a behavior is undesirable, the gear trajectory is observed with different weighting factors. Concurrently, the costs incurred by the remaining four objectives are observed to select the weighting factors for the study. Figure 9 shows the gear shift trajectory with a zero-weighting factor and the gear shift map for the selected Pareto point.

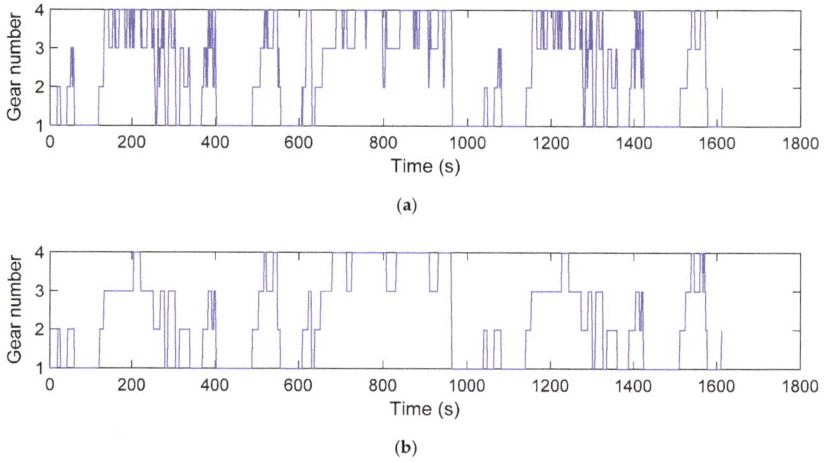

Figure 9. Gear shift trajectory: (**a**) Gear shift trajectory with zero-weighting factor; (**b**) Gear shift trajectory of the selected Pareto point.

Figure 10 represents the relationship of the Pareto points with the four objectives, i.e., fuel, HC, CO and NOx via a parallel coordinate plot. In the parallel coordinate plot, the "crossing lines" indicate a conflict between the two adjacent objectives. The intensity or degree of conflict factor present can be determined by the number of crossed lines. Instead, lines that do not cross demonstrate objectives which are in relative harmony [54]. Figure 10 shows that fuel, HC and CO aims are in harmony compared to NOx. NOx emissions have a high degree of conflict with CO. For further investigation, the best and worst Pareto points for fuel, CO, HC and NOx are plotted in Figure 11a,b, respectively.

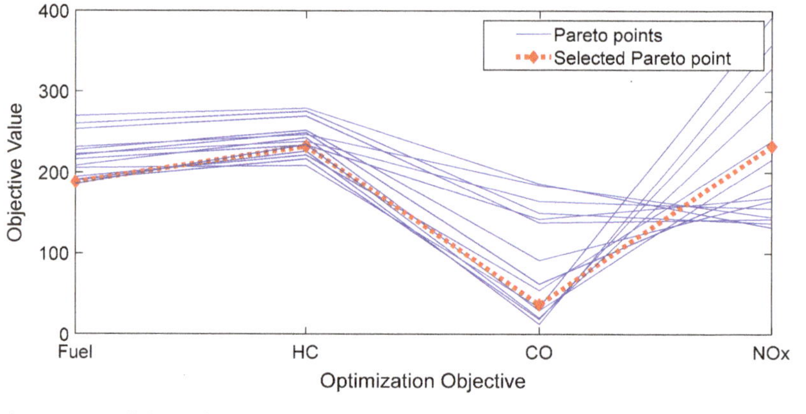

Figure 10. Parallel coordinate plots for Pareto points with varied weighting factors for optimization objectives (for clarity only 15 Pareto points are represented).

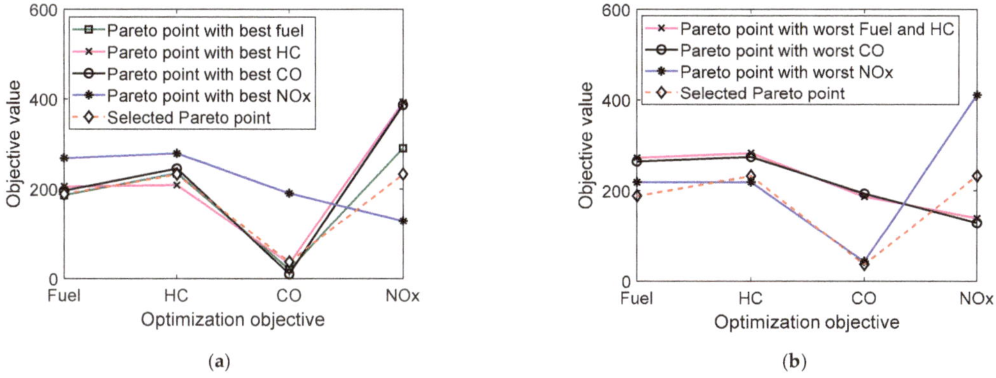

Figure 11. Parallel coordinate plots: (**a**) Pareto points with the best values for the objectives; (**b**) Pareto points with the worst values for the objectives.

As shown in both Figure 11a,b, the Fuel consumption, HC and CO objectives are consistent, while NOx behaves in conflict. Fuel and HC have a low degree of conflict; the best Pareto points overlap and worst values are observed from the same Pareto point. In addition, both the best and worst Pareto points of CO and HC show a higher degree of harmony. In contrast, the best Pareto point of NOx has a higher cost for fuel, HC and CO. Similar trend can be observed with the worst Pareto point of NOx. Selected Pareto point overlaps with the best fuel, HC and CO Pareto points. With the selected Pareto point, compared to the best Pareto point of each objective, fuel, HC, CO and NOx increased by 3%, 32%, 15% and 37%, respectively.

4. Rule-Based Power Management Strategy

The DP algorithm is based on knowing future drive cycle characteristics, but in a real-world implementation, the drive cycle is unknown. In addition, the DP procedure requires higher computational power to solve the energy management problem. However, the DP results can be used to design a power management strategy based on empirical rules. A rule-based control strategy is computationally effective for an embedded control system of a hybrid electric three-wheeler. However, the implemented rule-based strategy can generate driving patterns that may not be optimal, especially when the vehicle is operated outside the drive cycle characteristics to which the control strategy is optimized. In this section results obtained from the DP algorithm, are used to extract rules for a sub-optimal rule-based strategy. This approach, already known in the literature, is used to design a computationally effective rule-based supervisory power management strategy to operate a hybrid electric three-wheeler efficiently during real vehicle driving conditions. [15,19,24,25] used a similar methodology.

4.1. Rule Extraction from DP Based Control Strategy

4.1.1. Power Configuration Selection Strategy

The power configuration of the powertrain determines the energy flow direction and operating status of the main powertrain components according to the power demand from the driver. When power demand is positive, the powertrain of a hybrid electric vehicle can operate under two main configurations. First, motor only or EV configuration, the engine is turned off and the vehicle is propelled by the motor alone. Secondly, parallel hybrid configuration. Three power flow methods are present in the parallel hybrid configuration. First, the engine assist, the engine and motor are used to meet the power demand. Secondly, engine only, the vehicle is propelled by the engine alone. Finally, recharge, the engine is operated at a higher point than the power demand. Excess power is stored in the battery by operating the motor as a generator. Figure 12 shows the different power flow methods selected by the DP algorithm as a relationship of total power demand and transmission output speed of the powertrain.

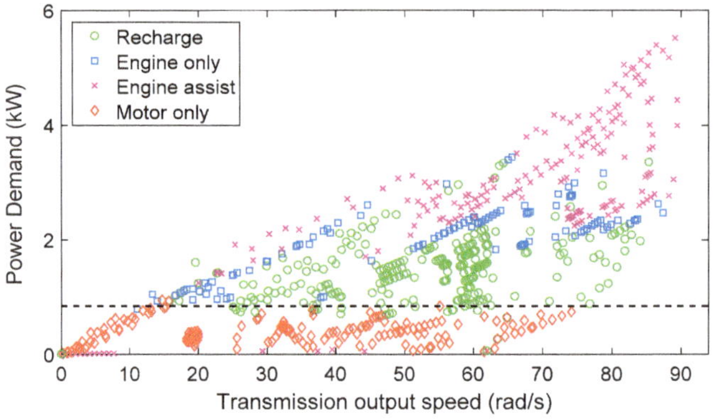

Figure 12. Operating points from DP optimization results when power demand is positive.

DP results show that a clear distinction is absent between recharging, engine only and parallel hybrid methods. However, between motor only and parallel hybrid configurations, a clear distinction is present. Below the power demand threshold shown by the black dashed line, DP algorithm gives priority to EV mode or motor only configuration. Results show a power demand threshold to select EV mode for the sub-optimal rule-based control strategy can be found using the power demand map of DP results.

4.1.2. Gear Shift Logic

Determining the gear shift schedule of the transmission is paramount for efficient control of the internal combustion engine. For a parallel hybrid powertrain, it is harder to obtain an efficient gear shift map using traditional methodologies. Therefore, the DP algorithm is used to determine the gear shift strategy. As mentioned above, in the DP procedure gear shift command is given as a control input to the powertrain model. Figure 13 shows the gear shift strategy from the DP optimization results as a relationship between transmission output speed, engine power demand and gear number.

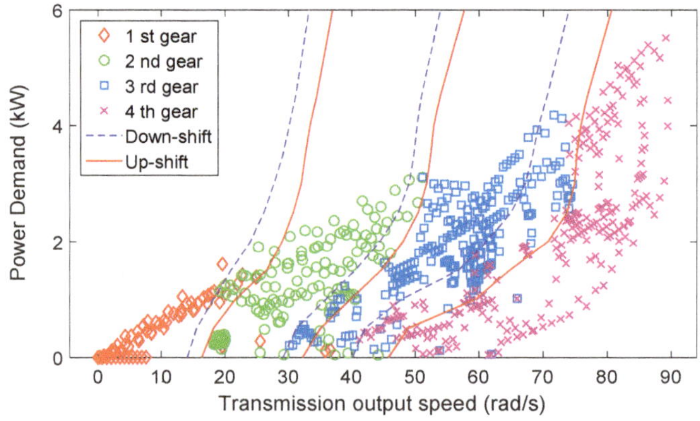

Figure 13. Transmission gear selection from DP algorithm when power demand is positive.

From the gear shift map of the DP results, four different regions can be found for each gear position. The boundaries between the regions can be used to determine gear shifting thresholds. Thresholds between each gear position are used to develop a gear shift map for the sub-optimal rule-based strategy. To avoid continuous shifting between gears, a hysteresis function is added to the shifting thresholds. Red solid lines and blue dashed lines show the up-shift and down-shift gear maps, respectively, for the proposed rule-based strategy.

4.1.3. Power Split Logic

During hybrid configuration, motor and engine both operate in harmony. DP results are studied to understand the power split strategy between the power sources. To quantify the power flow from the powertrain components, i.e., motor and engine, a Power Split Ratio (PSR) is defined as,

$$PSR = P_{eng}/P_{dem} \tag{42}$$

where P_{eng} is the power demanded from the engine and P_{dem} is the total power demanded at the transmission input of the powertrain. During vehicle operation, four distinct powertrain operating modes can be found through power split ratio:

- PSR = 0, only the motor provides the demanded power (EV mode);
- PSR = 1, only the engine provides the demanded power (engine only mode);
- 0 < PSR < 1, both engine and motor provide the demanded power (engine assist mode);
- PSR > 1, the engine provides surplus power and motor acts as a generator (recharge mode).

The engine operating points from the DP results is represented in Figure 14 as a relationship between the transmission input speed and power split ratio.

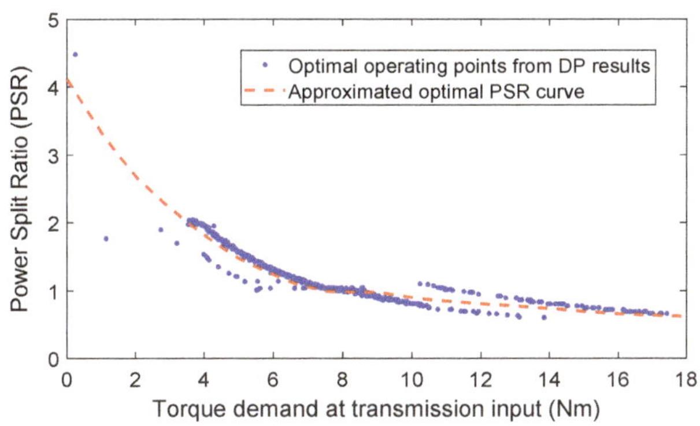

Figure 14. Engine operating points from DP results.

PSR map shows that engine operating points from DP results can be approximated. The red dashed line in the above figure shows the approximated PSR curve. The approximated PSR curve is used in the proposed sub-optimal rule-based strategy to determine the power split between engine and motor. Mathematical expression for the approximated PSR curve is defined as,

$$\begin{aligned} f(x) &= -0.0027x^3 + 0.0769x^2 - 0.845x + 4.1139, \quad T < 7.5 \\ f(x) &= 1, \quad 7.5 \leq T \leq 9 \\ f(x) &= -0.0002x^3 + 0.0055x^2 - 0.0909x + 1.0484, \quad T > 9 \end{aligned} \tag{43}$$

where T is the torque demand at transmission input. Note that similar to [23], PSR approximated line can be replaced with an optimal torque map. Where the relationship between the engine power demand is plotted with total power demand and transmission input speed to form optimal engine power surfaces. This method uses the transmission input speed axis for a better approximation of DP results.

4.2. Proposed Rule-Based Strategy

The extracted results from DP data indicate that power demand map, which determines P_{EV} (power demand boundary between EV and parallel configurations) and the gear shift map contribute significantly to achieve an efficient power management logic in the hybrid electric vehicle. However, the PSR map approximation method is not a universal procedure for the power split strategy. The PSR method does not consider the battery SoC level when deciding the power contribution and operational mode of the power sources, i.e., motor and the engine. This could lead to a scenario where battery is depleted or overcharged than the recommended levels. Therefore, additional empirical rules should be determined to prevent the battery from depleting or overcharging. Note that alternative approaches to link power split strategy with battery SoC are already known in the literature. For example, Lin et al. [24] use a neural network approach and Biasini et al. in [15] use a penalty function to change the motor to power demand ratio with battery SoC level.

The logic of the DP based; sub-optimal power management strategy proposed for the hybrid electric three-wheeler is summarized in Figure 15.

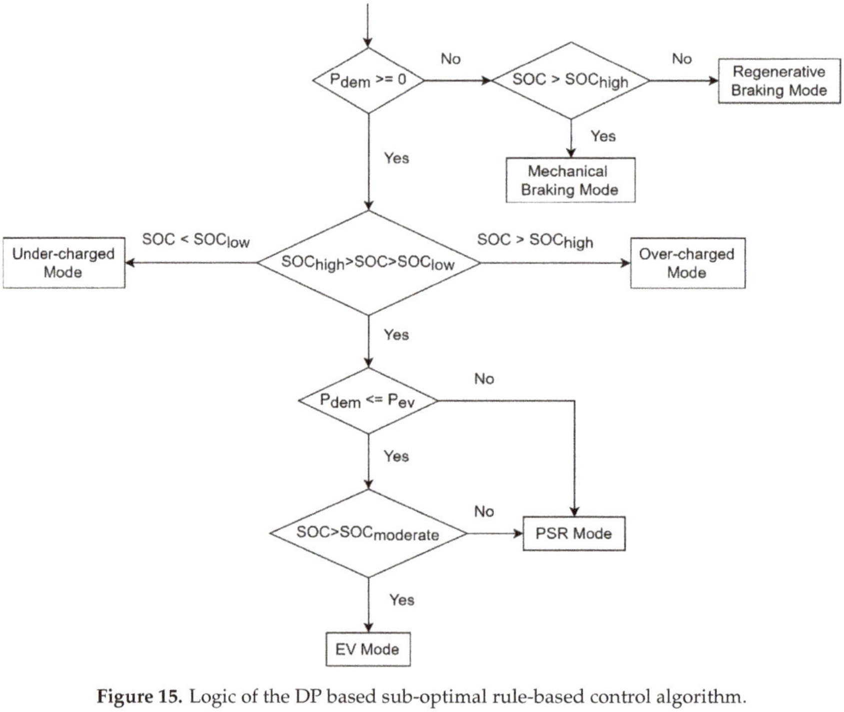

Figure 15. Logic of the DP based sub-optimal rule-based control algorithm.

Six main control modes are used for the proposed strategy. Depending on the battery SoC level and power demand, different modes are used to control the power flow. PSR mode propels the vehicle using both motor and engine in harmony. The power ratio is determined using the PSR map. For the charge sustaining policy, which assures that the SoC level is kept within predetermined levels, two SoC bounds are defined as SoC_{low} and SoC_{high}. In addition, a third SoC level is predetermined as $SoC_{moderate}$ to give priority to PSR mode. SoC thresholds are defined with a hysteresis function to avoid frequent shifting between different modes. Under charged and over-charged modes are used to ensure that SoC levels are kept within defined boundaries. During under-charged mode, priority is given to the IC engine and motor is operated as a generator. In contrast, during over-charged mode priority is given to the motor to propel the vehicle. EV mode is only used if SoC level is higher than $SoC_{moderate}$ and power demand is below P_{EV} values. In EV mode, vehicle is propelled by the motor alone. PSR mode is selected if SoC value is below $SoC_{moderate}$ but above SoC_{low}. If the power demand is negative, regenerative braking is applied. However, if the maximum brake torque supplied by the motor is below the torque demanded, additional torque is supplied by the mechanical brake system. In addition, if the SoC level is higher than SoC_{high}, only the mechanical brakes are used to decelerate the vehicle.

Core design parameters from the DP results used for the rule-based strategy are power demand map, which determines P_{EV}, gear shift map, which determines the gear shift strategy and PSR map, which determine power split strategy between the motor and engine. The selection of EV/PSR mode, gear shift and PSR decisions are determined sequentially because PSR command decision requires the knowledge of torque demand and speed at the transmission input of the powertrain. These parameters can only be calculated after determining the gear position. Figure 16 summarizes the logic proposed for the PSR mode.

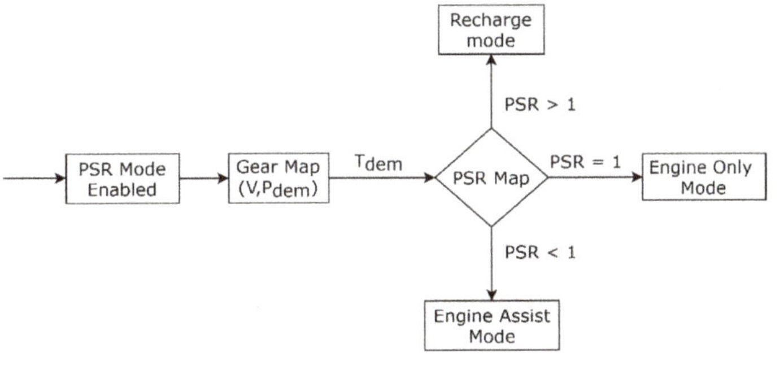

Figure 16. Flowchart for the DP based power split ratio (PSR) mode control algorithm.

Gear shift strategy during vehicle operation is crucial to improve overall powertrain efficiency. Under the proposed control strategy if the vehicle is operated by both engine and motor or by engine alone the gear shift strategy adopted from the DP results is used. However, this strategy cannot be used during EV mode or regenerative braking mode. The procedure to derive an effective gearbox strategy for these modes is already known in the literature [55]. In the present work, the heuristic method proposed by Zhu at el in [56], is used to develop the gear shift strategy during regenerative braking and EV modes of the vehicle. To achieve a desirable gear shift strategy, maximum torque curve of the motor and motor efficiency characteristics are considered.

4.3. Comparison of Proposed Rule-Based and DP Based Control Strategy

The sub-optimal rule-based power management strategy developed above is integrated into the implemented forward-facing quasi-static model. The results obtained from the DP results are compared with the simulation results of the forward-facing model. For both control strategies starting SoC is taken as 0.55. The Worldwide Harmonized Light Vehicles Test Cycle (WLTC) class 1 cycle and the New European Drive Cycle (NEDC) is selected for the simulation experiment. In addition, the vehicle is loaded to simulate the weight of the driver and two passengers. Note that DP and rule-based (RB) strategies are simulated in two different models (i.e., backward-facing and forward-facing models). A comparison of the models is presented in Section 2.2. Since the cumulative energy bought to the wheels in the two models are different, certain errors can be introduced to the validation results.

Depending on the power to mass ratio and maximum speed of the testing vehicle, the WLTC test cycle is selected. Three-wheelers fall into class 1 category established for vehicles with the power to mass ratio below or equal to 22 W/kg. Test cycle includes three-speed phases, i.e., low, medium and low speed. In general, three-wheelers operate at low speeds. For instance, the maximum legal speed limit for three-wheelers for all roads in Sri Lanka is only 40 km/h [57] and the maximum speed of the 2019 Bajaj RE 4 s model (IC engine three-wheeler) is only 65 km/h [58]. Since maximum speed of the test vehicle is below the WLTC class 1 test exception limit of 70 km/h, the medium speed section of the WLTC is replaced with a low-speed section. The used WLTC class 1 drive cycle is represented in Figure 17.

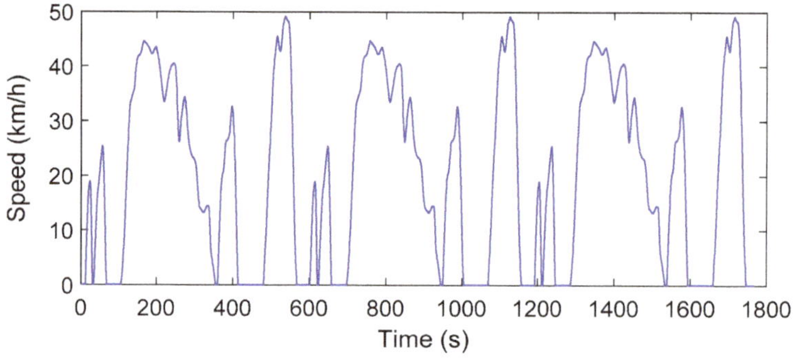

Figure 17. Speed profile of the WLTC Class 1 drive cycle.

The WLTC test cycle has a maximum speed of 49.1 km/h and an average speed without stops of 27.6 km/h. Overall results for the drive cycle are affected by the operating time of each control mode, i.e., EV, PSR, under-charged, over-charge, regenerative braking and mechanical braking. During the simulation, the rule-based strategy uses the EV mode for 43% and PSR mode for 42% of the time. Which contributes to 85% of the complete test. The regenerative mode was used for 15% of the time. Under-charged, over-charged, or mechanical braking modes are not used. Table 3 summarizes fuel and emission level change with the RB control strategy (sub-optimal rule-based strategy) compared to the DP results.

Table 3. Performance comparison of DP and rule-based strategies for the WLTC class 1 drive cycle.

Controller	Fuel (l/100 km)	HC (g/km)	CO (g/km)	NOx (g/km)
DP	1.32	0.142	1.096	0.405
Rule Based	1.38	0.145	1.151	0.408
Deviation	+4.92%	+2.09%	+5.02%	+0.68%

Figure 18 represents the SoC trajectory of the control strategies (i.e., RB and DP). Final SoC value difference is 0.004.

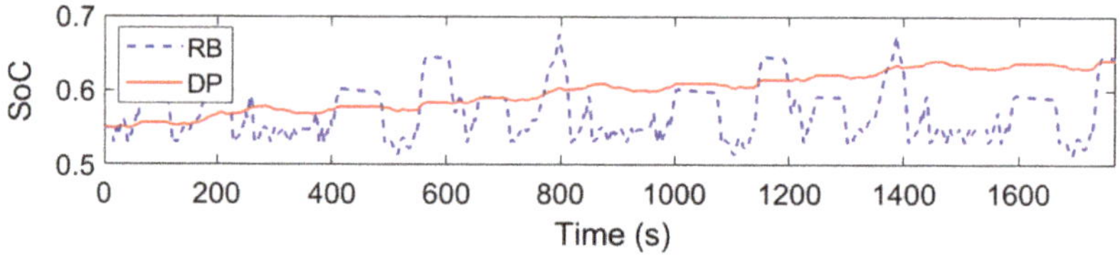

Figure 18. SoC trajectory of rule-based (RB) and DP strategies over WLTC.

To gauge the performance of the proposed rule-based strategy on a never-seen-before drive cycle, both DP and proposed RB strategies are simulated with the Urban Drive Cycle (UDC) segments of the NEDC. Although NEDC is replaced by the WLTC, still in countries such as Sri Lanka NEDC drive cycle is used in emission standards for petrol light-duty vehicles [59]. In the present study, a modified NEDC is used due to the low power characteristics of three-wheelers. Extra urban driving cycle (EUDC) segment of the NEDC drive cycle consists of a maximum speed of 120 km/h and average speed (without stops) of 69 km/h; therefore, this drive cycle segment has been excluded from the simulation. UDC-NEDC drive cycle has a maximum speed of 50 km/h and average speed (without stops) of 26 km/h. The used UDC-NEDC drive cycle is represented in Figure 19.

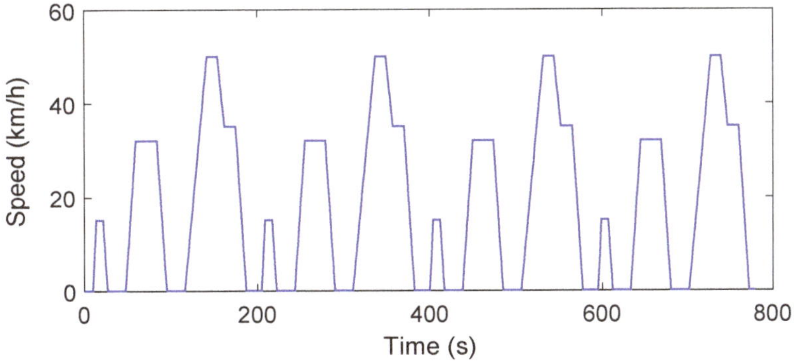

Figure 19. Speed profile of the UDC-NEDC drive cycle.

During the simulation, the rule-based strategy used the PSR mode for 34% and EV mode for 45% of the time. Regenerative mode and under-charged mode are used for 19% and 3% of the time, respectively. Remaining modes are not used. Table 4 summarizes fuel and emission level change with the proposed rule-based strategy compared to the DP results.

Table 4. Performance comparison of DP and rule-based strategies for the UDC-NEDC drive cycle.

Controller	Fuel (l/100 km)	HC (g/km)	CO (g/km)	NOx (g/km)
DP	1.53	0.162	2.264	0.438
Rule-Based	1.59	0.157	2.485	0.474
Deviation	+4.01%	−2.87%	+9.76%	+8.16%

Figure 20 represents the SoC trajectory of the control strategies (i.e., RB and DP). Final SoC value difference is 0.003.

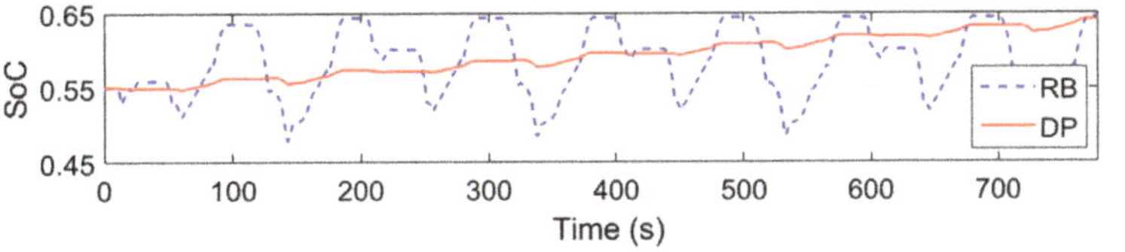

Figure 20. SoC trajectory of RB and DP strategies over UDC-NEDC.

To gauge the performance of the control strategy developed in real-life cycles, results are simulated in "Malabe" cycle from [14]. The "Malabe" real-life cycle is shown in Figure 21.

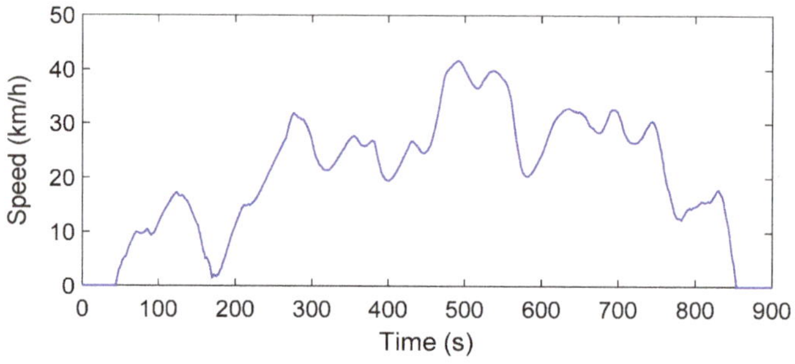

Figure 21. Speed profile of the real-life "Malabe" cycle.

During the simulation, the rule-based strategy used the PSR mode for 50% and EV mode for 39% of the time. Regenerative mode and over-charged mode are used for 6% and 5% of the time, respectively. Remaining modes are not used. Table 5 summarizes fuel and emission level change with the proposed rule-based control strategy compared to the DP results.

Table 5. Performance comparison of DP and rule-based strategies for the "Malabe" cycle.

Controller	Fuel (l/100 km)	HC (g/km)	CO (g/km)	NOx (g/km)
DP	1.15	0.124	0.801	0.309
Rule-Based	1.26	0.135	0.989	0.327
Deviation	+9.61%	+8.59%	+23.50%	+5.86%

Figure 22 represents the SoC trajectory of the control strategies (i.e., RB and DP). Final SoC value difference is 0.003.

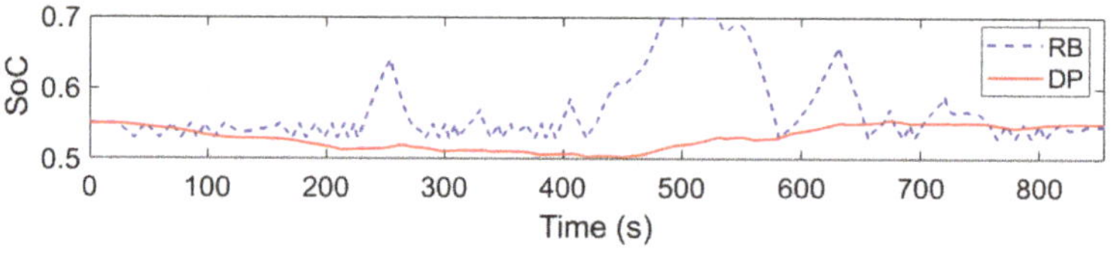

Figure 22. SoC trajectory of RB and DP strategies over "Malabe".

The ADVISOR parallel hybrid three-wheeler model and conventional powertrain model (internal combustion engine (ICE) based) developed in [14] are simulated over "Malabe" cycle with the same parameters of the present study and fuel consumption results of the DP and proposed rule-based strategy are compared. Findings show that compared to the conventional (ICE-based) powertrain, fuel consumption of the parallel hybrid power train reduced 37.4% with ADIVISOR unoptimized strategy, 52.9% with the proposed optimized rule-base strategy and 57.0% with the DP.

The present work reveals a methodology to establish an easy-to-implement, near-optimal and real-time control strategy for parallel hybrid electric three-wheelers. However, additional studies are necessary to investigate the performance of the proposed strategy in real-life conditions. The simulation models are developed with simplified models with ADVISOR maps and data. Therefore, extensive studies on evaluating the control strategy on detailed simulation models including dynamic characteristics of the powertrain models and hardware-in-loop experiments should follow this study. Furthermore, the role of uncertainty in the performance of the proposed strategy could be realized in a future study by investigating the sensitivity of the control parameters used.

5. Conclusions

In this article, a rule-based power management strategy that is based on an optimal control strategy was designed for a pre-transmission parallel hybrid electric three-wheeler through a sequential design methodology. The optimal control strategy (DP based) was simulated in a backward-facing model to optimize fuel consumption, emissions (i.e., HC, CO and NOx) and gear shift frequency. Results from the DP based control strategy were used to extract rules to develop a sub-optimal rule-based algorithm. The proposed rule-based strategy was assessed with a forward-facing model of the hybrid electric three-wheeler. The main conclusions are as follows:

1. Numerical issues were observed during the DP optimization procedure. Thus, the effect of discretization resolution of the state and control variables were investigated. Results show that computational effort and accuracy of the optimal result from the DP optimization procedure increase with higher discretization resolutions of the state and control variables.
2. The cost function of the optimization problem considered multiple objectives: fuel consumption, emissions and gear shift strategy. Results showed that gear shift strategy and remaining objectives behaved uniquely; contradictory to fuel and emission objectives, frequency of gear shift increased with lower weighting factors and vice versa. Moreover, within the engine model used in the present study, a higher degree of conflict was observed between NOx and remaining objectives (i.e., fuel, HC and CO).
3. Three main useful strategies were extracted from DP results, i.e., full EV mode on/off threshold, gear shift and power-split strategy to develop the rule-based algorithm. The rule-based strategy-maintained fuel consumption and emissions within 10% of the DP results for WLTC and NEDC drive cycles. The proposed control strategy is computationally less demanding, easy-to-implement on a vehicle and near-optimal; thus, a viable option to control a hybrid electric three-wheeler operating in densely populated urban roads.

As future research, the proposed control strategy could be assessed on a prototype hybrid-electric three-wheeler for performance and robustness. In addition, simulation models could be validated with experimental testing and improved further to include dynamic characteristics of the vehicle. With an advance simulation model, multi-objective optimization could be extended to include vehicle dynamics and drivability characteristics.

Author Contributions: Conceptualization, W.M., M.P. and P.K.; methodology, W.M.; software, W.M.; validation, W.M.; formal analysis, W.M.; investigation, W.M.; resources, W.M. and M.P.; data curation, W.M.; writing—original draft preparation, W.M.; writing—review and editing, W.M., M.P., R.A. and P.K.; visualization, W.M.; supervision, M.P. and R.A.; funding acquisition, M.P. All authors have read and agreed to the published version of the manuscript.

Funding: This research was funded by the Bentley CAE research unit of Sri Lanka Institute of Information Technology, Sri Lanka.

Institutional Review Board Statement: Not applicable.

Informed Consent Statement: Not applicable.

Data Availability Statement: Not applicable.

Acknowledgments: The authors would like to thank Kasun Abeyasinghe (Sri Lanka Institute of Information Technology, Sri Lanka), Dilini Gunathilake (University of Wollongong, Australia) and Sampath Liyanarachchi (Memorial University of Newfoundland, Canada) for their valuable contribution to the work. Although any errors are of our own and should not tarnish the reputation of those esteemed persons.

Conflicts of Interest: The authors declare no conflict of interest.

Abbreviations

BSFC	Brake Specific Fuel Consumption
CNG	Compressed Natural Gas
CPU	Central Processing Unit
DC	Direct Current
DP	Dynamic Programming
EV	Electric Vehicle
GPS	Global Positioning System
IC	Internal Combustion
LPG	Liquefied Petroleum Gas
NEDC	New European Drive Cycle
PSR	Power Split Ratio
RAM	Random Access Memory
RB	Rule-Based
SI	Spark Ignition
SoC	State of Charge
UDC	Urban Drive Cycle
WLTC	Worldwide Harmonized Light Vehicles Test Cycle

References

1. Ghosh, A. Possibilities and challenges for the inclusion of the electric vehicle (EV) to reduce the carbon footprint in the transport sector: A review. *Energies* **2020**, *13*, 2602. [CrossRef]
2. Saxena, S.N. Revolution in growth of three-wheeler electric vehicles in India. *J. Glob. Tour. Res.* **2019**, *4*, 117–126. [CrossRef]
3. Overview Information of the Bajaj RE 4s Three-Wheeler. Available online: https://www.globalbajaj.com/global/english/brands/intracity/re/re-4s/overview/ (accessed on 14 October 2020).
4. Kokate, V.L.; Uttekar, S.S.; Karandikar, P.B.; Holmukhe, R.M. Retrofitting of Auto Rickshaw to E-Rickshaw—A Feasibility Study. In Proceedings of the First International Conference on Power, Control and Computing Technologies (ICPC2T), Raipur, India, 3–5 January 2020; pp. 229–234.
5. Hofman, T.; van der Tas, S.G.; Ooms, W.; van Meijl, E.W.P.; Laugeman, B.M. Development of a Micro-Hybrid System for a Three-Wheeled Motor Taxi. *World Electr. Veh. J.* **2009**, *3*, 572–580. [CrossRef]
6. Majumdar, D.; Jash, T. Merits and Challenges of E-Rickshaw as An Alternative form of Public Road Transport System: A Case Study in the State of West Bengal in India. *Energy Procedia* **2015**, *79*, 307–314. [CrossRef]
7. Ahmadi, L.; Yip, A.; Fowler, M.; Young, S.B.; Fraser, R.A. Environmental feasibility of re-use of electric vehicle batteries. *Sustain. Energy Technol. Assess.* **2014**, *6*, 64–74. [CrossRef]
8. Rajper, S.Z.; Albrecht, J. Prospects of Electric Vehicles in the Developing Countries: A Literature Review. *Sustainability* **2020**, *12*, 1906. [CrossRef]
9. Khan, F.; Ali, Y.; Khan, A.U. Sustainable hybrid electric vehicle selection in the context of a developing country. *Air Qual. Atmos. Health* **2020**, *13*, 489–499. [CrossRef]
10. Vezzini, A.; Sharan, H.; Umanand, L. Low-pollution three-wheeler autorickshaw with power-assist series hybrid and novel variable DC-link voltage system. *J. Indian Inst. Sci.* **2005**, *85*, 105–118.
11. Amjad, S.; Rudramoorthy, R.; Neelakrishnan, S.; Gurusubramanian, S.; Raja, J.D.; Mathew, R.; Sundaravel, M. Plug-in hybrid conversion of three wheeler using a novel drive strategy. *Int. J. Altern. Propuls.* **2012**, *2*, 148. [CrossRef]
12. Jijith, R.V.; Indulal, S. Hybrid Electric Three-Wheeler with ANN Controller. In Proceedings of the International Conference on Circuits and Systems in Digital Enterprise Technology (ICCSDET), Kottayam, India, 21–22 December 2018; pp. 1–5.
13. Padmanaban, V.; Ramasubramanian, A.; Subramaniam, T. Investigation on use of plug—In hybrid electric vehicle (PHEV) technology using renewable energy for an autorickshaw. *J. KONES. Powertrain Transp.* **2012**, *19*, 383–394. [CrossRef]
14. Maddumage, W.U.; Abeyasighe, K.Y.; Perera, M.S.M.; Attalage, R.A.; Kelly, P. Comparing Fuel Consumption and Emission Levels of Hybrid Powertrain Configurations and a Conventional Powertrain in Varied Drive Cycles and Degree of Hybridization. *Sci. Technol.* **2020**, *19*, 20–33. [CrossRef]
15. Biasini, R.; Onori, S.; Rizzoni, G. A near-optimal rule-based energy management strategy for medium duty hybrid truck. *Int. J. Powertrains* **2013**, *2*, 232. [CrossRef]
16. Enang, W.; Bannister, C. Modelling and control of hybrid electric vehicles (A comprehensive review). *Renew. Sustain. Energy Rev.* **2017**, *74*, 1210–1239. [CrossRef]
17. Sudhakaran, S.; Indulal, S.; Lal Priya, P.S. Fuzzy Logic Controller Design for Retrofitted Three-wheeled Hybrid Electric Vehicle. In Proceedings of the 2020 First IEEE International Conference on Measurement, Instrumentation, Control and Automation (ICMICA), Kurukshetra, India, 24–26 June 2020; pp. 1–5.
18. Lin, C.C.; Peng, H.; Grizzle, J.W.; Kang, J.M. Power management strategy for a parallel hybrid electric truck. *IEEE Trans. Control Syst. Technol.* **2003**, *11*, 839–849. [CrossRef]

19. Kum, D.; Peng, H.; Bucknor, N.K. Supervisory Control of Parallel Hybrid Electric Vehicles for Fuel and Emission Reduction. *J. Dyn. Syst. Meas. Control* **2011**, *133*, 1–10. [CrossRef]
20. Lin, C.-C.; Filipi, Z.; Wang, Y.; Louca, L.; Peng, H.; Assanis, D.N.; Stein, J. Integrated, Feed-Forward Hybrid Electric Vehicle Simulation in SIMULINK and its Use for Power Management Studies. *SAE Tech. Pap.* **2001**, *1*. [CrossRef]
21. Zou, Y.; Shi-jie, H.; Dong-ge, L.; Wei, G.; Hu, X. Optimal Energy Control Strategy Design for a Hybrid Electric Vehicle. *Discret. Dyn. Nat. Soc.* **2013**, *2013*, 1–8. [CrossRef]
22. Sundström, O.; Guzzella, L.; Soltic, P. Optimal Hybridization in Two Parallel Hybrid Electric Vehicles using Dynamic Programming. *IFAC Proc. Vol.* **2008**, *41*, 4642–4647. [CrossRef]
23. Wang, R.; Lukic, S.M. Dynamic programming technique in hybrid electric vehicle optimization. In Proceedings of the IEEE International Electric Vehicle Conference, Greenville, SC, USA, 4–8 March 2012; pp. 1–8.
24. Lin, C.-C.; Filipi, Z.; Louca, L.; Peng, H.; Assanis, D.; Stein, J. Modelling and control of a medium-duty hybrid electric truck. *Int. J. Heavy Veh. Syst.* **2004**, *11*, 349. [CrossRef]
25. Asghar, M.; Bhatti, A.I.; Izhar, T. Benchmark fuel economy for a parallel hybrid electric three-wheeler vehicle (rickshaw). *Adv. Mech. Eng.* **2018**, *10*. [CrossRef]
26. Huang, X.; Wang, Y.; Xing, Z.; Du, K. Emission factors of air pollutants from CNG-gasoline bi-fuel vehicles: Part II. CO, HC and NOx. *Sci. Total Environ.* **2016**, *565*, 698–705. [CrossRef] [PubMed]
27. Wallington, T.J.; Sullivan, J.L.; Hurley, M.D. Emissions of CO_2, CO, NOx, HC, PM, HFC-134a, N2O and CH4 from the global light duty vehicle fleet. *Meteorol. Z.* **2008**, *17*, 109–116. [CrossRef]
28. Ram, N.S.M. Trends in Automotive Electronics in Two and Three Wheeled Vehicles. *SAE Tech. Pap.* **2003**. [CrossRef]
29. Huang, Y.; Wang, H.; Khajepour, A.; Li, B.; Ji, J.; Zhao, K.; Hu, C. A review of power management strategies and component sizing methods for hybrid vehicles. *Renew. Sustain. Energy Rev.* **2018**, *96*, 132–144. [CrossRef]
30. Arsie, I.; Graziosi, M.; Pianese, C. Optimization of supervisory control strategy for parallel hybrid vehicle with provisional load estimate. In Proceedings of the 7th International Symposium on Advanced Vehicle Control (AVEC), Arnhem, The Netherlands, 23–27 August 2004; pp. 483–488.
31. Borhan, H.; Vahidi, A.; Phillips, A.M.; Kuang, M.L.; Kolmanovsky, I.V.; Di Cairano, S. MPC-Based Energy Management of a Power-Split Hybrid Electric Vehicle. *IEEE Trans. Control Syst. Technol.* **2012**, *20*, 593–603. [CrossRef]
32. Liu, Y.; Li, J.; Lei, Z.; Li, W.; Qin, D.; Chen, Z. An Adaptive Equivalent Consumption Minimization Strategy for Plug-In Hybrid Electric Vehicles Based on Energy Balance Principle. *IEEE Access* **2019**, *7*, 67589–67601. [CrossRef]
33. Cheng, R.; Rodemann, T.; Fischer, M.; Olhofer, M.; Jin, Y. Evolutionary Many-Objective Optimization of Hybrid Electric Vehicle Control: From General Optimization to Preference Articulation. *IEEE Trans. Emerg. Top. Comput. Intell.* **2017**, *1*, 97–111. [CrossRef]
34. Tulpule, P.; Marano, V.; Rizzoni, G. Effects of different PHEV control strategies on vehicle performance. In Proceedings of the American Control Conference, St. Louis, MO, USA, 10–12 June 2009; pp. 3950–3955.
35. Wei, X. Modeling and Control of a Hybrid Electric Drivetrain for Optimum Fuel Economy, Performance and Driveability. Ph.D. Thesis, The Ohio State University, Columbus, OH, USA, 2004.
36. Wang, X.; He, H.; Sun, F.; Sun, X.; Tang, H. Comparative Study on Different Energy Management Strategies for Plug-In Hybrid Electric Vehicles. *Energies* **2013**, *6*, 5656–5675. [CrossRef]
37. He, H.; Sun, F.; Zhang, J. Application Study on the Dynamic Programming Algorithm for Energy Management of Plug-in Hybrid Electric Vehicles. *Energies* **2015**, *8*, 3225–3244. [CrossRef]
38. Chen, Z.; Mi, C.C.; Xu, J.; Gong, X.; You, C. Energy Management for a Power-Split Plug-in Hybrid Electric Vehicle Based on Dynamic Programming and Neural Networks. *IEEE Trans. Veh. Technol.* **2014**, *63*, 1567–1580. [CrossRef]
39. Markel, T.; Simpson, A. Cost-Benefit Analysis of Plug-In Hybrid Electric Vehicle Technology. *World Electr. Veh. J.* **2007**, *1*, 294–301. [CrossRef]
40. Markel, T.; Smith, K.; Pesaran, A.A. Improving Petroleum Displacement Potential of PHEVs Using Enhanced Charging Scenarios. In *Electric and Hybrid Vehicles*; Elsevier: Amsterdam, The Netherlands, 2010; pp. 211–225. ISBN 9780444535658.
41. Chan, C.C.; Bouscayrol, A.; Chen, K. Electric, Hybrid, and Fuel-Cell Vehicles: Architectures and Modeling. *IEEE Trans. Veh. Technol.* **2010**, *59*, 589–598. [CrossRef]
42. Denis, N.; Dubois, M.R.; Trovao, J.P.F.; Desrochers, A. Power Split Strategy Optimization of a Plug-in Parallel Hybrid Electric Vehicle. *IEEE Trans. Veh. Technol.* **2018**, *67*, 315–326. [CrossRef]
43. Nguyễn, B.-H.; Trovão, J.P.F.; German, R.; Bouscayrol, A. Real-Time Energy Management of Parallel Hybrid Electric Vehicles Using Linear Quadratic Regulation. *Energies* **2020**, *13*, 5538. [CrossRef]
44. Karaoğlan, M.U.; Kuralay, N.S.; Colpan, C.O. The effect of gear ratios on the exhaust emissions and fuel consumption of a parallel hybrid vehicle powertrain. *J. Clean. Prod.* **2019**, *210*, 1033–1041. [CrossRef]
45. Liu, Y.; Liu, J.; Zhang, Y.; Wu, Y.; Chen, Z.; Ye, M. Rule learning based energy management strategy of fuel cell hybrid vehicles considering multi-objective optimization. *Energy* **2020**, *207*, 118212. [CrossRef]
46. Assadian, F.; Mohan, G.; Longo, S. Comparative analysis of forward-facing models vs backward-facing models in powertrain component sizing. In Proceedings of the Hybrid and Electric Vehicles Conference (HEVC), London, UK, 6–7 November 2013; Volume 2013, pp. 1–6.
47. Elbert, P.; Ebbesen, S.; Guzzella, L. Implementation of Dynamic Programming for n-Dimensional Optimal Control Problems with Final State Constraints. *IEEE Trans. Control Syst. Technol.* **2013**, *21*, 924–931. [CrossRef]

48. Sundstrom, O.; Guzzella, L. A generic dynamic programming Matlab function. In Proceedings of the IEEE International Conference on Control Applications, St. Petersburg, Russia, 8–10 July 2009; pp. 1625–1630.
49. Sundström, O.; Ambühl, D.; Guzzella, L. On Implementation of Dynamic Programming for Optimal Control Problems with Final State Constraints. *Oil Gas Sci. Technol. Rev. l'IFP* **2010**, *65*, 91–102. [CrossRef]
50. Chen, X.; Du, W.; Qian, F. An Adaptive Multi-Objective Differential Evolution Algorithm for Solving Chemical Dynamic Optimization Problems. In *Computer Aided Chemical Engineering*; Elsevier: Amsterdam, The Netherlands, 2015; Volume 37, pp. 821–826.
51. Marler, R.T.; Arora, J.S. Survey of multi-objective optimization methods for engineering. *Struct. Multidiscip. Optim.* **2004**, *26*, 369–395. [CrossRef]
52. Jakob, W.; Blume, C. Pareto Optimization or Cascaded Weighted Sum: A Comparison of Concepts. *Algorithms* **2014**, *7*, 166–185. [CrossRef]
53. Yang, X.-S. Multi-Objective Optimization. In *Nature-Inspired Optimization Algorithms*; Elsevier: Amsterdam, The Netherlands, 2014; pp. 197–211. ISBN 9780124167438.
54. Fleming, P.J.; Purshouse, R.C.; Lygoe, R.J. Many-Objective Optimization: An Engineering Design Perspective. In *Lecture Notes in Computer Science*; Springer: Berlin, Heidelberg, 2005; Volume 3410, pp. 14–32.
55. Liang, J.; Walker, P.D.; Ruan, J.; Yang, H.; Wu, J.; Zhang, N. Gearshift and brake distribution control for regenerative braking in electric vehicles with dual clutch transmission. *Mech. Mach. Theory* **2019**, *133*, 1–22. [CrossRef]
56. Zhu, B.; Zhang, N.; Walker, P.; Zhou, X.; Zhan, W.; Wei, Y.; Ke, N. Gear shift schedule design for multi-speed pure electric vehicles. *Proc. Inst. Mech. Eng. Part D J. Automob. Eng.* **2015**, *229*, 70–82. [CrossRef]
57. Minister of Transport, S.L. Motor Traffic (Speed Limits) Regulations, No. 1 of 2012. In *Gazette Extraordinary No. 1763/26 of June 22, 2012*; Department of Government Printing: Colombo, Sri Lanka, 2012; pp. 1A–2A.
58. Specifications—Bajaj RE 4S Twin Spark 4-Stroke DTSi Three-Wheeler. Available online: https://www.globalbajaj.com/global/english/brands/intracity/re/re-4s/specifications/ (accessed on 13 October 2020).
59. President, S.L. Vehicular exhaust emission standards for importation, manufacturing or assembling of vehicles. In *Gazette Extraordinary No. 2079/42 of July 12, 2018*; Department of Government Printing: Colombo, Sri Lanka, 2018; p. 3A.

energies

MDPI

Article

Real-Time Energy Management of Parallel Hybrid Electric Vehicles Using Linear Quadratic Regulation

Bảo-Huy Nguyễn [1,2,3], João Pedro F. Trovão [1,4,*], Ronan German [2,3] and Alain Bouscayrol [2,3]

1 e-TESC Lab., University of Sherbrooke, Sherbrooke, QC J1K 2R1, Canada; Bao.Huy.Nguyen@USherbrooke.ca
2 Centrale Lille, Arts et Métiers Institute of Technology, Univ. Lille, Yncrea Hauts-de-France, ULR 2697-L2EP, F-59000 Lille, France; Ronan.German@univ-lille.fr (R.G.); Alain.Bouscayrol@univ-lille.fr (A.B.)
3 MEGEVH, French Scientific Network on Hybrid and Electric Vehicles, F-59000 Lille, France
4 Canada Research Chair in Efficient Electric Vehicles with Hybridized Energy Storage Systems, University of Sherbrooke, Sherbrooke, QC J1K 2R1, Canada
* Correspondence: Joao.Trovao@USherbrooke.ca

Received: 20 September 2020; Accepted: 19 October 2020; Published: 22 October 2020

Abstract: Optimization-based methods are of interest for developing energy management strategies due to their high performance for hybrid electric vehicles. However, these methods are often complicated and may require strong computational efforts, which can prevent them from real-world applications. This paper proposes a novel real-time optimization-based torque distribution strategy for a parallel hybrid truck. The strategy aims to minimize the engine fuel consumption while ensuring battery charge-sustaining by using linear quadratic regulation in a closed-loop control scheme. Furthermore, by reformulating the problem, the obtained strategy does not require the information of the engine efficiency map like the previous works in literature. The obtained strategy is simple, straightforward, and therefore easy to be implemented in real-time platforms. The proposed method is evaluated via simulation by comparison to dynamic programming as a benchmark. Furthermore, the real-time ability of the proposed strategy is experimentally validated by using power hardware-in-the-loop simulation.

Keywords: energy management strategy (EMS); parallel hybrid electric vehicle (HEV); energetic macroscopic representation (EMR); hardware-in-the-loop (HIL)

1. Introduction

The world is facing critical issues of environmental pollution and scarcity of fossil fuel resources. Meanwhile, transportation systems play an important role in environmental care and fossil fuel consumption [1]. Hybrid electric vehicles (HEVs) are among the current promising solutions for these problems [2–4]. HEVs can be classified as series, parallel, and series-parallel configurations [5], in which parallel HEVs have their engines directly connected to the drivetrains. Both internal combustion engine (ICE) and electrical machines produce torque to propel the vehicle. This torque assistance ability makes parallel HEVs appropriate for heavy-duty vehicles such as trucks [6–8]. Since the mechanical power can be shared between the engine and the machine, the parallel configuration does not require bulky machines as with series HEVs. Moreover, the power coupling in the parallel configuration can be realized by simple mechanical devices, e.g., a belt. This sort of HEVs do not need complex and large components like its series-parallel counterpart.

However, parallel HEVs suffer from some disadvantages. Firstly, the electrical drive must ensure most of the transient phases; hence, high peak currents are imposed to the batteries. That can notably degrade the batteries [9,10]. To reduce the aging stress on batteries, supercapacitors (SCs) can be added to support the transients [11–13]. With a proper power distribution strategy, SCs can significantly reduce the batteries rms current (see, e.g., [14,15]), which is directly related to their life-time [16,17].

Secondly, in the parallel configuration, the engine is mechanically connected to the drivetrain subsystem. Thus, the engine speed is strongly related to the vehicle velocity. That causes the difficulty to operate the ICE within its optimal region. Fuel consumption is therefore a critical issue of parallel HEVs. An energy management strategy (EMS) is essential to improve the fuel saving by sharing the power between the engine and the machine. In the case of parallel HEVs, it is the torque distribution strategy.

EMS development methods have been classified into rule-based and optimization-based approaches [18], in which the latter group has attracted numerous efforts from both academic fields and industry [19–21]. In [19], an intelligent EMS for a power-split plug-in HEV is developed using dynamic programming (DP) and an artificial neural network (ANN). DP deduces the optimal solutions for six standard driving cycles that are used to train the neural networks. Two ANNs are developed with and without specific trip information. The proposed EMS is validated by simulation. The lack of experimental results in this paper implies the difficulty to realize the ANNs in a real-time platform. DP is computed via off-line simulation to generate data for training the ANNs. The off-line training processes do not affect the real-time implementation but increase the complexity of the method.

In [20], an approximate Pontryagin's minimum principle (PMP) with the restricted five Hamiltonian candidates is proposed for energy management of a plug-in parallel HEV. A reduction of calculating time from six hours to four minutes is reported. However, the method still required the driving cycle known in advance. Thus, it is considered an off-line method, even though the authors claim the feasibility of real-time implementation.

In [21], a plug-in HEV is managed using multi-objective stochastic DP with varied weighting factors to trade-off the batteries' life-time and fuel consumption. Electrical consumption is also taken into account by a fixed penalty factor. Due to the essence of random-process simulated disturbances, stochastic DP strategies are suitable to be applied for fixed-routine vehicle such as trains, buses, and delivery trucks.

These strategies often offer high performance because they are based on optimization techniques. Furthermore, they can be developed through organized approaches, i.e., the developer can follow a systematic procedure to obtain the strategy. Nevertheless, optimization-based methods are often complicated and computationally expensive. Hence, it is still difficult to use these EMSs for real-world applications. Sub-optimal strategies are generally deduced.

The objective of this paper is to propose a novel real-time optimization-based strategy for torque distribution of a parallel hybrid truck. A simple but efficient EMS is thus proposed for an easy real-time implementation while ensuring a low energy consumption. In order to overcome the complexity of the EMS development, Energetic Macroscopic Representation (EMR) is employed. This graphical formalism allows control-oriented decompositions of the system model, so that the control scheme can be systematically deduced [22]. The EMR of the system is a guide for model simplification in order to develop a simple EMS. Then, linear quadratic regulation (LQR) is applied to deduce a feedback control law of the batteries state-of-charge (SoC). This closed-loop control serves as a real-time strategy warranting both fuel saving and batteries charge sustaining. It is noteworthy that this is the first time LQR is employed as a real-time EMS for HEVs. LQR has been used in [23] to tune a model predictive controller (MPC), which leads to a complex EMS. In one article [24], the authors applied LQR to deduce an off-line EMS without proposing a real-time application. This current paper aims to use LQR to develop a real-time EMS for torque distribution of a parallel hybrid vehicle.

The proposed EMS is of interest due to its explicit simple control law. The obtained strategy is a single analytical equation containing fundamental system parameters and measurable variables. Hence, the strategy is easy for real-time implementation and so realistic for real-world applications. This helps to avoid high computational cost such as in the issue of model predictive control (MPC) [25]. Moreover, the development method is straightforward. It does not require any initial guess, iterative simulations, or additional adaptation such as adaptive Pontryagin's minimum principle (A-PMP, also called λ-control) like in [15,26].

The effectiveness of the new EMS is examined by comparison to an off-line optimal benchmark achieved by using DP [27] in term of fuel consumption, as used by many relevant works [15,28,29]. Furthermore, the real-time ability of the proposed strategy is demonstrated via experiment by using reduced-scale power hardware-in-the-loop (HIL) simulation [30].

Section 2 presents the modeling of the studied system for developing EMS. The proposed strategy is developed in Section 3. The comparative evaluation via simulation is given in Section 4. Section 5 addresses the experimental validation of the novel EMS.

2. Modeling for Energy Management Strategy Development

2.1. Modeling of the Studied System

The studied system is a parallel hybrid delivery truck (Figure 1). The electrical drive is mechanically coupled with the ICE via a belt. The engine and the machine propel the truck via a gearbox. The studied system is modeled by the equations given in Equations (1)–(10) and graphically organized by using EMR (Figure 2). The batteries are modeled by Equation (1). The open-circuit voltage $u_{\text{bat OC}}$ and the equivalent series resistance (ESR) r_{bat} are the non-linear functions of the SoC. These functions are given by look-up tables. The electrical drive is modeled by Equation (2). Here, a static model is used by assuming that its torque T_{ED} perfectly follows its reference $T_{\text{ED ref}}$ [31]. The efficiency η_{ED} is given by its average values in traction and regenerative modes.

$$\begin{cases} u_{\text{bat}} = u_{\text{bat OC}}(SoC_{\text{bat}}) - r_{\text{bat}}(SoC_{\text{bat}})i_{\text{bat}} \\ SoC_{\text{bat}} = SoC_{\text{bat init}} - \frac{1}{3600C_{\text{bat}}}\int_0^t i_{\text{bat}}dt \end{cases} \tag{1}$$

$$\begin{cases} T_{\text{ED}} = T_{\text{ED ref}} \\ i_{\text{trac}} = \frac{T_{\text{ED}}\Omega_{\text{ED}}}{u_{\text{bat}}\eta_{\text{ED}}} \end{cases} \tag{2}$$

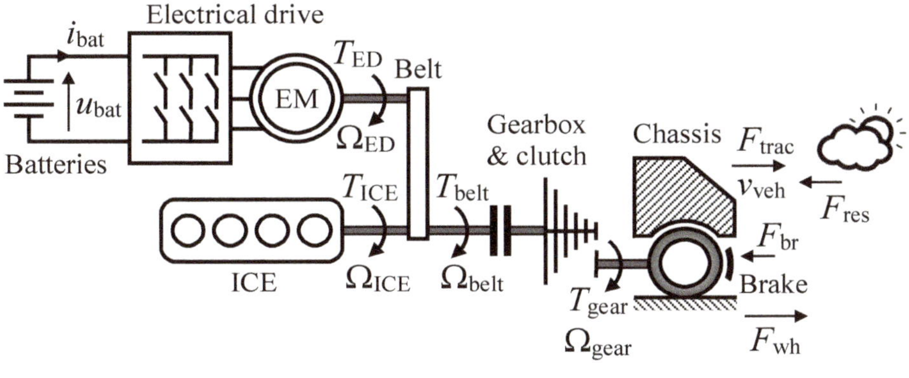

Figure 1. Studied system: a parallel hybrid truck.

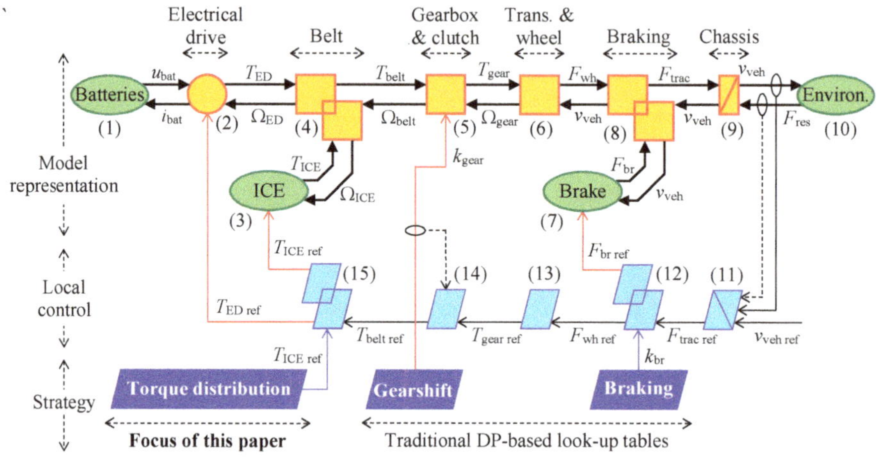

Figure 2. EMR and control of the studied system.

The static model of the ICE is addressed by Equation (3) with the assumed perfect torque control. The fuel consumption rate $\dot{m}_{\text{fuel}}$, in g/s, is given by a fuel consumption map of the engine torque and speed (look-up table). Here, it should be noted that the fuel consumption map is only used for modeling in order to practically examine the control and strategy. It is not required in the EMS development step using the novel method proposed in this paper, which will be presented in the next section. The belt is modeled by Equation (4). The engine speed Ω_{ICE} and the drive speed Ω_{ED} are fixed due to the constant value of the belt ratio k_{belt}. The belt ensures the distribution of the mechanical power between the engine and the machine.

$$\begin{cases} T_{\text{ICE}} = T_{\text{ICE ref}} \\ m_{\text{fuel}} = \int_0^t \dot{m}_{\text{fuel}}(T_{\text{ICE}}, \Omega_{\text{ICE}})dt \end{cases} \quad (3)$$

$$\begin{cases} T_{\text{belt}} = T_{\text{ICE}} + T_{\text{ED}} k_{\text{belt}} \eta_{\text{belt}} \\ \Omega_{\text{ICE}} = \Omega_{\text{belt}} \\ \Omega_{\text{ED}} = \Omega_{\text{belt}} k_{\text{belt}} \end{cases} \quad (4)$$

The gearbox (with clutch) is modeled by Equation (5). The clutch engagement is associated with the gearshift. The gear ratio k_{gear} is shifted discontinuously. The transmission and the wheel are given by Equation (6) with a global ratio k_{tran}. Equation (7) expresses the mechanical brake model. Here, the assumption is no response delay of the braking force. The chassis and the braking coupling are modeled by Equations (8) and (9), respectively. The vehicle mass M_{veh} includes goods and driver. The environment, including the air and the road, is modeled by Equation (10) with three resistive force components, in which g is the gravitational acceleration (m/s^2), ρ the air drag coefficient, $c_x A$ the aerodynamic standard (m^2), and α the slope angle (rad).

$$\begin{cases} T_{\text{gear}} = T_{\text{belt}} k_{\text{gear}} \eta_{\text{gear}} \\ \Omega_{\text{belt}} = \Omega_{\text{gear}} k_{\text{gear}} \end{cases} \quad (5)$$

$$\begin{cases} F_{\text{wh}} = T_{\text{gear}} k_{\text{tran}} \eta_{\text{tran}} \\ \Omega_{\text{gear}} = v_{\text{veh}} k_{\text{tran}} \end{cases} \quad (6)$$

$$F_{\text{br}} = F_{\text{br ref}} \quad (7)$$

$$F_{\text{trac}} = F_{\text{wh}} + F_{\text{br}} \text{ with } F_{\text{br}} \leq 0 \quad (8)$$

$$v_{\text{veh}} = \frac{1}{M_{\text{veh}}} \int_0^t (F_{\text{trac}} - F_{\text{res}}) dt \tag{9}$$

$$F_{\text{res}} = k_{\text{roll}} M_{\text{veh}} g + 0.5 \rho c_x A (v_{\text{veh}} + v_{\text{wind}})^2 + M_{\text{veh}} g \sin \alpha \tag{10}$$

2.2. Local Control of the System

EMR follows the principle of inversion which allows "automatic" deduction of the control scheme from the model organization. The control is deduced by functionally inverting the model representation [22]. If the element contains no dynamical delay, i.e., there are only algebraic operators, it should be directly inverted. The direct inversion is realized by algebraic manipulations. If there is dynamical delay, i.e., the element contains differential equations reorganized by the integral causality, it should be indirectly inverted by using a closed-loop controller.

The control of the hybrid traction subsystem with equation given in Equations (11)–(15) is deduced as follows. The vehicle dynamics (Equation (9)) is indirectly inverted by a closed-loop velocity controller (Equation (11)) (crossed parallelogram). Here, the IP control structure is employed to cancel the zero of the closed-loop transfer function. Then, the braking coupling relationship (Equation (8)) is inverted by Equation (12) (overlapped parallelogram) as a braking force distribution. The ratio k_{br} is imposed from a braking strategy. Next, Equations (13) and (14) (parallelograms) are the direct inversion of Equations (6) and (5) to calculate the $T_{\text{gear ref}}$ and the $T_{\text{belt ref}}$, respectively. Finally, the belt model (Equation (4)) is inverted to deduce the torque distribution (Equation (15)) (overlapped parallelogram). It computes the electrical drive torque reference $T_{\text{ED ref}}$ from the ICE torque reference $T_{\text{ICE ref}}$ imposed by the torque distribution strategy.

$$F_{\text{trac ref}} = F_{\text{res meas}} + k_{\text{I chas}} \int_0^t (v_{\text{veh ref}} - v_{\text{veh meas}}) dt - k_{\text{P chas}} v_{\text{veh meas}} \tag{11}$$

$$\begin{cases} F_{\text{wh ref}} = F_{\text{trac ref}} k_{\text{br}} \\ F_{\text{br ref}} = F_{\text{trac ref}} (1 - k_{\text{br}}) \end{cases} \tag{12}$$

$$T_{\text{gear ref}} = \frac{F_{\text{wh ref}}}{k_{\text{tran}}} \tag{13}$$

$$T_{\text{belt ref}} = \frac{T_{\text{gear ref}}}{k_{\text{gear}}} \tag{14}$$

$$T_{\text{ED ref}} = \frac{T_{\text{belt ref}} - T_{\text{ICE ref}}}{k_{\text{belt}}} \tag{15}$$

2.3. Model Reduction for Energy Management Strategy

To develop EMS, the system model should be reduced. Firstly, the full dynamical model is complex, which is often a multi-variable high-order model. That can make the EMS development complicated and hard to be realized. Hierarchical schemes are therefore often of interest to organize the strategies of complex system, such as [32]. Second, the dynamics at the higher level (strategy) are slower than the ones at the lower level (control) [33,34]. Addressing all the fast dynamics can cause huge computation for the EMS, which could be difficult for real-time implementation. Finally, when the local control is properly developed, the controlled variables can be considered as perfect response to the references.

This work focuses on developing the torque distribution strategy. In this study, the gearshift and braking strategies are developed by using look-up tables deduced from DP for ensuring fair comparisons. Hence, the drivetrain subsystem model can be represented as an equivalent mechanical source that imposes the belt speed Ω_{belt} to the system. A reduced EMR (Figure 3) is obtained for the development of the torque distribution strategy.

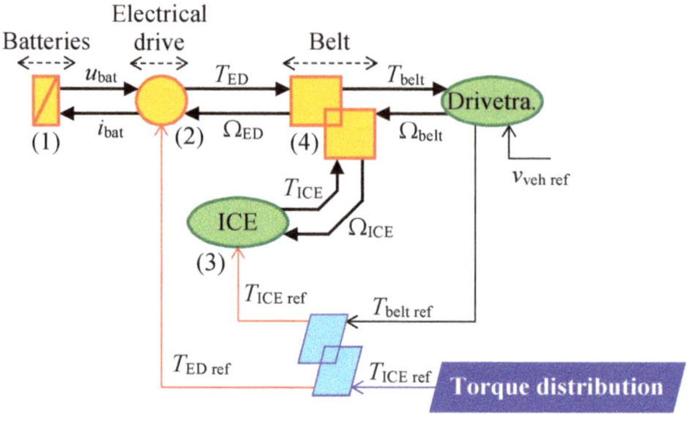

Figure 3. Reduced EMR for torque distribution strategy development.

In EMR, the source element (oval) is the terminal of the considered model. By using a source element to represent the drivetrain subsystem, the considered model for EMS development is reduced to (1)–(4) (see Figure 3).

The SoC of the batteries is the state variable to be controlled at the strategy level. Thus, the batteries are depicted as an accumulation element (crossed rectangle), which represents the considered dynamics of the studied model.

The ICE torque reference $T_{ICE\ ref}$ is imposed by the strategy block; thus, it is the control variable. The belt torque reference $T_{belt\ ref}$ is imposed by the drivetrain subsystem as the disturbance to be compensated.

A mathematical model can then be deduced from the reduced EMR of the system. By neglecting the efficiencies, the reduced mathematical model of the system can be written as follows:

$$\frac{d}{dt}SoC_{bat} = \frac{k_{belt}\Omega_{ICE}}{C_{bat}u_{bat}}(T_{ICE\ ref} - T_{belt\ ref}); \quad (16)$$

where Ω_{ICE} and u_{bat} are measurable disturbances; $T_{belt\ ref}$ indirectly determined by the driver request; k_{belt} and C_{bat} the given parameter; SoC_{bat} the state variable; and $T_{ICE\ ref}$ the control variable. This is a linear model which is convenient for applying optimal control.

3. Proposed Real-time Energy Management Strategy

3.1. Approach

3.1.1. Original Problem Statement

The energy management problem of the parallel hybrid truck can be stated as follows. Find an optimal control law $T^{opt}_{ICE\ ref}$ for the model (16) to minimize the cost function J:

$$J = \int_0^t \dot{m}_{fuel}(T_{ICE}, \Omega_{ICE})dt. \quad (17)$$

Applying directly the optimal control theory to solve this problem leads to a trivial solution in which $T^{opt}_{ICE\ ref} = 0$ [35]. This solution can be intuitively explained as: the best way to save fuel is not to use the ICE. However, it is not an expected solution because the batteries will be fully discharged very quickly.

To overcome this issue, two approaches are often used in literature. The first one is to add the final constrain of the batteries SoC:

$$SoC_{bat\ final} = SoC_{bat\ init}. \quad (18)$$

Then, optimal control techniques such as DP [36,37] or PMP [35,38] are applied. This approach, however, is suitable only for off-line strategies when the driving cycles are known in advance. In act, there is no way to ensure this final state constraint in real-time since the final time is unknown. To develop real-time EMSs, λ-control [26,39] is often applied to adapt the PMP-based optimal solution to real-time operations. The strategies, therefore, become sub-optimal.

The second approach is to charge the cost of SoC variation by adding to the cost function a penalty such as [25]:

$$J = \int_0^t \left[a\left(\dot{m}_{\text{fuel}}\right)^2 + b(SoC - SoC_{\text{ref}})^2 \right] dt. \tag{19}$$

where a and b are weighting factors. The problem is then solved by using MPC [25] or PMP [40] with λ-control scheme [26].

There is a common drawback of these approaches that the fuel consumption rate $\dot{m}_{\text{fuel}}(T_{\text{ICE}}, \Omega_{\text{ICE}})$ must be considered to develop the strategies. For numerical methods such as DP, a look-up table of the fuel consumption rate is directly used. For analytical methods such as PMP, the fuel consumption map is approximated by polynomial functions of which derivatives can be analytically calculated, such as [41]. This is because the PMP method requires analytical expression of the partial derivative. In both cases, a fuel consumption map of ICE is required. This is a drawback in real-world applications since such data are not often available.

Strategy development using these approaches is also often complicated. Moreover, methods such as MPC require strong computational efforts that leads to difficulties to implement them in real-time.

As a consequence, the problem could be reformulated for simplification without requirement of any data map and low computational efforts.

3.1.2. Problem Reformulation

This study proposes an alternative approach, firstly by reformulating the problem. It is known that the fuel consumption rate is mainly proportional to the ICE power [26,42]. In other words, for a given speed, the higher the engine torque is, the higher the fuel consumption rate $\dot{m}_{\text{fuel}}$ is. The data of the engine (Figure 4) show an almost linear behavior of the fuel consumption rate as a function of the torque and the speed.

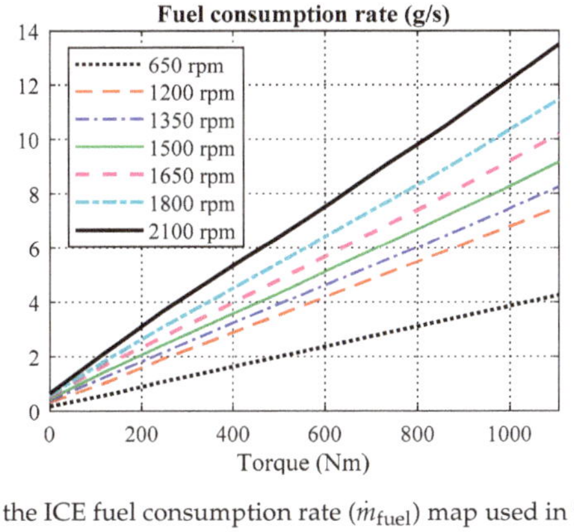

Figure 4. An example of the ICE fuel consumption rate ($\dot{m}_{\text{fuel}}$) map used in this study (Detroit Diesel Engine 205 kW).

Hence, it can be considered that minimizing the engine torque will minimize the fuel consumption. That means instead of using the original cost function (Equation (17)), it can be:

$$J = \int_0^t T_{\text{ICE ref}} dt. \tag{20}$$

Here, Equation (20) can be considered as an approximation of the cost (Equation (17)). This approximation offers the following advantages: (i) it does not require any data set of fuel consumption rate of the engine; and (ii) it allows simple analytical calculations that are useful for developing real-time strategy.

Additionally, as discussed above, a term of SoC variation should be used with penalty in order to ensure the charge sustaining of the energy storage as follows:

$$J = \int_0^t \left[T^2_{\text{ICE ref}} + Q(SoC_{\text{bat}} - SoC_{\text{bat ref}})^2\right]dt; \tag{21}$$

where Q is an equivalent conversion factor to convert the SoC variation to the engine torque.

Since the reduced model (Equation (16)) is linear while the cost function (Equation (21)) is in a quadratic form of the state and control variables, the LQR method can be employed [43]. That can deduce a simple analytical control law that is suitable for on-board real-time implementation.

3.2. Strategy Development

3.2.1. Linear Quadratic Regulation (LQR)

The LQR control law is developed for the linear system in the following form:

$$\frac{d}{dt}x = Ax + Bu; \tag{22}$$

where x generally denotes state variables vector; u the control variables vector; A the dynamical matrix; B the control matrix; with the cost function in the form given by:

$$J = \int_0^\infty \left(x^T Qx + u^T Ru\right)dt \tag{23}$$

where Q and R are weighting matrices. In real-world real-time applications, the final time of the driving cycle is unknown. Thus, the formulation with an infinite time horizon is suitable.

By applying PMP to this linear system and quadratic cost function, the control law is obtained as follows:

$$u = -R^{-1}B^T P(x - x_{\text{ref}}); \tag{24}$$

where P is the solution of the algebraic Riccati equation:

$$PA + A^T P - PBR^{-1}B^T P + Q = 0 \tag{25}$$

.

3.2.2. Proposed LQR-Based Strategy

Applying the reduced mathematical model (Equation (16)) and the reformulated cost function (Equation (21)) to their general forms (Equations (22) and (23)), respectively, the system coefficients are defined by:

$$\begin{cases} A = 0 \\ B = \frac{k_{\text{belt}}\Omega_{\text{ICE}}}{C_{\text{bat}}u_{\text{bat}}} \\ Q \text{ remain} \\ R = 1 \end{cases} \tag{26}$$

.

Replacing Equation (26) in Equation (25), it leads to:

$$P = \frac{\sqrt{Q}}{B}. \tag{27}$$

By replacing Equation (27) in the general form (Equation (24)) of the control law, the ICE torque reference can be deduced as follows:

$$T_{\text{ICE ref}} = \sqrt{Q}(SoC_{\text{bat ref}} - SoC_{\text{bat}}). \quad (28)$$

It is seen that, eventually, the LQR-based control law has a form of a proportional (P) controller of the batteries SoC (see Appendix A).

3.2.3. Weighting Factor Determination

The square-root of the weighting factor Q can be determined by using the well-known pole-placement technique for the P controller. The reduced model (16) can be rewritten in Laplace domain as follows:

$$SoC_{\text{bat}} = \frac{k_{\text{belt}}\Omega_{\text{ICE}}}{C_{\text{bat}}u_{\text{bat}}s}(T_{\text{ICE ref}} - T_{\text{belt ref}}); \quad (29)$$

where s is the Laplace operator. Since the belt torque reference $T_{\text{belt ref}}$ is the disturbance of the studied control loop, it is neglected while synthesizing the controller. Thus, the closed-loop transfer function from $SoC_{\text{bat ref}}$ to SoC_{bat} is:

$$SoC_{\text{bat}} = \frac{1}{\frac{C_{\text{bat}}u_{\text{bat}}}{k_{\text{belt}}\Omega_{\text{ICE}}\sqrt{Q}}s + 1}SoC_{\text{bat ref}}. \quad (30)$$

The closed-loop system is a first-order transfer function, in which its response time (to 95% of the step reference) is three times of the time constant, as given by:

$$t_{\text{res}} = 3\frac{C_{\text{bat}}u_{\text{bat}}}{k_{\text{belt}}\Omega_{\text{ICE}}\sqrt{Q}} \quad (31)$$

where t_{res} is the response time of the closed-loop system. Hence, the square-root of the weighting factor can be determined as:

$$\sqrt{Q} = 3\frac{C_{\text{bat}}u_{\text{bat}}}{k_{\text{belt}}\Omega_{\text{ICE}}t_{\text{res}}}. \quad (32)$$

Eventually, by replacing Equation (32) in Equation (28), the LQR-based torque distribution strategy for the studied parallel hybrid truck is obtained as follows:

$$T_{\text{ICE ref}} = 3\frac{C_{\text{bat}}u_{\text{bat}}}{k_{\text{belt}}\Omega_{\text{ICE}}t_{\text{res}}}(SoC_{\text{bat ref}} - SoC_{\text{bat}}). \quad (33)$$

where the response time t_{res} is the only parameter needed to be defined by the strategy developer. In fact, it is a trade-off between the fuel saving (longer t_{res}) and charge sustaining of the batteries (shorter t_{res}). In order to guarantee the charge-sustaining condition of the energy management problem for parallel HEVs (see Equation (18)), the battery SoC reference $SoC_{\text{bat ref}}$ should be set as $SoC_{\text{bat init}}$.

The proposed real-time LQR-based strategy is implemented as illustrated in Figure 5. The limitations of the electrical drive torque and the batteries SoC are treated by using the conventional switching method. When the system reaches its limitations, the ICE must provide all the demanded traction power. The strategy is simple and straightforward without requiring complex data such as an engine fuel consumption map. Thus, it is suitable for on-board real-time implementation for real-world applications.

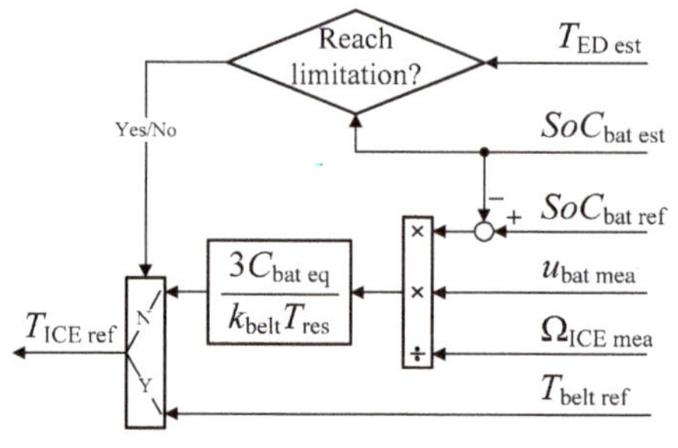

Figure 5. Implementation of the proposed LQR-based EMS.

4. Comparative Evaluations by Simulation

4.1. Simulation Conditions

The studied vehicle is based on a parallel hybrid truck designed in [44] with main parameters given in Table 1. A traditional six-level gearbox is employed. A 205-kW diesel engine is coupled with a 58-kW permanent magnet synchronous machine (PMSM) drive via a belt with the 1:1 ratio. The electrical drive is supplied by a 300-V 62-Ah batteries pack. The PMSM electrical drive is modeled by using nominal efficiency in traction and regenerative modes which are given in [44]. It could be of interest if an efficiency map is available that may compute the power consumption of the drive with higher accuracy. However, the use of the electrical drive efficiency data only affects the comparison between different traction models rather than between the different EMSs which is the objective of this paper. Hence, the nominal efficiency parameters are sufficient for this study.

Table 1. Examined system parameters for simulation of the parallel hybrid truck.

Parameters		Values
Vehicle (Based on the Hybrid Delivery Truck Designed in [44])		
Vehicle total mass	M_{veh}	7514 kg
Aerodynamic standard	c_xA	0.73×6.9 m^2
Rolling coefficient	k_{roll}	0.008
Final drive ratio	k_{FD}	3.33
Wheel radius	R_{wh}	0.397 m
Gearbox		
Gearbox ratio	k_{gear}	(7.14 4.17 2.50 1.59 1.00 0.78)
Efficiency	η_{gear}	(0.94 0.95 0.9 0.95 0.91 0.91)
Belt		
Belt ratio	k_{belt}	1
Efficiency	η_{belt}	0.95
ICE (Detroit Diesel Corp. Series 50 8.5 Diesel Engine)		
Maximal power	$P_{ICE\ max}$	205 kW
Maximal speed	$\Omega_{ICE\ max}$	2100 rpm
Idle speed	$\Omega_{ICE\ idle}$	650 rpm
Maximal torque	$T_{ICE\ max}$	1100 Nm
Mass density of diesel	M_{vol}	850 g/L
Electrical Drive (PMSM)		
Maximal power	$P_{ED\ max}$	58 kW
Maximal torque	$T_{ED\ max}$	400 Nm
Nominal speed	$\Omega_{ED\ nom}$	1500 rpm
Maximal speed	$\Omega_{ED\ max}$	4000 rpm
Nominal efficiency in traction mode	$\eta_{ED\ trac}$	96%
Nominal efficiency in regenerative mode	$\eta_{ED\ regen}$	90%
Batteries (LiPho A123 20Ah 2010 Cells)		
Battery bank capacity	C_{bat}	62 Ah
Battery bank resistance (at 70% SoC)	r_{bat}	26 mΩ
Battery bank OCV (at 70% SoC)	$u_{bat\ OC}$	300 V

This study examines the system with two standard driving cycles: the Urban Dynamometer Driving Schedule (UDDS) and the New European Driving Cycle (NEDC). UDDS was developed for testing heavy-duty vehicles, which is the case of the studied hybrid truck. Meanwhile, NEDC is for passenger cars and light-duty trucks. However, in this study, NEDC was employed, due to its simple profile, to better analyze the results. DP was employed to ensure the global optimal solution as a benchmark.

4.2. Results and Discussions

The objective of the torque distribution strategies is to minimize the engine fuel consumption. Hence, with different driving cycles, fuel consumption per 100 km is the criteria to evaluate and compare the different EMSs. Figure 6 shows a comparative evaluation of the proposed LQR-based strategy with the DP-based optimal solution and the conventional ICE truck. The hybrid truck can save up to 7.8% the fuel consumption with DP in the case of driving with UDDS cycle. It should be noted that DP is the off-line benchmark and only gives the theoretical optimal result, whereas the proposed real-time strategy saves 6.3% without knowing the driving cycle in advance. The fuel saving is 7.9% (DP) and 5.6% (proposed EMS) with NEDC.

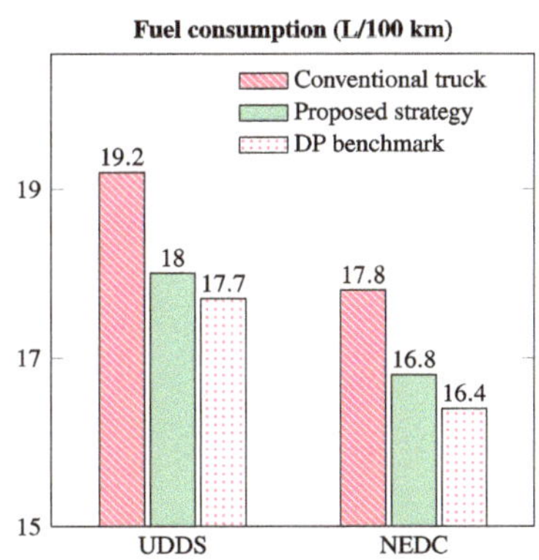

Figure 6. Comparison of fuel consumptions via simulation.

To have better understanding of the system behavior with the proposed strategy, the results with NEDC are presented here in detail. This driving cycle is simple; thus, the behavior is easier to analyze and understand. The examined NEDC contains four repeated urban cycles and a highway part (Figure 7). The controller acts to force the vehicle velocity to respond well to its reference without any overshoot. The torque distribution results are given in Figure 8. During every urban cycle, the ICE torque and the electrical drive torque perform in the same patterns. That confirms the consistency of the proposed EMS. The strategy let the machine support the engine as much as possible during the accelerations, then the ICE produce power to recharge the batteries. The electrical drive torque T_{ED} is kept within the torque constraints due to the drive power limitation. The electrical drive works to support the ICE to reduce the engine torque production as much as possible. When the machine reaches its torque limitation, the engine has to compensate for the remaining part to follow the torque reference needed to drive the vehicle. It should be noted that the electrical drive torque limitations reduce when its speed is higher than the machine nominal speed. Furthermore, it is seen that the drive torque is balanced in traction and regenerative braking mode. It indirectly shows the charge sustaining of the battery, which is often a requirement of the non-plug-in HEVs.

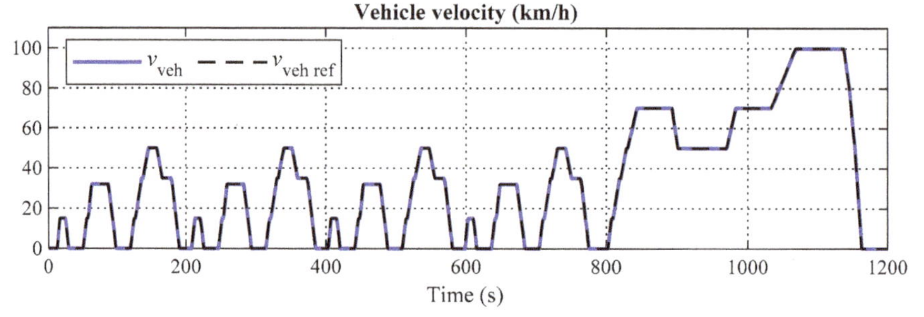

Figure 7. (Simulation) vehicle velocity following the test profile NEDC.

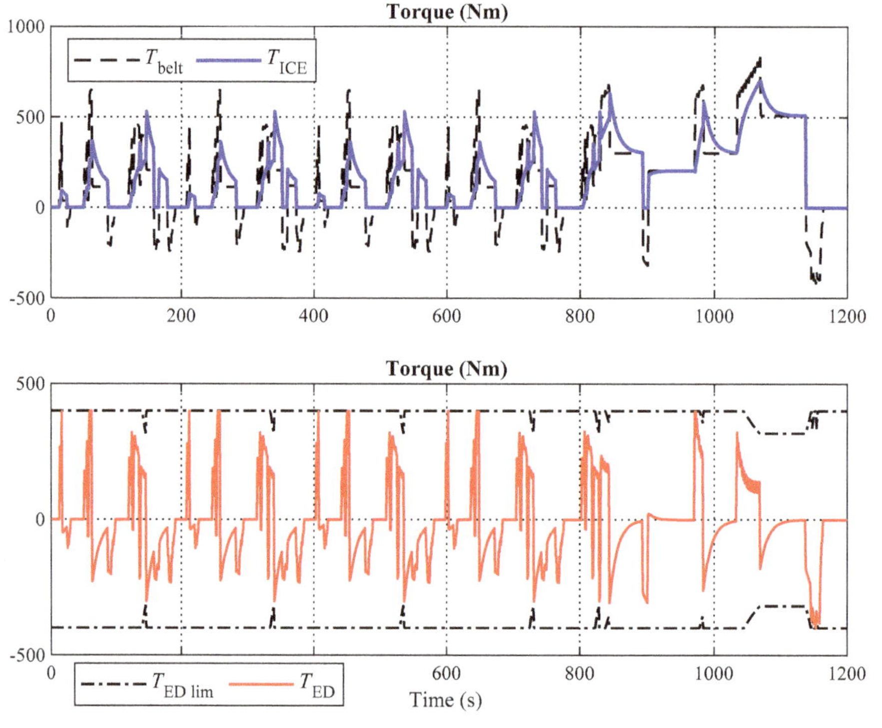

Figure 8. (Simulation) torque distribution results of belt, ICE, and ED torques with NEDC.

The batteries charge sustaining can be observed via the SoC evolutions, in a comparison with the result from DP (Figure 9). DP strategy can anticipate that there is a huge amount of regenerative energy at the end of the driving. Thus, it allows the batteries SoC to continuously reduce after each urban cycle. Meanwhile the real-time EMS does not "know" the driving condition in advance. Hence, it "tries" to ensure the charge sustaining after every urban one. The proposed LQR-based EMS perform well in both fuel saving and batteries charge sustaining. Yet, the DP strategy can save more fuel by a priori knowing the cycle.

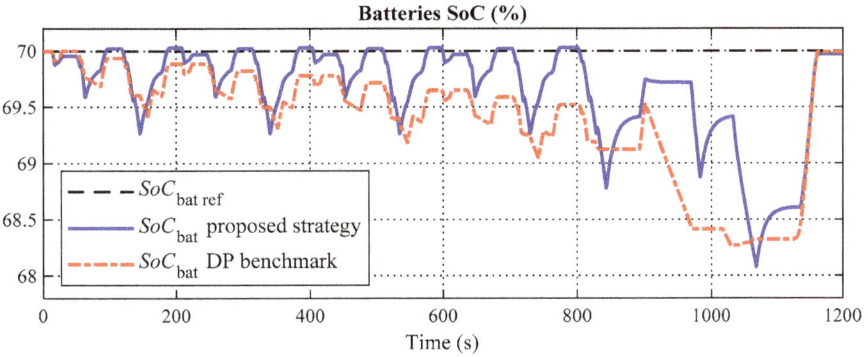

Figure 9. (Simulation) batteries SoC evolutions with NEDC.

It can be seen that the batteries SoC variation is kept within a narrow range of about 2%. This is because the batteries are often over-sized for non-plug-in HEVs in order to ensure the charge-sustaining condition. It is noteworthy that if the batteries are more deeply discharged, there will be a need of more energy to recharge them to the initial SoC. Working around this SoC range is optimal for this vehicle, as proven by the off-line optimal solution deduced by DP. The agreement between DP and the LQR-based strategy in term of batteries SoC range confirms the close-to-optimal performance of the proposed method. The batteries can be downsized by using advanced design methods; however, the sizing problem is not considered here because this paper focuses on EMS development.

5. Experimental Validation of the Proposed Strategy

5.1. Experimental Setup

The experiments are carried out to validate the proposed real-time torque distribution strategy by using reduced-scale power HIL simulation [30]. Figure 10 illustrates the experimental setup configuration and Figure 11 shows the test bench in the laboratory.

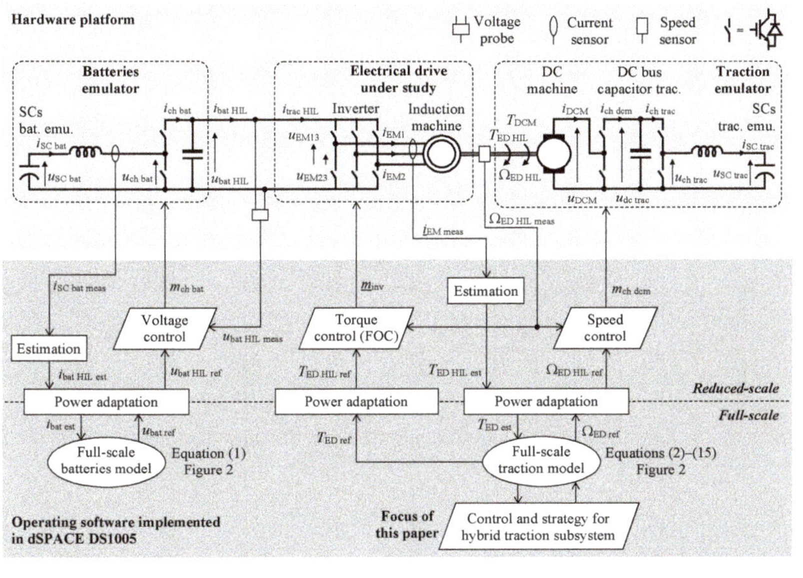

Figure 10. Experimental setup configuration.

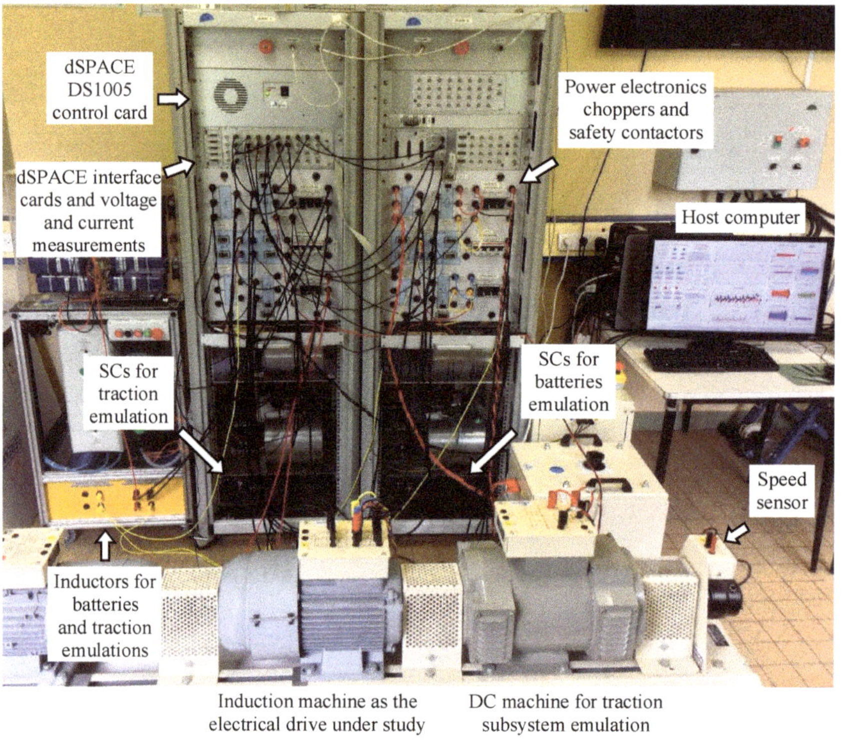

Figure 11. Experimental test bench.

The electrical drive is realized by a three-phase wound rotor induction machine (IM) and a voltage-source inverter. The DC bus of the inverter is connected in parallel with the DC bus capacitor of the battery emulator. The IM is connected in delta-connection scheme to increase the speed range since the emulated batteries voltage $u_{\text{bat HIL}}$ is much lower than the nominal DC bus voltage of the machine. The traction emulation is realized by using a DC machine, of which the armature is connected to a chopper to control the armature current i_{DCM}. That DC machine drive is supplied by a voltage source composed by a SCs pack, an inductor, a chopper, and a DC bus capacitor. The SCs pack charges and discharges to consume and to provide the emulated traction power, respectively. The traction DC bus voltage $u_{\text{dc trac}}$ is controlled to be constant. The electrical drive and the traction emulator are mechanically connected via a single shaft with a speed sensor. The batteries emulator is realized in the same way as described in [15], in which a controllable voltage source is composed by using a bidirectional DC/DC converter and SCs are employed to emulate the batteries.

The model and control program are implemented in a dSPACE controller board. As there is no full-scale machine available in the laboratory, a reduced-scale HIL testing is achieved. The mechanical emulation of the traction subsystem is realized in a similar way to [45]. The only difference in this work is that a SC pack, connected via a DC/DC converter, is employed to exchange the emulated traction energy. Via the power adaptation blocks, the full-scale traction model imposes the drive speed and torque references to the system. The DC machine is controlled by a speed controller to follow the speed reference, whereas the IM is controlled by a torque controller to follow the torque reference. The emulated torque is estimated from the measured currents of the IM. This torque is then imposed to the full-scale traction model via a power adaptation block.

5.2. Results and Discussions

During the driving cycle, the gearbox is shifted by a DP-based look-up table, which forms the electrical drive speed, and therefore also the ICE speed, as shown in Figure 12. The rotational speed is scaled with a ratio of 1.8 due to the limitations of the electrical machines used for the experiments. The lower boundary is the ICE idle speed. The left axis indicates the experimental results, while the right axis shows the simulation ones. It can be seen that the experimental speed matches well with the simulation. These results confirm that the emulator emulates well the behaviors of the traction subsystem. Some small differences appear when the engine is shifted too fast. This is because the DP-based gearshift strategy is developed with the assumption that the gearbox can be shifted without any delay.

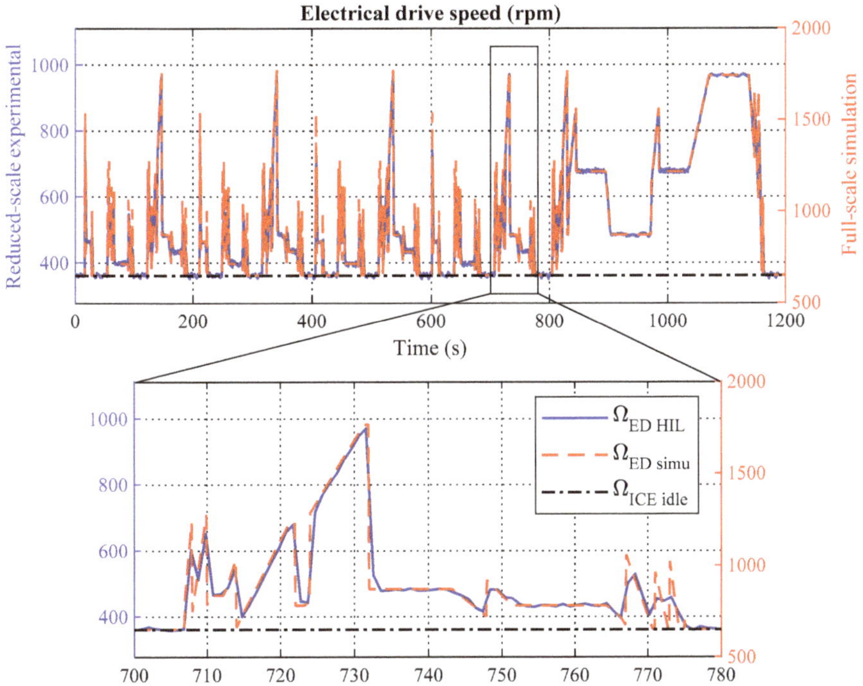

Figure 12. (Experimental) electrical drive speed (also emulated ICE speed) with NEDC.

The results of electrical drive torque with the proposed LQR-based torque distribution strategy are plotted in Figure 13. The torque is scaled 80 times by considering the power limitation of the experimental IM drive. The experimental and simulated torques are kept within the boundaries of the electrical drive torque limitations. Here, it should be noted that the electrical drive torque limitation varies with time. It is reduced when the machine speed is higher than its rated value.

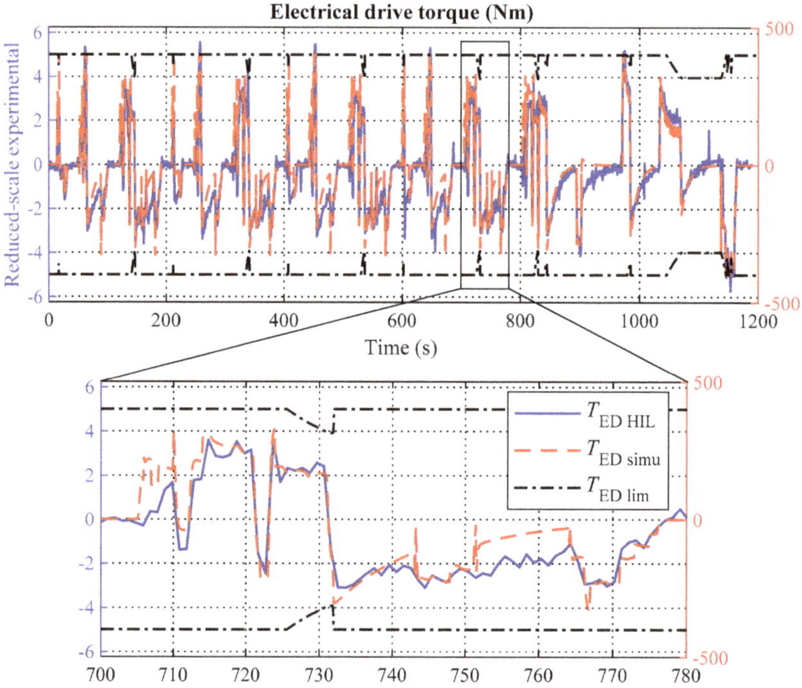

Figure 13. (Experimental) electrical drive torque obtained by the proposed LQR-based strategy with NEDC.

The software program of the experiments is implemented in the dSPACE DS1005 card with a sampling time of 0.2 ms. In fact, most of the computational resource of the card is devoted for the complex control schemes of the batteries and traction emulators and for the full-scale model of the traction subsystem. Very little computational effort is required to perform the LQR-based strategy, which composed of a simple Equation (33). That demonstrates the real-time capability of the proposed EMS which is the main objective of the experiments despite the limitation of the experimental set-up in terms of power.

6. Conclusions

In this paper, a novel real-time energy management strategy has been proposed and validated for a parallel hybrid truck. Firstly, the truck has been modeled and controlled considering its full dynamical behaviors. To overcome the complexity of the system, the model, represented by using EMR, has been reduced to obtain a suitable mathematical model. Then, optimal control theory has been applied to this reduced mathematical model to deduce the real-time EMS.

To avoid the requirement of using ICE fuel consumption data, which is hard to achieve in real-world applications, the cost function has been reformulated. Furthermore, the variation of the batteries SoC has been added to the function as a penalty. Eventually, the cost is in form of a quadratic function of the control and the state variables. Since the reduced model is linear, the LQR method has been applied to obtain the optimization-based control law of the ICE torque reference. The obtained strategy is a closed-loop control scheme of the batteries SoC, which is suitable to be implemented in real-time. The development procedure and the strategy implementation require only basic system parameters such as the batteries capacity and the belt ratio. The proposed EMS is, therefore, realistic for real-world applications.

The new strategy has been compared to the DP-based off-line optimal solution and the case of the conventional ICE-only truck. Simulation results have verified the performances of the LQR-based strategy. Moreover, reduced-scale power HIL simulation has been used to validate the proposed EMS in experiments. It has been shown that the LQR-based strategy works properly in real-time.

Author Contributions: Conceptualization and methodology B.-H.N., J.P.F.T., and A.B.; software, B.-H.N.; validation, B.-H.N., J.P.F.T., R.G., and A.B.; formal analysis, B.-H.N.; investigation, B.-H.N. and R.G.; resources, J.P.F.T. and A.B.; data curation, B.-H.N.; writing—original draft preparation, B.-H.N.; writing—review and editing, B.-H.N., J.P.F.T., R.G., and A.B.; visualization, B.H.N.; supervision, J.P.F.T. and A.B.; project administration, J.P.F.T. and A.B.; funding acquisition, J.P.F.T. and A.B. All authors have read and agreed to the published version of the manuscript.

Funding: This work is supported by the Canada Research Chair Program (950-230672), by the Mitacs Accelerate Program (IT09159), by the C2EI IDEAL program within the framework of the European Union by FEDER funding of the "Hauts-de-France" region, and also under the support of the *"Service de Coopération et d'Action Culturelle du Consulat Général de France à Québec"*.

Conflicts of Interest: The authors declare no conflict of interest.

Appendix A

The algebraic Riccati Equation (25) is available for the linear time-invariant (LTI) system; moreover, the model (16) is time-varying. For that, a variable substitution can be done:

$$u_{dum} = \frac{k_{belt}\Omega_{ICE}}{C_{bat}u_{bat}}(T_{ICE\ ref} - T_{belt\ ref}) \tag{A1}$$

where u_{dum} is a dummy control variable. The model is now linear. The corresponding quadratic cost function is, therefore:

$$J_{dum} = \int_0^t \left[u_{dum}^2 + Q_{dum}(SoC_{bat} - SoC_{bat\ ref})^2\right]dt. \tag{A2}$$

Applying the LQR method and then replacing the dummy control variable u_{dum} by the $T_{ICE\ ref}$, the same control law as Equation (33) is eventually obtained. In this paper, these steps are neglected to use the variables with physical meaning only.

References

1. Bauer, N.; Mouratiadou, I.; Luderer, G.; Baumstark, L.; Brecha, R.J.; Edenhofer, O.; Kriegler, E. Global fossil energy markets and climate change mitigation—An analysis with REMIND. *Clim. Change* **2016**, *136*, 69–82. [CrossRef]
2. Liu, P.; Jin, Z.; Hua, Y.; Zhang, L. Development of Test-Bed Controller for Powertrain of HEV. *Energies* **2020**, *13*, 3372. [CrossRef]
3. Lee, W.; Kim, T.; Jeong, J.; Chung, J.; Kim, D.; Lee, B.; Kim, N. Control Analysis of a Real-World P2 Hybrid Electric Vehicle Based on Test Data. *Energies* **2020**, *13*, 4092. [CrossRef]
4. Kim, H.; Wi, J.; Yoo, J.; Son, H.; Park, C.; Kim, H. A Study on the Fuel Economy Potential of Parallel and Power Split Type Hybrid Electric Vehicles. *Energies* **2018**, *11*, 2103. [CrossRef]
5. Chan, C.C.; Bouscayrol, A.; Chen, K. Electric, Hybrid, and Fuel-Cell Vehicles: Architectures and Modeling. *IEEE Trans. Veh. Technol.* **2010**, *59*, 589–598. [CrossRef]
6. Van Mullem, D.; Van Keulen, T.; Kessels, J.; De Jager, B.; Steinbuch, M. Implementation of an Optimal Control Energy Management Strategy in a Hybrid Truck. In Proceedings of the 6th IFAC Symposium Advances in Automotive Control, Munich, Germany, 12–14 July 2010; pp. 61–66.
7. Biasini, R.; Onori, S.; Rizzoni, G. A near-optimal rule-based energy management strategy for medium duty hybrid truck. *Int. J. Powertrains* **2013**, *2*, 232–261. [CrossRef]
8. Mayet, C.; Welles, J.; Bouscayrol, A.; Hofman, T.; Lemaire-Semail, B. Influence of a CVT on the fuel consumption of a parallel medium-duty electric hybrid truck. *Math. Comput. Simul.* **2019**, *158*, 120–129. [CrossRef]

9. Barré, A.; Deguilhem, B.; Grolleau, S.; Gérard, M.; Suard, F.; Riu, D. A review on lithium-ion battery ageing mechanisms and estimations for automotive applications. *J. Power Sources* **2013**, *241*, 680–689. [CrossRef]
10. Uddin, K.; Moore, A.D.; Barai, A.; Marco, J. The effects of high frequency current ripple on electric vehicle battery performance. *Appl. Energy* **2016**, *178*, 142–154. [CrossRef]
11. Erdinc, O.; Vural, B.; Uzunoglu, M. A wavelet-fuzzy logic based energy management strategy for a fuel cell/battery/ultra-capacitor hybrid vehicular power system. *J. Power Sources* **2009**, *194*, 369–380. [CrossRef]
12. Song, Z.; Hofmann, H.; Li, J.; Hou, J.; Han, X.; Ouyang, M. Energy management strategies comparison for electric vehicles with hybrid energy storage system. *Appl. Energy* **2014**, *134*, 321–331. [CrossRef]
13. Schaltz, E.; Khaligh, A.; Rasmussen, P.O. Influence of Battery/Ultracapacitor Energy-Storage Sizing on Battery Lifetime in a Fuel Cell Hybrid Electric Vehicle. *IEEE Trans. Veh. Technol.* **2009**, *58*, 3882–3891. [CrossRef]
14. Hredzak, B.; Agelidis, V.G.; Demetriades, G. Application of explicit model predictive control to a hybrid battery-ultracapacitor power source. *J. Power Sources* **2015**, *277*, 84–94. [CrossRef]
15. Nguyen, B.-H.; German, R.; Trovao, J.P.F.; Bouscayrol, A. Real-Time Energy Management of Battery/Supercapacitor Electric Vehicles Based on an Adaptation of Pontryagin's Minimum Principle. *IEEE Trans. Veh. Technol.* **2019**, *68*, 203–212. [CrossRef]
16. Groot, J.; Swierczynski, M.; Stan, A.I.; Kær, S.K. On the complex ageing characteristics of high-power LiFePO4/graphite battery cells cycled with high charge and discharge currents. *J. Power Sources* **2015**, *286*, 475–487. [CrossRef]
17. Savoye, F.; Venet, P.; Millet, M.; Groot, J. Impact of Periodic Current Pulses on Li-Ion Battery Performance. *IEEE Trans. Ind. Electron.* **2011**, *59*, 3481–3488. [CrossRef]
18. Salmasi, F.R. Control Strategies for Hybrid Electric Vehicles: Evolution, Classification, Comparison, and Future Trends. *IEEE Trans. Veh. Technol.* **2007**, *56*, 2393–2404. [CrossRef]
19. Chen, Z.; Mi, C.C.; Xu, J.; Gong, X.; You, C. Energy Management for a Power-Split Plug-in Hybrid Electric Vehicle Based on Dynamic Programming and Neural Networks. *IEEE Trans. Veh. Technol.* **2014**, *63*, 1567–1580. [CrossRef]
20. Hou, C.; Ouyang, M.; Xu, L.; Wang, H. Approximate Pontryagin's minimum principle applied to the energy management of plug-in hybrid electric vehicles. *Appl. Energy* **2014**, *115*, 174–189. [CrossRef]
21. Moura, S.J.; Stein, J.L.; Fathy, H.K. Battery-Health Conscious Power Management in Plug-In Hybrid Electric Vehicles via Electrochemical Modeling and Stochastic Control. *IEEE Trans. Control. Syst. Technol.* **2012**, *21*, 679–694. [CrossRef]
22. Bouscayrol, A.; Hautier, J.-P.; Lemaire-Semail, B. Graphic Formalisms for the Control of Multi-Physical Energetic Systems: COG and EMR. In *Systemic Design Methodologies for Electrical Energy Systems*; Roboam, X., Ed.; ISTE Ltd.: London, UK, 2013; pp. 89–124.
23. Unger, J.; Kozek, M.; Jakubek, S. Nonlinear model predictive energy management controller with load and cycle prediction for non-road HEV. *Control. Eng. Pr.* **2015**, *36*, 120–132. [CrossRef]
24. Lu, B.; Natarajan, B.; Schulz, N. Optimal control based power management in hybrid military vehicle. In *Proceedings of the 2012 IEEE International Electric Vehicle Conference*; Institute of Electrical and Electronics Engineers (IEEE): Greenville, SC, USA, 2012; pp. 1–7.
25. Borhan, H.; Vahidi, A.; Phillips, A.M.; Kuang, M.L.; Kolmanovsky, I.V.; Di Cairano, S. MPC-Based Energy Management of a Power-Split Hybrid Electric Vehicle. *IEEE Trans. Control. Syst. Technol.* **2011**, *20*, 593–603. [CrossRef]
26. Kessels, J.T.B.A.; Koot, M.W.T.; Bosch, P.V.D.; Kok, D.B. Online Energy Management for Hybrid Electric Vehicles. *IEEE Trans. Veh. Technol.* **2008**, *57*, 3428–3440. [CrossRef]
27. Sundstrom, O.; Guzzella, L. A generic dynamic programming Matlab function. In *Proceedings of the 2009 IEEE International Conference on Control Applications*; Institute of Electrical and Electronics Engineers (IEEE): St. Petersburg, Russia, 2009; pp. 1625–1630.
28. Sciarretta, A.; Serrao, L.; Dewangan, P.; Tona, P.; Bergshoeff, E.; Bordons, C.; Charmpa, L.; Elbert, P.; Eriksson, L.; Hofman, T.; et al. A control benchmark on the energy management of a plug-in hybrid electric vehicle. *Control. Eng. Pract.* **2014**, *29*, 287–298. [CrossRef]
29. Pisu, P.; Rizzoni, G. A Comparative Study of Supervisory Control Strategies for Hybrid Electric Vehicles. *IEEE Trans. Control. Syst. Technol.* **2007**, *15*, 506–518. [CrossRef]
30. Bouscayrol, A. Hardware-in-the-Loop Simulation. In *The Industrial Electronics Handbook: Control and Mechatronics*, 2nd ed.; Wilamowski, B.M., Irwin, J.D., Eds.; CRC Press: Boca Raton, FL, USA, 2011.

31. Mayet, C.; Delarue, P.; Bouscayrol, A.; Chattot, E.; Verhille, J.-N. Comparison of Different EMR-Based Models of Traction Power Substations for Energetic Studies of Subway Lines. *IEEE Trans. Veh. Technol.* **2016**, *65*, 1021–1029. [CrossRef]
32. Kamal, E.; Adouane, L. Hierarchical Energy Optimization Strategy and Its Integrated Reliable Battery Fault Management for Hybrid Hydraulic-Electric Vehicle. *IEEE Trans. Veh. Technol.* **2018**, *67*, 3740–3754. [CrossRef]
33. Bouscayrol, A.; Delarue, P.; Guillaud, X. Power strategies for maximum control structure of a wind energy conversion system with a synchronous machine. Renew. *Energy* **2005**, *30*, 2273–2288. [CrossRef]
34. Trovao, J.P.F.; Pereirinha, P.G.; Jorge, H.M.; Antunes, C.H. A multi-level energy management system for multi-source electric vehicles—An integrated rule-based meta-heuristic approach. *Appl. Energy* **2013**, *105*, 304–318. [CrossRef]
35. Delprat, S.; Lauber, J.; Guerra, T.; Rimaux, J. Control of a Parallel Hybrid Powertrain: Optimal Control. *IEEE Trans. Veh. Technol.* **2004**, *53*, 872–881. [CrossRef]
36. Sundström, O.; Ambühl, D.; Guzzella, L. On Implementation of Dynamic Programming for Optimal Control Problems with Final State Constraints. *Oil Gas Sci. Technol. Rev. IFP* **2009**, *65*, 91–102. [CrossRef]
37. Pérez, L.V.; Bossio, G.R.; Moitre, D.; García, G.O. Optimization of power management in an hybrid electric vehicle using dynamic programming. *Math. Comput. Simul.* **2006**, *73*, 244–254. [CrossRef]
38. Serrao, L.; Onori, S.; Sciarretta, A.; Guezennec, Y.; Rizzoni, G. Optimal energy management of hybrid electric vehicles including battery aging. In *Proceedings of the 2011 American Control Conference*; Institute of Electrical and Electronics Engineers (IEEE): San Francisco, CA, USA, 2011; pp. 2125–2130.
39. Onori, S.; Tribioli, L. Adaptive Pontryagin's Minimum Principle supervisory controller design for the plug-in hybrid GM Chevrolet Volt. *Appl. Energy* **2015**, *147*, 224–234. [CrossRef]
40. Kim, N.; Cha, S.; Peng, H. Optimal Control of Hybrid Electric Vehicles Based on Pontryagin's Minimum Principle. *IEEE Trans. Control. Syst. Technol.* **2011**, *19*, 1279–1287. [CrossRef]
41. Mura, R.; Utkin, V.; Onori, S. Energy Management Design in Hybrid Electric Vehicles: A Novel Optimality and Stability Framework. *IEEE Trans. Control. Syst. Technol.* **2015**, *23*, 1307–1322. [CrossRef]
42. Koot, M.; Kessels, J.; Dejager, B.; Heemels, W.; VandenBosch, P.; Steinbuch, M.M. Energy Management Strategies for Vehicular Electric Power Systems. *IEEE Trans. Veh. Technol.* **2005**, *54*, 771–782. [CrossRef]
43. Bryson, A.E.; Ho, Y.-C. *Applied Optimal Control: Optimization, Estimation, and Control*; Taylor & Francis: New York, NY, USA, 1975.
44. Hofman, T.T.; Steinbuch, M.M.; Van Druten, R.; Serrarens, A. Hybrid component specification optimisation for a medium-duty hybrid electric truck. *Int. J. Heavy Veh. Syst.* **2008**, *15*, 356. [CrossRef]
45. Allegre, A.-L.; Bouscayrol, A.; Verhille, J.-N.; Delarue, P.; Chattot, E.; El-Fassi, S. Reduced-Scale-Power Hardware-in-the-Loop Simulation of an Innovative Subway. *IEEE Trans. Ind. Electron.* **2010**, *57*, 1175–1185. [CrossRef]

energies

MDPI

Article

Impact of Powertrain Components Size and Degradation Level on the Energy Management of a Hybrid Industrial Self-Guided Vehicle

Amin Ghobadpour *, Ali Amamou, Sousso Kelouwani *, Nadjet Zioui and Lotfi Zeghmi

Hydrogen Research Institute, Université du Québec à Trois-Rivières, Trois-Rivières, QC G9A 5H7, Canada; ali.amamou@uqtr.ca (A.A.); nadjet.zioui@uqtr.ca (N.Z.); lotfi.zeghmi@uqtr.ca (L.Z.)

* Correspondence: amin.ghobadpour@uqtr.ca (A.G.); sousso.kelouwani@uqtr.ca (S.K.); Tel.: +1-819-979-6956 (A.G.); +1-819-376-5011 (ext. 3948) (S.K.); Fax: +1-819-376-5152 (S.K.)

Received: 18 August 2020; Accepted: 19 September 2020; Published: 24 September 2020

Abstract: This paper deals with the design of an energy management strategy (EMS) for an industrial hybrid self-guided vehicle (SGV), considering the size of a fuel cell (FC) stack and degradation of a battery pack. In this context, first, a realistic energy model of the SGV was proposed and validated, based on experiments. This model provided a basis for individual components analysis, estimating energy requirements, component sizing, and testing various EMSs, prior to practical implementation. Second, the performance of the developed FC/battery SGV powertrain was validated under three EMS modes. Each mode was studied by considering four different FC sizes and three battery degradation levels. The final results showed that a small FC as a range extender is recommended, to reduce system cost. It is also important to maintain the FC in its high efficiency zones with a minimum ON/OFF cycle, leading to efficiency and lifetime enhancement of FC system. Battery *SOC* have to be kept at a high level during SGV operation, to support the FC during SGV acceleration. In order to improve the SGV's overall autonomy, it is also important to minimize the stop and go and rotational SGV motion with appropriate acceleration and deceleration rate.

Keywords: industrial self-guided vehicle; differential drive mobile robots; energy management strategy; fuel cell

1. Introduction

In the context of industry 4.0, the issue of indoor material handling and transportation by low-speed vehicles, such as mobile robots, is a frequently discussed topic [1]. Such vehicles can play an important role by providing mobility and reducing the cost of production and material transport [2]. Vehicle-based transport systems using self-guided vehicles (SGVs) are widely used in facilities such as manufacturing plants, warehouses, distribution centers, and transshipment terminals. Differential drive mobile robots (DDMRs) are the most common and popular way to drive SGVs, due to their simplicity and zero-radius turning [3]. In this context, the electric powertrains and especially the battery based powertrains were proposed to replace the internal combustion engines (ICEs) in industrial vehicle applications [4]. However, limited autonomy, as well as the time-consuming process of recharging the batteries, are still challenging. To extend the autonomy, a primary method is to choose a bigger battery in terms of capacity, which in turn would increase the cost and the recharge time. Another notable drawback of the battery is its limited life cycle. For instance, the lifespan of a lithium battery varies from 500 to 1500 charge/discharge cycles, depending on the utilization pattern [5]. Another solution is to extend the SGV autonomy by adding a second power source like a proton exchange membrane fuel cell (PEMFC) system. Unlike the batteries, a PEMFC can be charged in just a few minutes and has the potential to provide power over a long period of time [6]. In addition, PEMFC has a high operating efficiency (around 50–70%) compared to ICEs [7]. The PEMFC has been proposed for various robotic applications, such as unmanned underwater vehicle [6,8], humanoid robot [9], and construction machinery [10,11]. Regarding the indoor industry hybrid vehicles, some progress has been made in case of forklifts [12,13], however, other types of indoor vehicles have not been considered as such.

These studies showed promising results in terms of autonomy enhancement of an industrial SGV, by employing a hybrid fuel cell/battery powertrain. However, the PEMFC has some fundamental problems, such as slow dynamics and hydrogen storage issues, for practical robotic applications [14,15]. The slow dynamic characteristic of the PEMFC causes a problem for a DDMR application, where the power requirements can vary considerably during a mission [6,8]. These variable power peaks might degrade the fuel cell (FC) and reduce its efficiency [16]. Concerning the hydrogen storage problem, it is important to note that SGVs are characterized by their mobility and compactness because of the importance of transferability in small spaces. Thus, this can lead to a challenge in the hybridization, due to the limited available space for components of an industrial SGV. For example, the size of the FC stack in terms of power, and the size of the hydrogen tank in terms of capacity and volume should be appropriate to avoid oversizing.

In this respect, when vehicle power sources are hybridized by using big batteries and small PEMFCs as a range extender, the size of the PEMFC stack can be reduced and power transients become negligible as they are absorbed by the battery pack. This enables both capital and volume savings in a vehicle design process. Moreover, the less cyclic operation of the PEMFC increases its lifetime and reduces the system control challenges [17]. Therefore, an energy management strategy (EMS) is needed to satisfy the energy needs of the SGV, the dynamic response of the system, the state of charge (SOC) of the battery, the level of hydrogen, and the energy efficiency of the system [18–20]. The fuel economy and performance of a hybrid SGV can be affected by the component size, EMS, etc. [21]. Nevertheless, based on the author's knowledge, there is no standard method for analyzing the energy and components sizing of such applications, and it has not been discussed in the literature. Therefore, the design process of an EMS for a hybrid SGV system is a challenging task, given the limited durability and energy efficiency of FC and batteries, as well as the operational constraints of the SGV.

Typically, model-based design is a technique that is applied to design an embedded software by engineers in different applications. However, there is a lack of a well-defined energy model that can act as a real-world indoor vehicle. In [22], a simplified energy equation for a DDMR is derived in terms of the robot's velocity vector. In [23], various motion profiles for automatic floor cleaning mobile robots are presented and their efficiency is discussed. However, a solid energy model is not yet presented. The work presented in [24] is the first step towards the construction of a realistic energetic model, but it does not include the losses within the DC motors. In addition, in [25], the design and efficiency mapping of an electric drive are investigated for a mobile robotic container platform used in industrial halls. However, it is based on a basic battery model that can negatively affect the accuracy of the energetic model.

This paper proposes an EMS for an SGV, by taking the FC size and battery pack degradation into account. In light of the above-discussed papers, modeling of energy sources (batteries and FC) is essential for the design of a hybrid powertrain, which is not thoroughly inspected in the literature. Therefore, a detailed model, including the vehicle dynamic, battery, FC, and motors, is developed in the first place. An appropriate energy model is worthwhile for tuning the EMS and component sizing, before the implementation in a real platform. By using a realistic energy model, it is possible to calculate how much energy is needed to work in a given path or for a specific period (e.g., during an 8 h working shift). Consequently, the range-extender capacity (required power and fuel tank capacity) can be sized by considering the available energy of the onboard battery and the EMS. Subsequently, an EMS composed of three modes is proposed for the studied SGV. The EMS between the battery and the secondary source (FC) might affect the battery lifespan and the overall energy efficiency of the hybrid DDMR. Hence, the objective of this study was the evaluation of the impact of the FC sizing and battery degradation on an industrial SGV to operate for an entire 8 h shift time.

The rest of this paper proceeds as follows. The project background, modeling, developed method, and EMS are described in Section 2. In Section 3, real-world tests are described; Section 4 presents the simulation results and their experimental validation by empirical data. Finally, the conclusions are drawn in Section 5.

2. Materials and Methods

2.1. Project Background

In this research, an industrial battery-powered SGV was considered as a case study for developing an FC/battery hybrid electric SGV (Figure 1) for warehouse applications. The driving speed was limited to 1.8 m s^{-1} to avoid dangerous collisions between the mobile vehicle and staff. The DDMR powertrain consists of the two electric motors with single-speed gearboxes for each driven wheel. The nominal power range of each motor was between zero and 180 W. In this context, the manufacturer neglects the regenerative braking system due to the low-speed application. Based on the experimental tests, the maximum operation time of the SGV without load was around 3.5 h at 1 m s^{-1} average speed. The SGV is equipped with two lasers sensors, two encoders, and an inertial measurement unit (IMU) for autonomous navigation. In addition, an onboard industrial computer was utilized to compute localization and navigation algorithms.

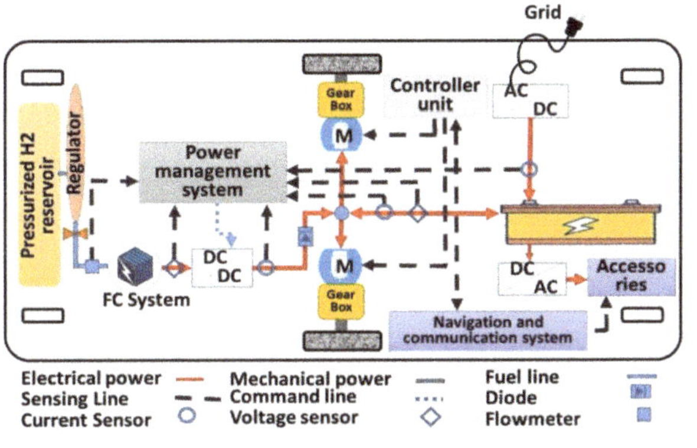

Figure 1. Hybrid self-guided vehicle (SGV) schematic diagram.

Different levels of modeling such as physical models, lookup table data, and efficiency maps were considered to model the battery pack, the FC, the electric machines, etc. Each model was based on the manufacturer's datasheets and the experimental test results. The methodology for each component modeling is described in the following sections.

2.2. The DDMR Powertrain Modeling

Figure 2 shows a schematic view of DDMR with two drive wheels and four castors that were added for balance. Each drive wheel could independently be driven either forward or backward. Subsequently, the robot's trajectories could be varied by applying different speeds on the drive wheels.

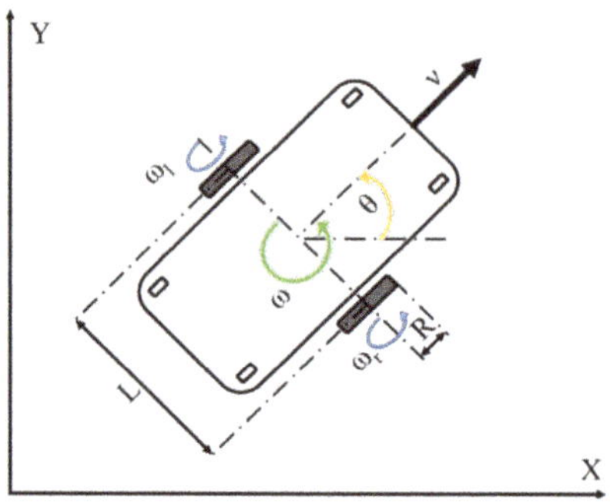

Figure 2. Kinematic model of a differential drive mobile robot (DDMR).

Regarding Figure 2, ω_r and ω_L present the rotational velocities of the right and left wheels, respectively, which were related to the linear v, and rotational ω velocities of the robot, by the following equations:

$$v = \frac{R(\omega_r + \omega_l)}{2} \tag{1}$$

$$\omega = \frac{R(\omega_r - \omega_l)}{L} \tag{2}$$

where R represents the wheel's radius, and L symbolizes the distance between the two wheels. The total energy consumption ($E_{Tot.}$) related to the DDMR motion could be formulated as the following equations:

$$E_{Tot.} = E_{Kin.} + E_{roll} + E_{Others}, \tag{3}$$

where $E_{Kin.}$, E_{roll}, and E_{Others} denote the energy needed to overcome the kinetic, rolling resistance losses, and other accessories, such as sensors consumption energies, respectively. The kinetic energy equation was obtained by [24]:

$$E_{Kin.} = \int_{t0}^{tf} \frac{1}{2}\, mv(t)^2 + \frac{1}{2}\, I\, \omega(t)^2\, dt \tag{4}$$

where $v(t)$ and $\omega(t)$ represent the linear and angular velocities of the robot from initial time (t_0) to a final time (t_f), m is the robot mass, I signifies the moment of inertia of the robot. Rolling resistance, which is caused by the slight deformation of the ground or the wheels at the contact point with the ground, depends on the weight of the robot (m) and the coefficient of rolling friction (μ). The E_{roll} is expressed by:

$$\mathrm{E}_{Roll} = \int_{t0}^{tf} \frac{1}{2}\, \mu\, m\, g\, (r\omega_R(t)) + \frac{1}{2}\, \mu\, m\, g\, (r\omega_L(t))\, dt, \tag{5}$$

In order to estimate the rolling resistance coefficient, the coast-down method is commonly used in vehicle applications [26]. According to this technique, the DC motors are disengaged from the wheels, then the robot is freely rolled with an initial velocity of 1 m s^{-1}. The robot slowly comes to rest, which takes almost five seconds.

$$\mu = \frac{v_i}{gt} \tag{6}$$

where g represents gravity acceleration. Accordingly, μ is estimated to be 0.02, which makes sense, since the robot was moved on a flat cement surface.

In order to model the DDMR movement, an equivalent total mass M_{tot} (mass of the vehicle and the equivalent mass of the rotating parts) was considered. The vehicle velocity v_{ev} was obtained using Newton's second law of motion, with the traction and resistive forces, F_{tr} and F_{res}:

$$M_{tot.} \frac{d}{dt} v_{ev} = F_{tr} - F_{res} \tag{7}$$

$$F_{res} = C_{roll} M_{tot.} g \cos \alpha + 0.5\, \rho\, C_d\, A\, (v_{ev} + v_w) + M_{tot.}\, g \sin \alpha \tag{8}$$

where C_{roll} denotes the rolling resistance coefficient, g is the acceleration due to gravity, α is the slope angle, C_d is the air drag coefficient, A represents the frontal area of the vehicle. However, the air drag coefficient and wind velocity (v_w) are considered to be zero due to the low-speed indoor application of the DDMR. The main parameters of the case study DDMR are listed in Table 1. These parameters were defined based on experimental tests and components datasheet from the manufacturer. Furthermore, the main subsystem models of the hybrid DDMR are described in the following sections.

Table 1. The simulation parameters of the main components.

Parameter	Symbol	Value
Rolling resistance coefficient	C_{roll}	0.02
Drive wheel radius	R	0.197 m
Wheelbase	L	0.72 m
Total mass	m	100 kg
Moment of inertia for motor's rotor	I	0.3 kg m^2
Linear acceleration	a	1 m s^{-2}
Rotational acceleration	ω	1 Rad. s^{-2}

The total efficiency (η_{tot}) of the DDMR powertrain could be calculated as the ratio between the total net kinematic energy ($E_{Kin.}$) from Equation (4) and the vehicle total energy ($E_{Tot.}$) from Equation (3). Thus, the η_{Tot} is expressed as follows:

$$\eta_{tot} = \frac{E_{Kin.}}{E_{Tot.}} \times 100 \tag{9}$$

2.3. Traction Subsystem

In the case of the DDMR, there are two similar motors that were independently modeled in the current study. Considering the speed (v_{ev}) and the required traction force (F_{tr}), the vehicle requested power (P_m) from the electric motor side could be expressed as:

$$P_m = \frac{F_{tr}\, v_{ev}}{\eta_m \eta_t}, \tag{10}$$

where η_{m} and η_{t} denote the motor and transmission efficiency, respectively. For simplification purposes, a static model was considered for the traction subsystem, according to the efficiency map. The electric machines and power electronic components, such as converters and motor drives, were modeled based on the experimental lookup table data provided by the manufacturer. The look-up table determined the electric motor torque, speed, and related efficiency.

Demanded current (I_{TS}) by the traction system was related to the traction force F_{tr} from the wheels speed (v_{wheel}), which could be derived from the following expression:

$$I_{TS} = \frac{F_{tr}\, v_{wheel}}{V_{Batt.}\, (\eta_{tr})^k} \; With\; k = \begin{cases} 1\; if\; P_{tr} > 0 \\ -1\; if\; P_{tr} < 0 \end{cases} \tag{11}$$

where V_{Batt}. and η_{tr} denote battery voltage and traction system efficiency, respectively. k is the coefficient that is influenced by traction power depending on the vehicle acceleration or deceleration.

2.4. Energy Storage Subsystem (Battery)

SOC is one of the most important battery parameters. This parameter needs to be carefully monitored to avoid damaging the battery pack, owing to overcharging/discharging. Several studies were done to estimate the battery *SOC* level [27–29]. Among the various methods, coulomb counting based on current integration remains one of the most commonly used methods, due to its reasonable accuracy and implementation simplicity [28,30]. The *SOC* of the battery ($SOC_{Batt.}$) was calculated by the following expression:

$$SOC_{Batt} = SOC_{Init.} - \frac{100}{3600\ Q_{Batt.}} \int_0^t I_{Batt.}\ dt \tag{12}$$

where $SOC_{Init.}$ is the initial *SOC* of the battery, $Q_{Batt.}$ and $I_{Batt.}$ denote battery capacity and currents, respectively.

The battery was modeled using an open-circuit voltage V_{OC} with a series resistance (R_S), and a parallel combination of resistance capacitance ($R_c\ C_c$), as presented by the following equation [31].

$$I_{Batt.} = \frac{V_{OC} - R_S.I_{Batt.} - V_{Batt.}}{R_C} + C_c \frac{d}{dt}\left(V_{OC} - R_S I_{Batt.} - V_{Batt.}\right) \tag{13}$$

Battery total power ($P_{Batt.}$) was determined by using resistance ($R_{Batt.}$) and open-circuit voltage (V_{OC}) from the following equation [32]:

$$P_{Batt.} = V_{OC}.I_{Bat.} - I_{Bat.}^2 \cdot R_{Batt.} \tag{14}$$

According to a recent research on battery degradation, reported in reference [5], the SOH level of a lithium-ion battery reached a low of 75 percent after 1500 charging and discharging cycles at standard conditions. Therefore, a battery degradation model from [5] was considered to be a typical reference for the battery *SOH* level in this study. A summary of battery specifications is listed in Table 2.

Table 2. Summary of battery specifications [33].

Parameter	Value
Battery nominal voltage and capacity	25.6 V, 40 Ah
Charging Voltage	29.2 VDC
Charging Current	4–8 A
Open Circuit Voltage Range	29.2 VDC
Maximum Continuous Discharge Current	40 A
Maximum Peak Discharge	80 A (3 Sec.)
Operating Temperature	−10 °C to 60 °C

2.5. Fuel Cell as Range Extender

The FC is considered to be a voltage source, using its static polarization curve, based on a semi-empirical model that was developed by Squadrito et al. [34]. The cell voltage (V_{cell}) was determined by subtracting activation ($\xi_{act.}$), Ohmic (ξ_{Ohm}.), and concentration ($\xi_{conc.}$) overvoltage from the reversible cell potential (E_{nernst}). The detailed explanation of the PEMFC model could be found in [35]:

$$V_{FC} = E_{nernst} - \xi_{act} - \xi_{ohm} - \xi_{conc} \tag{15}$$

$$V_{FC} = V_0 - blog\left(i_{fc}\right) - ri_{fc} + \alpha \left(i_{fc}\right)^k log\left(1 - \beta i_{fc}\right) \tag{16}$$

where V_0 denotes the thermodynamic voltage of the FC. The activation overvoltage is the sum of the cathode and anode overvoltage presented by $blog\ (i_{fc})$. The Ohmic overvoltage, indicated by ri_{fc}, is characterized as the resistance of the membrane and the electrodes. It mainly depends on the resistance of the internal membrane (r) and the FC current. The concentration overvoltage occurs when the FC is faced with a high current density. α and k variables are the parameters related to the diffusion process and the dimensionless number related to the phenomenon of water flooding, respectively. β is the inverse of the limiting density current [35]. In addition, a static characteristic represents the hydrogen (H_2) mass flow versus the current linear function that describes the FC system consumption. The hydrogen flow was determined on the basis of experimental data, by a first-order function approximation, where a and b represent fitting parameters [36].

$$\dot{m}_{H2} = a + bi_{fc} \tag{17}$$

In order to take into account the added onboard energy source, the SOC of the H_2 tank (SOC_{Tank}) was calculated by the following equation:

$$SOC_{\ Tank} = \frac{m_{H2-init} - \int \dot{m}_{H2} i_{fc}}{m_{H2-init}} \tag{18}$$

where $m_{H2\text{-}init}$ is the initial mass of H_2 [g], $\dot{m}_{H2}$ is the H_2 mass flow [g s^{-1}], and i_{fc} is the FC current. H_2 consumption is supposed to have a constant rate when the FC is turned ON. The FC system is composed of the DC–DC converter, a smoothing inductor, and a boost chopper for its current control [37]. Nevertheless, their energetic performances are included in the FC static characteristics. The following equations are used for the converter unit model:

$$L\frac{d}{dt}I_{fc} = V_{fc} - V_{hfc} - r_L I_{fc} \tag{19}$$

$$\begin{cases} V_{hfc} = m_{hfc}V_{bat} \\ I_{hfc} = m_{hfc}I_{fc}\ \eta_{hfc} \end{cases} \quad With\ k\ = \begin{cases} 1\ if\ P > 0 \\ -1\ if\ P < 0 \end{cases} \tag{20}$$

where L is the converter inductance, V_{fc} is the output voltage of FC, V_{hfc} is the input voltage in the chopper, r_L is the converter resistance, m_{hfc} is the modulation ratio, and η_{hfc} = 95% is the average efficiency of the converter.

The power of the FC (P_{FC}) as a range extender could be computed from the H_2 lower heating value (LHV_{H2}), as follows:

$$P_{FC} = LHV_{H2} \cdot \dot{m}_{H2} \cdot \eta_{FC} \tag{21}$$

It should be noted that the SGV system has other components such as onboard electronics, sensors, microcontrollers, which are the accessory parts of the hybrid SGV powertrain. These components are extremely efficient nowadays, but they still consume a portion of the battery's current. Therefore, the current loss due to the electronics was presented as I_{Others}.

The Kirchhoff's current law used to model the parallel connection between the battery pack, traction motors (I_{TS}), FC (I_{FC}), and I_{Others} is given by Equation (22):

$$I_{Batt.} = I_{TS} - I_{FC} + I_{Others} \tag{22}$$

2.6. Energy Management Strategy (EMS)

An EMS with three modes of operation is proposed in this study (Figure 3). The first mode is known as charge-sustaining (CS), which maintains the battery SOC at the desired level such that the FC starts to supply its maximum power when the battery SOC drops to the minimum threshold of 40%. For the second mode, called charge-depleting (CD), the battery is the main power source until the FC starts providing its maximum power as the battery SOC reaches its minimum threshold (20%). The third mode is called charge-blending (CB), which emphasizes the FC efficiency by turning ON the FC when the battery SOC reaches the threshold value of 60%. For the CB mode, the FC supplies constant power corresponding to its maximum efficiency. For the three modes, the FC is turned OFF when the battery SOC reaches 85%. Since each strategy employs the FC within a certain range, it is important to size energy resources appropriately, to avoid energy shortages. The following design considerations are established for the proposed EMS:

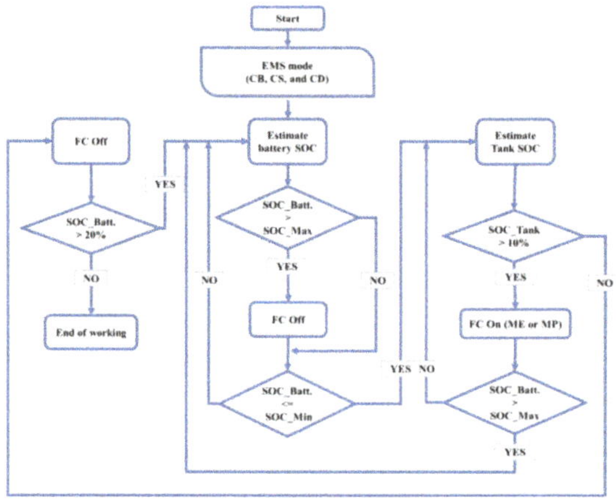

EMS mode	SOC_Min	FC mode
CS	40%	Maximum power (MP)
CD	20%	Maximum power (MP)
CB	60%	Maximum efficiency (ME)

Figure 3. Flow chart of the proposed energy management strategy (EMS).

(a) According to [38], the minimum battery degradation can be achieved if the *SOC* is maintained over 40%.

(b) When the SOC_{Tank} is less than 10%, the tank does not have enough pressure to provide H_2 to the FC.

A MATLAB Simulink (Version 2017, MathWorks company, Massachusetts, USA) model was developed to present the mathematical models of the different SGV components. The MATLAB Simulink model, presented in Figure 4, includes the different subsystems such as the SGV powertrain, traction system, battery pack, fuel cell, and energy management strategies.

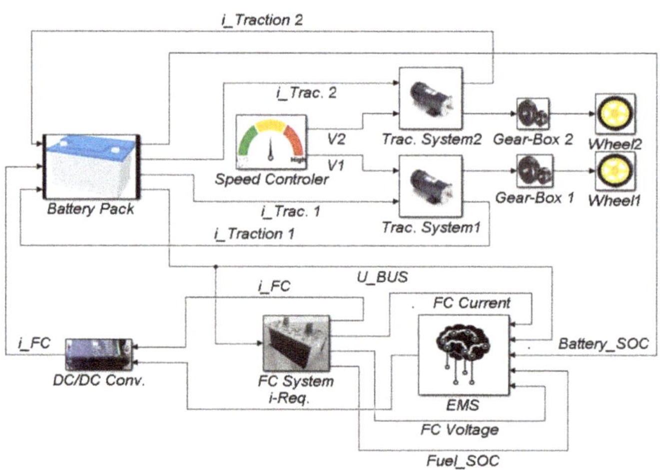

Figure 4. The overall Matlab Simulink model of the SGV.

3. Experimental Tests

Figure 5 presents the studied SGV. For measuring the actual required power and energy consumption, the SGV was moved using different velocity profiles, including the pure transition in the X-direction, forward and backward, trapezoid speed profile, circular movement, and rotation around the center of gravity, as demonstrated in Table 3.

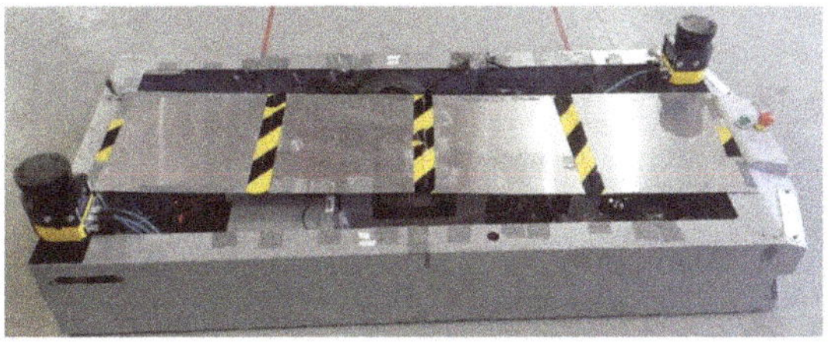

Figure 5. The studied SGV.

Table 3. Experimental working cycle parameters.

Measured Parameters	Motion Type				
	Transition in the X-Direction			Rotational	
	Trapezoid	Backward	Forward	Rotation Around the Center	Circular Pathway
	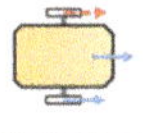	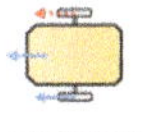	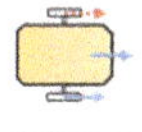	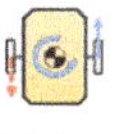	
Wheels speed profile (Solid line: right wheel speed Dashed line: left wheel speed)					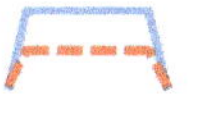
Average speed of the right wheel (Rad s^{-1})	1.4	1.5	1.5	0.57	1.3
Average speed of left wheel (Rad s^{-1})	1.4	1.5	1.5	−0.57	2
Average linear speed (m s^{-1})	0.85	0.9	0.9	0	0.32
Average linear acceleration (m s^{-2})	0.7	−0.9	0.9	0	0.9
Average rotational velocity of the vehicle (Rad s^{-1})	0	0	0	0.31	0.19

Each motion test was conducted for 11 s and repeated three times in the same condition at flat cement surface. In addition, each experiment included three sections, acceleration from stationary, constant velocities, and deceleration to stationary. For example, in the trapezoidal velocity profile scenario, maximum linear velocity was considered to be 1.4 m s^{-1}, to achieve the high efficiency of the electric motors. Similarly, the average linear acceleration from rest to maximum speed and vice versa was adjusted to 0.7 m s^{-2}, to prevent high mechanical and electrical stresses. The maximum linear acceleration for forward, backward, and circular pathway were fixed to 0.9 m s^{-2}.

Moreover, in the industrial environment, the SGVs were usually employed in stop-and-go loop working conditions. Therefore, a typical pathway was designed by considering the mixture of transitional and rotational movements, as shown in Figure 6. Measured data from this pathway was used to evaluate the SGV model. In this context, the battery voltage and current, motor power, and the velocity of the wheels were recorded by the developed data logger with a 0.1 s sample rate.

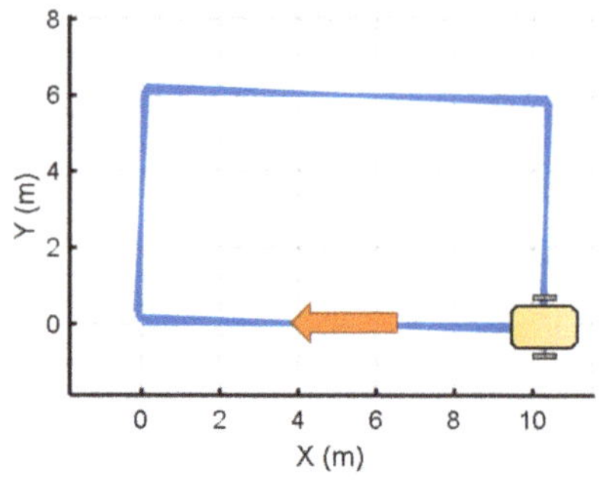

Figure 6. The real-world mixed working cycle pathways.

4. Result and Discussions

4.1. Model Validation

The proposed FC model was experimentally validated in our previous work [36,39,40]. The performance of the Squadrito model was thoroughly investigated with regards to the estimation of the polarization curve for different FCs, under various operating temperatures [36]. A current profile was applied to the FCs, and the measured data (current, temperature, and voltage) were transferred to an online identification algorithm to tune the parameters of the models in real-time, by minimizing the error between the estimated voltage and the measured one. Once the parameters of the model were estimated, the model was used for predicting the polarization curve of the stack. The conducted study showed that the proposed semi-empirical FC model was accurate regarding the polarization curve prediction for different nominal power FC in the above and below zero temperature conditions [36,39].

In order to validate the dynamic battery model, a test bench was developed at the UQTR Hydrogen Research Institute (Figure 7). The test bench was composed mainly of a 24 V Lithium battery. The charging/discharging process and the acquisition of data (temperature, current, voltage) were performed through an embedded computer NI CompactRIO 9022 (Version 2017, National Instruments, Texas, USA). A charger and a programmable load were used in order to charge and discharge the battery. The battery being tested was placed inside an environmental chamber, which was used to set the ambient temperature.

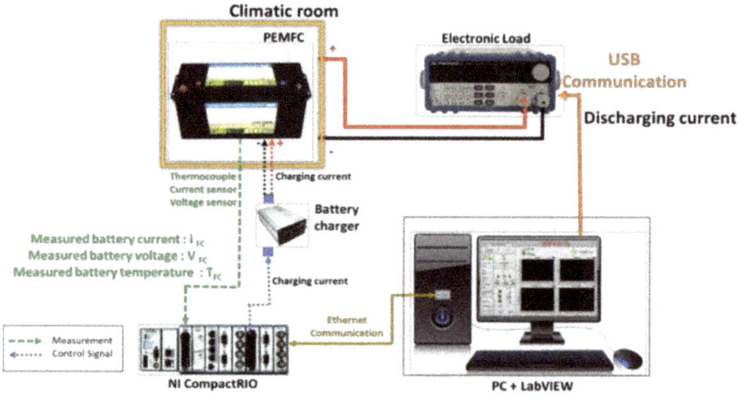

Figure 7. Battery test bench.

The proposed test bench was used to validate the dynamic battery model by comparing experimental and simulation results during the charging and discharging process. Figure 8 presents battery voltage, charging/discharging current, battery *SOC*, and error between experimental and simulation data. Results showed that the proposed battery model was accurate regarding the charging and discharging curves.

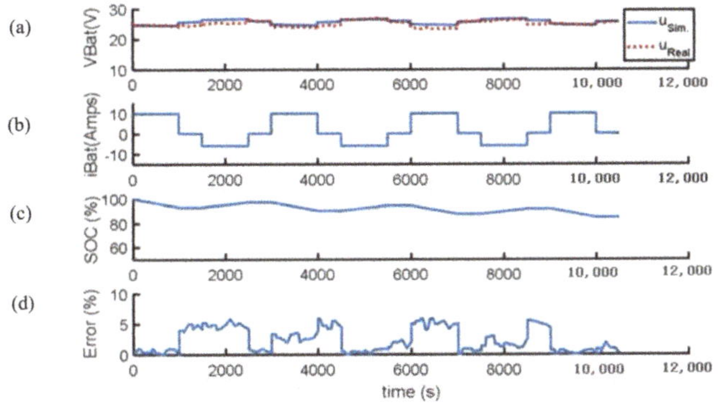

Figure 8. Battery model validation during the charging and discharging process. (**a**) battery voltage; (**b**) battery current; (**c**) battery *SOC*; (**d**) calculated error between measured data and simulation.

To evaluate the overall SGV model, a comparative study was performed between the simulation results and the extracted experimental data. In a first step, the SGV was tested without a load while the wheels were off the ground. The experimental data and simulation results are shown in Figure 9. Figure 9a compares the battery voltage from the real-world test and simulation, and Figure 9b compares the consumed current by the motors. These results confirmed a good synergy between the energy storage system model and the real battery-powered SGV performance.

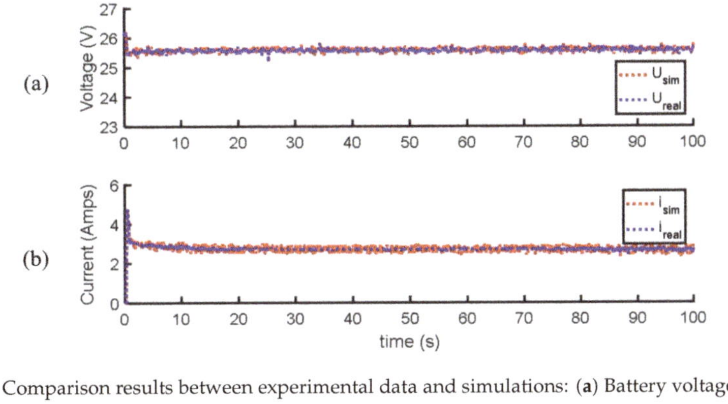

Figure 9. Comparison results between experimental data and simulations: (**a**) Battery voltage, and (**b**) current by traction motors.

Table 4 presents the experimental and simulation results for the different proposed motion (trapezoid, backward, forward, rotation around the center, and circular pathway). Table 4 shows that the average requested power, maximum demanded power, consumed energy, and average efficiency calculated by the SGV energetic model agreed with the experimental results. Consequently, the highest requested power (509.5 W) was obtained for the backward and forward movements caused by the high acceleration rate of 0.9 m s^{-2}. However, the lowest requested power (235.1 W) was obtained for the trapezoid movement scenario, which was due to the low acceleration rate. Therefore, the acceleration rate affected the requested power and subsequently the size of the energy source.

Table 4. Average value of the experimental and simulation results of the defined scenarios.

Motions Condition	Movement Type	Ave. Demanded Power (W)	Max. Demanded Power (W)	Energy Consumption (Wh)	Overall Efficiency (%)
Transitional in X-direction	Forward	292.7 * 283.5 **	508.8 * 497.3 **	0.322 * 0.312 **	34.3 * 33.5 **
	Backward	293.6 * 283.6 **	509.5 * 498.5 **	0.323 * 0.312 **	33.9 * 32.4 **
	Trapezoid speed profile	256.3 * 246.1 **	235.1 * 242.2 **	0.282 * 0.271 **	37.4 * 35.9 **
Rotational	Rotation around the center	531.8 * 509.7 **	264.4 * 274.8 **	0.585 * 0.561 **	19.3 * 20.2 **
	Circular pathway	1053.6 * 1088.1 **	457.7 * 450.3 **	1.159 * 1.069 **	38.07 * 36.9 **

* Experimental results. ** Simulation results.

On the other hand, the highest energy consumption (1.159 Wh) was related to the circular movement, compared to the trapezoid movement scenario with the lowest energy consumption (0.282 Wh), which might happen because of the combined transitional and rotational movements and variation of electric motors' efficiency in different speeds. The comparison of results between the rotation around the center of gravity and the transitional movements showed that the rotational movement required almost two times more average power than the transitional one. From these results, it could be concluded that the use of the SGV in the range of the operating speed with the maximum efficiency of the electric motors is important. It also highlights the importance of component sizing and optimal path planning of the SGV, which can affect the overall efficiency of the energy system. Hence, it is important to provide a practical working cycle to estimate the required energy and power for selecting an appropriate FC range extender.

4.2. Mixed Working Cycle Analysis

In order to evaluate the performance of the SGV in a more realistic condition, the mixed working cycle was analyzed in this section. The measured wheel speed from the experiment, which was used as input of the SGV model, is shown in Figure 10a. In this regard, the linear velocity and the angular speed, battery voltage, battery current, and traction power of the developed model of the vehicle were compared to the experimental results, as shown in Figure 10b–f. Hence, the validity of the developed model was investigated by an adequate accuracy between the measured values and the simulation results.

Furthermore, the traction power from the experimental working cycle (Figure 10f) showed that the power peak (+470 W) was reached during acceleration from 0 to 0.9 m s^{-1}, while the average requested power was around 160 W. In fact, the demanded power of the SGV to accelerate from 0 to 0.9 m s^{-1} mainly corresponded to the high starting currents by the electric motors. It was obvious that due to acceleration, the maximum required current (+23 Amps) was obtained in the bingeing, which was around five times higher than the constant speed. Moreover, in the brake mode, a reverse current was generated (sent back to the battery), which might be wasted if it was higher than the battery charging current (Table 2). Therefore, the movement pattern (stop and go) could affect the powertrain efficiency of the SGV. As a result, considering the minimum stop and go situation along with appropriate acceleration and deceleration rate, is recommended, to decrease SGV power requirement while designing the working cycle for an industrial application.

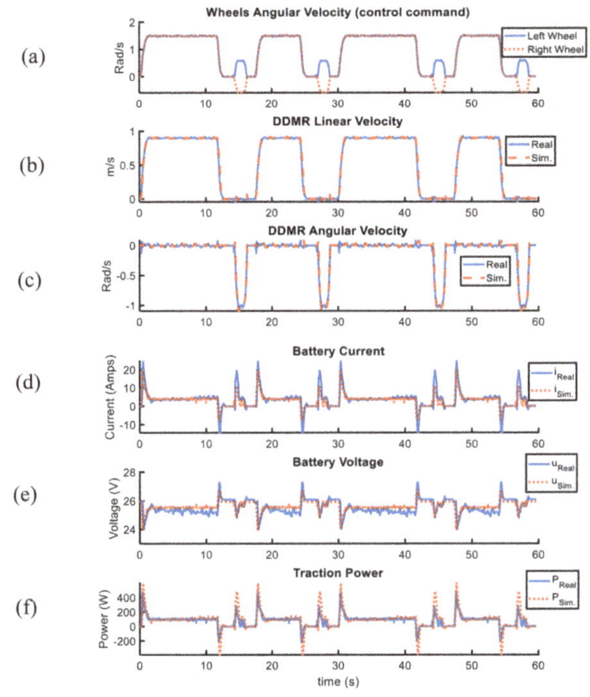

Figure 10. Comparison results of experimental data from physical DDMR versus the model. (**a**) Wheels speed (**b**) linear speed, (**c**) angular speed, (**d**) battery voltage, (**e**) battery current, and (**f**) traction power.

Furthermore, Figure 11 illustrates the comparison results between the transitional and rotational movement of the mixed work cycles. Figure 11a,b show that the amount of 0.396 Wh (94,73%) and 0.022 Wh (5.26%) were consumed for the transitional and rotational movements, respectively. It was found that a significant portion of the vehicle's stored energy was spent for one working cycle with only 360-degree rotational movement. The rotational movement might be increased in crowded working environments where humans and SGVs work together. In fact, the vehicle will need more alternate rotation and stop-and-go mode to avoid obstacles, by finding an appropriate path to reach the desired position. It is important to note that the overall rotational efficiency was around 20% compared to 33% of transitional motion (Figure 11a). Therefore, in order to increase the system's overall autonomy, it was important to minimize SGV rotational motion.

Motions condition	Energy Consumption (Wh)	Overall Efficiency (%)
Transitional	0.396	35.8
Rotational	0.022	19.8

(**a**)

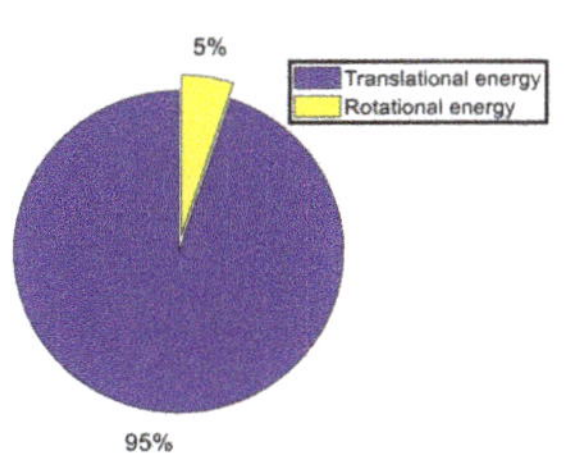

(**b**)

Figure 11. (**a**) Comparison of the transitional and rotational movement using experimental data. (**b**) Percentage of consumed energy for transitional and rotational movement.

Regarding the limited autonomy of the battery-powered system, the FC was suggested as a range extender. The selection of an appropriate FC in terms of the nominal power size and onboard hydrogen tank capacity are important. For instance, if the FC nominal power selected was smaller than the requirement, the system might face a lack of energy. On the other hand, a system with an oversized FC might lead to a higher system cost and fuel consumption. Moreover, the EMS might affect the size and lifetime of the energy sources.

4.3. Impact of FC Sizing and Battery Degradation on EMS Performance

Industrial SGVs usually work on a relatively specific path in a controlled indoor environment, which makes energy consumption estimation possible by employing a realistic model during the specified work hours. Therefore, by knowing the required power and energy and considering the available energy from the battery, the requisite energy from the range extender could be calculated. In addition, the FC system lifetime was enhanced if its load fluctuations were reduced and if frequent starts and stops of the FC were avoided [41].

In order to evaluate the impact of FC sizing on EMS performance, the mixed working cycle was repeated many times to reach an 8 h working shift. In this context, the measured velocity profile for the mixed working cycle was imposed to the proposed model as an input. After this, the performance of the proposed FC/battery SGV powertrain was validated under the three EMS modes, including the CD, CS, and CB. Then, each EMS mode was investigated by considering four different nominal FC power (300 W, 200 W, 160 W, and 100 W).

For instance, Figure 12a indicates the impact of different EMS modes on the *SOC* of the battery with a 160 W FC as a range extender. Figure 12b,d,f show the power distribution between power sources in the CB, CS, and CD modes (respectively) with a 160 W FC for the entire 8 h work shift under the mixed working cycles. Furthermore, Figure 12c,e,g present the total supplied energy from each source for the entire working shift in the CB, CS, and CD modes, respectively.

Results show that the suggested hybrid system would fail in the CD mode. As mentioned before, in the CD mode, the battery pack supplies the required power without the assistance of the FC until the battery *SOC* reaches the low battery threshold level (20%). Therefore, the system failure occurs after 5 h in the CD mode, due to the insufficient available power of the battery and FC, which cannot alone support the power peak requirement of the SGV (+470 W). This meant that an FC system with a higher power (i.e., 500 W) was needed to satisfy the power peak of the SGV during the acceleration with the CD mode, to accomplish the 8 h working shift.

On the other hand, in the CS mode, the FC was turned ON at its maximum power when the battery *SOC* was around 40%, to ensure battery *SOC* sustenance until the end of work. Indeed, CS strategy acted similar to the CD mode in the beginning, if the battery was fully charged. Nevertheless, due to the higher battery *SOC* threshold level, the CD mode kept the battery *SOC* higher than the minimum authorized *SOC* (20%) during the 8 h working shift, to supply the required peak power. It was important to note that when the FC was ON, it charged the battery during low power demand periods. Since the FC operates at its maximum power rate during the CD and CS modes, it leads to high fuel consumption and FC degradation.

In the CB mode, when the *SOC* reaches the threshold value of 60%, the FC supplies a constant power while the battery supplies the required peak power. Accordingly, the CB strategy allows turning ON the FC stack in its maximum fuel efficiency with a minimum ON/OFF cycle, to minimize hydrogen consumption. Moreover, it prevents the battery *SOC* to reach the minimum threshold before the end of the working shift and avoids frequent power source fluctuations that could affect the FC lifespan. Therefore, it might increase the battery and FC lifetime by reducing their degradation. It should be noted that if a regenerative braking system is considered, up to 45 Wh of energy could be recovered (around 3.5% of the total consumed energy).

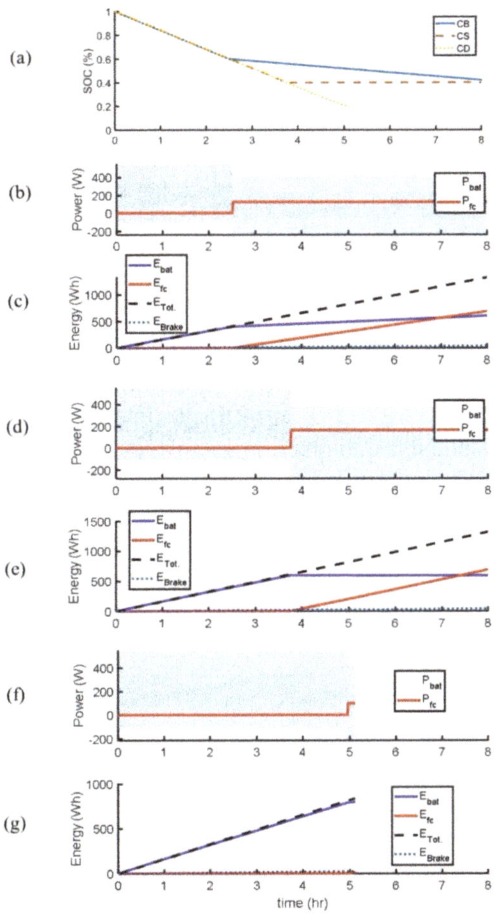

Figure 12. (**a**) Battery *SOC* levels with the three strategies, (**b**) battery and FC power in the CB strategy, (**c**) provided energy by each energy sources during CB strategy, (**d**) battery and FC power in the CS strategy, (**e**) provided energy by each energy sources during CS strategy, (**f**) battery and FC power in the CD strategy, and (**g**) provided energy by each energy sources during the CD strategy.

As a general outcome, Table 5 lists the comparison results for each EMS with the four different nominal FC power and three battery charge and discharge cycles (0, 750, and 1500), in terms of the working time, H_2 consumption, energy provided by sources, and final *SOC* of the battery. These results showed that the system could not meet an 8 h working shift in the CD mode with different FC power sizes, due to the insufficient power supplied by both energy sources for the power peak requirement (+470 W). Moreover, an FC with 100 W nominal power was not able to supply the energy needed to operate the SGV for the entire 8 h shift, with all the proposed EMSs.

Table 5. Comparison results of energy sources with different EMS and battery lifetime.

EMS Mode	FC Nominal Power (Max Power) (W)	Battery Charge and Discharge Cycle	Working Time (s) (h)	H_2 Cons. (gr.)	Provided Energy by FC (Wh)	Provided Energy by Battery (Wh)	Batt. SOH (%)	Batt. Final *SOC* (%)
Charge Depleting (CD)	100	0 [a]	18,411 (5.114)	+0.57	16	796	100	20
		750	16,040 (4.455)	+0.57	16	691	90	20
		1500	12,767 (3.546)	+0.56	15	347	75	20
	160	0	18,615 (5.120)	+0.72	26	798	100	20
		750	16,125 (4.459)	+0.71	25	693	90	20
		1500	12,753 (3.548)	+0.68	24	352	75	20
	200	0	18,658 (5.128)	+1.03	33	802	100	20
		750	16,153 (4.467)	+0.98	32	697	90	20
		1500	12,823 (3.555)	+0.86	30	355	75	20
	300	0	18,658 (5.132)	+1.53	50	810	100	20
		750	16,153 (4.477)	+1.46	48	703	90	20
		1500	12,823 (3.562)	+1.29	45	359	75	20
Charge sustaining (CS)	100	0	25,460 (7.072)	+12	330	798	100	20
		750	23,088 (6.413)	+12	330	693	90	20
		1500	19.813 (5.505)	+12	330	349	75	20
	160	0	28,800 (8)	+24	689	591	100	39.5
		750	28,800 (8)	+28	780	502	90	38.5
		1500	28,800 (8)	+33	925	359	75	38.3
	200	0	28,800 (8)	+30	844	440	100	54.7
		750	28,800 (8)	+35	975	310	90	56.9
		1500	28,800 (8)	+44	1156	132	75	60.1
	300	0	28,800 (8)	+42	1186	105	100	85.5
		750	28,800 (8)	+42	1184	106	90	75.2
		1500	28,800 (8)	+42	1185	104	75	60.7
Charge Blending (CB)	100	0	26,629 (7.396)	+10	381	800	100	20
		750	24,258 (6.738)	+10	380	695	90	20
		1500	20,986 (5.829)	+10	382	551	75	20
	160	0	28,800 (8)	+19	683	598	100	40
		750	28,800 (8)	+22	845	422	90	37.6
		1500	28,800 (8)	+26	878	406	75	34.3
	200	0	28,800 (8)	+23	853	430	100	56.3
		750	28,800 (8)	+26	955	329	90	59.9
		1500	28,800 (8)	+32	942	342	75	40
	300	0	28,800 (8)	+28	1032	245	100	72.1
		750	28,800 (8)	+31.5	1192	96	90	77.1
		1500	28,800 (8)	+36.7	1414	−121	75	84

([a] means new battery).

For the CD strategy, the proposed FC power needs to be greater than the SGV power picks during acceleration, to reach 8 h of working time. However, for the CS and CB strategies, an FC with a minimum nominal power of 160 W is mandatory to attain 8 h of working time with minimum hydrogen consumption. A lower FC nominal power could be a good choice if the objective was to reduce system cost and energy cost because hydrogen energy is costly, compared to the battery charging station energy. It is also important to note that FC presents a lower efficiency (around 55%) compared to battery efficiency (around 90%) and the slow dynamic characteristics of the FC cause a problem for an SGV application, where the power requirements can vary considerably during a mission. In order to increase the system efficiency and to reduce hydrogen consumption, it is better to consider an EMS that uses the FC at its maximum efficiency similar to the CB mode, instead of using the FC at its maximum power as the CS and CD modes. As shown in Table 5, the simulation result illustrates a difference of around 26% in hydrogen consumption for the same provided energy by the 160 W FC system, during the typical 8 h shift working cycles between the CS and CB modes.

Furthermore, Table 5 confirms that it is necessary to consider a higher nominal power of the FC system if the objective is to minimize the energy provided by the battery, in order to reduce battery degradation and battery charging time. In addition, the FC system tank unlike the batteries could be filled in a few minutes and has the potential to provide power over long periods of time, as required for long-term missions.

The battery degradation should be considered by the EMS because it can affect the FC sizing. Accordingly, three battery charge and discharge cycles (0, 750, and 1500) are considered, as shown in Table 5. Results showed that battery degradation affected the capabilities of the EMS, as well as vehicle performance, by reducing the energy storage system capacity. By comparing the results, it could be seen that a 200 W FC system in the CB mode was able to maintain the battery *SOC* at the desired level (40%), even until the end of the useful life of the battery (up to 1500 battery charge and discharge cycles). Moreover, the results showed that battery degradation impact on the vehicle performance was more drastic in the system with a smaller secondary power source because the final *SOC* was usually lower than the desired level. However, it should also be noted that in a system with a larger range extender, a part of the generated power by the FC system might be wasted due to the insufficient capacity of the energy storage system at the end of its useful life. For instance, in a hybrid system with a 300 W FC and the CB mode strategy, around 121 Wh of the generated energy might be wasted (3.14 g of H_2) due to battery aging. Therefore, a tradeoff between FC size and battery capacity, which was affected by the degradation, must be considered on EMS design. To summarize, the FC and hydrogen tank sizing of the hybrid SGV strongly depended on the EMS and the desired autonomy. It was also important to consider the degradation of power sources and the maximum load that could be transported by the SGV during the FC and hydrogen tank sizing.

5. Conclusions

This study analyzed the performance of an industrial SGV EMS for different FC sizes and battery degradation levels. In this regard, a realistic energy model, which was the core of the proposed analysis, was developed for the studied SGV. Different levels of modeling such as physical models, lookup table data, and efficiency maps were considered to model the battery pack, the FC, and the electric machines, etc. The validity of the developed model was confirmed by an adequate accuracy between the measured values from the real-world tests and the simulation results during a specific working cycle. The proposed model was then used to investigate the effect of three EMS modes (CD, CS, and CB) on FC sizing, by considering different nominal FC power (300 W, 200 W, 160 W, and 100 W) and different battery degradation levels.

In industrial environment like warehouses, the SGV execute a predefined cycle with known velocity. Therefore, it is possible to accurately estimate the size of the FC system as a range extender, depending on the EMS and the desired working shift. It is important that the EMS of a hybrid SGV system considers the cost, the limited durability, and energy efficiency of FC and batteries, as well as the operational constraints of the SGV. In this context, it is recommended to avoid oversized sources by considering a small FC as a range extender, to reduce system cost and to maintain the FC in its high efficiency zones with a minimum ON/OFF cycle, leading to efficiency and lifetime enhancement of FC system. Battery *SOC* have to be kept at a high level during SGV operation, to support the FC during SGV acceleration. In order to improve the SGV's overall autonomy, it is also important to minimize the stop and go and rotational SGV motion with appropriate acceleration and deceleration rate.

Some prospects for extending the scope of this study remain as follows:

- Exploring the effect of FC degradation on EMS performance and FC sizing.
- Proposing an online EMS, based on the online identification of the maximum efficiency of the fuel cell system that changes over time and an online battery management system.

Author Contributions: Conceptualization, A.G.; Data curation, A.G. and L.Z.; Formal analysis, A.G.; Funding acquisition, S.K.; Investigation, A.G.; Methodology, A.G., A.A.; Project administration, S.K.; Resources, A.A. and S.K.; Software, A.G.; Supervision, S.K., N.Z. and A.A.; Validation, A.G., A.A., S.K. and N.Z.; Visualization, A.G.; Writing–original draft, A.G.; Writing–review & editing, A.A. and N.Z. All authors have read and agreed to the published version of the manuscript.

Funding: This research received no external funding.

Acknowledgments: The authors would like to thank the Noovelia Research Chair on the Development of Intelligent Navigation Systems for Industrial Guided Vehicles, the Natural Sciences and Engineering Research Council of Canada and the Fondation de l'Université du Québec à Trois-Rivières for their supports.

Conflicts of Interest: The authors declare no conflict of interest.

References

1. Hermann, M.; Pentek, T.; Otto, B. Design principles for industrie 4.0 scenarios. In Proceedings of the 2016 IEEE 49th Hawaii International Conference on System Sciences (HICSS), Koloa, HI, USA, 5–8 January 2016; pp. 3928–3937.
2. Watzenig, D.; Horn, M. *Automated Driving: Safer and More Efficient Future Driving*; Springer: Berlin/Heidelberg, Germany, 2016.
3. Martins, F.N.; Sarcinelli-Filho, M.; Carelli, R. A velocity-based dynamic model and its properties for differential drive mobile robots. *J. Intell. Robot. Syst.* **2017**, *85*, 277–292. [CrossRef]
4. Ehsani, M.; Gao, Y.; Longo, S.; Ebrahimi, K. *Modern Electric, Hybrid Electric, and Fuel Cell Vehicles*; CRC Press: Boca Raton, FL, USA, 2018.
5. Li, X.; Wang, Z.; Zhang, L.; Zou, C.; Dorrell, D. State-of-health estimation for Li-ion batteries by combing the incremental capacity analysis method with grey relational analysis. *J. Power Sources* **2019**, *410*, 106–114. [CrossRef]
6. Thangavelautham, J.; Strawser, D.; Cheung, M.Y.; Dubowsky, S. Lithium hydride powered PEM fuel cells for long-duration small mobile robotic missions. In Proceedings of the 2012 IEEE International Conference on Robotics and Automation, Saint Paul, MN, USA, 14–18 May 2012; pp. 415–422.
7. Eriksson, E.; Gray, E.M. Optimization and integration of hybrid renewable energy hydrogen fuel cell energy systems–A critical review. *Appl. Energy* **2017**, *202*, 348–364. [CrossRef]
8. Thangavelautham, J.; Gallardo, D.; Strawser, D.; Dubowsky, S. Hybrid fuel cell power for long duration robot missions in field environments. In *Field Robotics*; World Scientifi: Singapore, 2012; pp. 471–478.
9. Joh, H.-I.; Ha, T.J.; Hwang, S.Y.; Kim, J.-H.; Chae, S.-H.; Cho, J.H.; Prabhuram, J.; Kim, S.-K.; Lim, T.-H.; Cho, B.-K.; et al. A direct methanol fuel cell system to power a humanoid robot. *J. Power Sources* **2010**, *195*, 293–298. [CrossRef]
10. Kesner, S.B.; Plante, J.-S.; Boston, P.J.; Fabian, T.; Dubowsky, S. Mobility and power feasibility of a microbot team system for extraterrestrial cave exploration. In Proceedings of the 2007 IEEE International Conference on Robotics and Automation, Roma, Italy, 10–14 April 2007; pp. 4893–4898.
11. Lee, S.-Y.; Min, I.-G.; Kim, H.-J.; Nam, S.W.; Lee, J.; Kim, S.J.; Jang, J.H.; Cho, E.; Song, K.H.; Hong, S.-A.; et al. Development of a 600 w proton exchange membrane fuel cell power system for the hazardous mission robot. *J. Fuel Cell Sci. Technol.* **2010**, *7*, 3. [CrossRef]

12. He, X.; Jiang, Y. Review of hybrid electric systems for construction machinery. *Autom. Constr.* **2018**, *92*, 286–296. [CrossRef]
13. Hosseinzadeh, E.; Rokni, M.; Advani, S.G.; Prasad, A.K. Performance simulation and analysis of a fuel cell/battery hybrid forklift truck. *Int. J. Hydrog. Energy* **2013**, *38*, 4241–4249. [CrossRef]
14. Amamou, A.A.; Kelouwani, S.; Boulon, L.; Agbossou, K. A comprehensive review of solutions and strategies for cold start of automotive proton exchange membrane fuel cells. *IEEE Access* **2016**, *4*, 4989–5002. [CrossRef]
15. Vivas, F.; de las Heras, A.; Segura, F.; Andújar, J. A review of energy management strategies for renewable hybrid energy systems with hydrogen backup. *Renew. Sustain. Energy Rev.* **2018**, *82*, 126–155. [CrossRef]
16. Lü, X.; Qu, Y.; Wang, Y.; Qin, C.; Liu, G. A comprehensive review on hybrid power system for PEMFC-HEV: Issues and strategies. *Energy Convers. Manag.* **2018**, *171*, 1273–1291. [CrossRef]
17. Hannan, M.; Azidin, F.; Mohamed, A. Hybrid electric vehicles and their challenges: A review. *Renew. Sustain. Energy Rev.* **2014**, *29*, 135–150. [CrossRef]
18. Garcia, P.; Fernandez, L.M.; Garcia, C.A.; Jurado, F. Energy management system of fuel-cell-battery hybrid tramway. *IEEE Trans. Ind. Electron.* **2009**, *57*, 4013–4023. [CrossRef]
19. Jiang, W.; Fahimi, B. Active Current Sharing and Source Management in Fuel Cell–Battery Hybrid Power System. *IEEE Trans. Ind. Electron.* **2010**, *57*, 752–761. [CrossRef]
20. Segura, F.; Andújar, J.M.; Durán, E. Analog current control techniques for power control in PEM fuel-cell hybrid systems: A critical review and a practical application. *IEEE Trans. Ind. Electron.* **2010**, *58*, 1171–1184. [CrossRef]
21. Torres, J.; Gonzalez, R.; Gimenez, A.; Lopez, J. Energy management strategy for plug-in hybrid electric vehicles. A comparative study. *Appl. Energy* **2014**, *113*, 816–824. [CrossRef]
22. Kim, C.H.; Kim, B.K. Minimum-energy translational trajectory generation for differential-driven wheeled mobile robots. *J. Intell. Robot. Syst.* **2007**, *49*, 367–383. [CrossRef]
23. Mei, Y.; Lu, Y.-H.; Hu, Y.C.; Lee, C.G. Energy-efficient motion planning for mobile robots. In Proceedings of the IEEE International Conference on Robotics and Automation, ICRA'04, New Orleans, LA, USA, 26 April–1 May 2004; Volume 5, pp. 4344–4349.
24. Liu, S.; Sun, D. Minimizing energy consumption of wheeled mobile robots via optimal motion planning. *IEEE/ASME Trans. Mechatron.* **2013**, *19*, 401–411. [CrossRef]
25. Krcmar, L.; Mach, O.; Cernohorsky, J. Design and Efficiency Mapping of an Electric Drive for Mobile Robotic Container Platform for Use in Industrial Halls. In Proceedings of the 2018 IEEE International Symposium on Power Electronics, Electrical Drives, Automation and Motion (SPEEDAM), Amalfi, Italy, 20–22 June 2018; pp. 963–967.
26. Andersen, L.G.; Larsen, J.K.; Fraser, E.S.; Schmidt, B.; Dyre, J.C. Rolling resistance measurement and model development. *J. Transp. Eng.* **2014**, *141*, 04014075. [CrossRef]
27. Hannan, M.A.; Lipu, M.H.; Hussain, A.; Mohamed, A. A review of lithium-ion battery state of charge estimation and management system in electric vehicle applications: Challenges and recommendations. *Renew. Sustain. Energy Rev.* **2017**, *78*, 834–854. [CrossRef]
28. Xiong, R.; Cao, J.; Yu, Q.; He, H.; Sun, F. Critical review on the battery state of charge estimation methods for electric vehicles. *IEEE Access* **2017**, *6*, 1832–1843. [CrossRef]
29. Sone, Y.; Ooto, H.; Eguro, T.; Yoshida, T.; Kubota, M.; Yoshida, H.; Yamamoto, M.; Sakai, S.; Ogawa, K.; Takeda, Y.; et al. Charge and discharge performance of over-discharged lithium-ion secondary battery—Lessons learned from the operation of the interplanetary spacecraft HAYABUSA. *Electrochemistry* **2007**, *75*, 950–957.
30. Ng, K.S.; Moo, C.-S.; Chen, Y.-P.; Hsieh, Y.-C. Enhanced coulomb counting method for estimating state-of-charge and state-of-health of lithium-ion batteries. *Appl. Energy* **2009**, *86*, 1506–1511.
31. Thirugnanam, K.; TP, E.R.J.; Singh, M.; Kumar, P. Mathematical modeling of Li-ion battery using genetic algorithm approach for V2G applications. *IEEE Trans. Energy Convers.* **2014**, *29*, 332–343.
32. Dhameja, S. *Electric Vehicle Battery Systems*; Elsevier: Amsterdam, The Netherlands, 2001.
33. Power, B. Bioenno Power Lithium Iron Phosphate (LiFePO4) Battery Model BLF-2440A. Bioenno Power Co. Available online: https://www.bioennopower.com/products/24v-40ah-lfp-battery-pvc-blf-2440a (accessed on 27 July 2020).
34. Squadrito, G.; Maggio, G.; Passalacqua, E.; Lufrano, F.; Patti, A. An empirical equation for polymer electrolyte fuel cell (PEFC) behaviour. *J. Appl. Electrochem.* **1999**, *29*, 1449–1455. [CrossRef]

35. Amamou, A.; Kandidayeni, M.; Boulon, L.; Kelouwani, S. Real time adaptive efficient cold start strategy for proton exchange membrane fuel cells. *Appl. Energy* **2018**, *216*, 21–30. [CrossRef]
36. Ettihir, K.; Boulon, L.; Agbossou, K. Optimization-based energy management strategy for a fuel cell/battery hybrid power system. *Appl. Energy* **2016**, *163*, 142–153. [CrossRef]
37. Delarue, P.; Bouscayrol, A.; Semail, E. Generic control method of multileg voltage-source-converters for fast practical implementation. *IEEE Trans. Power Electron.* **2003**, *18*, 517–526.
38. Young, K.; Wang, C.; Wang, L.Y.; Strunz, K. Electric vehicle battery technologies. In *Electric Vehicle Integration into Modern Power Networks*; Springer: Berlin/Heidelberg, Germany, 2013; pp. 15–56.
39. Amamou, A.; Kandidayeni, M.; Boulon, L.; Kelouwani, S.; Macia, A. Efficient Model Selection for Real-Time Adaptive Strategy of a Fuel Cell System on Vehicular Applications. *Int. J. Hydrog. Energy* **2020**, *45*, 19664–19675.
40. Kandidayeni, M.; Macias, A.; Amamou, A.A.; Boulon, L.; Kelouwani, S.; Chaoui, H. Overview and benchmark analysis of fuel cell parameters estimation for energy management purposes. *J. Power Sources* **2018**, *380*, 92–104. [CrossRef]
41. Keränen, T.; Karimäki, H.; Viitakangas, J.; Vallet, J.; Ihonen, J.; Hyötylä, P.; Uusalo, H.; Tingelöf, T. Development of integrated fuel cell hybrid power source for electric forklift. *J. Power Sources* **2011**, *196*, 9058–9068. [CrossRef]

energies

MDPI

Article

Integrated Optimization of Routing and Energy Management for Electric Vehicles in Delivery Scheduling

Lixing Wang [1,2,*], Zhenning Wu [1] and Changyong Cao [2,*]

1 School of Computers and Engineering, Northeastern University, Shenyang 110000, China; wuzhenning@ise.neu.edu.cn
2 Laboratory for Soft Machines & Electronics, School of Packaging, Michigan State University, East Lansing, MI 48824, USA
* Correspondence: wanglixing@mail.neu.edu.cn (L.W.); ccao@msu.edu (C.C.); Tel.: +86-185-2444-6209 (L.W.); +1-517-353-9504 (C.C.)

Abstract: At present, electric vehicles (EVs) are attracting increasing attention and have great potential for replacing fossil-fueled vehicles, especially for logistics applications. However, energy management for EVs is essential for them to be advantageous owing to their limitations with regard to battery capacity and recharging times. Therefore, inefficiencies can be expected for EV-based logistical operations without an energy management plan, which is not necessarily considered in traditional routing exercises. In this study, for the logistics application of EVs to manage energy and schedule the vehicle route, a system is proposed. The system comprises two parts: (1) a case-based reasoning subsystem to forecast the energy consumption and travel time for each route section, and (2) a genetic algorithm to optimize vehicle routing with an energy consumption situation as a new constraint. A dynamic adjustment algorithm is also adopted to achieve a rapid response to accidents in which the vehicles might be involved. Finally, a simulation is performed to test the system by adjusting the data from the vehicle routing problem with time windows. Solomon benchmarks are used for the validations. The analysis results show that the proposed vehicle management system is more economical than the traditional method.

Keywords: electric vehicle; energy consumption; energy management; logistics; supply chain; vehicle routing problem

Citation: Wang, L.; Wu, Z.; Cao, C. Integrated Optimization of Routing and Energy Management for Electric Vehicles in Delivery Scheduling. *Energies* **2021**, *14*, 1762. https://doi.org/10.3390/en14061762

Academic Editor: João Pedro Trovao

Received: 5 March 2021
Accepted: 19 March 2021
Published: 22 March 2021

1. Introduction

Fossil energy consumption by vehicles has increased significantly over the past decades, leading to ever-worsening environmental pollution [1]. Some densely populated cities, such as Tokyo, Beijing, and Shanghai have strict policies to control the increase in fossil-fueled vehicles. City managers are also establishing suitable policies to support electrical vehicles (EVs) in urban freight transport [2,3]. EVs play an important role in replacing traditional fossil-fueled vehicles worldwide, particularly in logistics applications. For example, Amazon is procuring 100,000 EVs and plans to deploy them in their package delivery system by 2021 [4]. Mathematical modeling has also been proposed for EVs to explore the relationship between the delivery costs and sustainability impact [5].

Energy management is essential for EVs because of their limited battery capacity and specific recharging times [6]. Fully charging an EV takes much longer than refueling a traditional vehicle, and fully charged EVs cannot travel as far as fossil-fueled trucks with full fuel tanks. Therefore, recharging stops for EVs should be incorporated into route planning as an additional consideration [7]. Thus, it is necessary to develop a new vehicle routing model for EVs that determines both the shortest possible route and the best energy management strategy.

The vehicle routing problem (VRP) involves planning for vehicles to deliver and collect goods or people. The classical VRP is defined as a single depot with route length constraints [8]. Several variants of this classical problem have been studied, including the vehicle routing problem with time windows (VRPTW), because real-life cases are more complex than theoretical problems [9]. VRPTW is a problem in which every customer must start within a given time window (a, b), with the vehicle arriving before ***a*** and waiting until the customer becomes available. However, arrivals after ***b*** are prohibited. In the case of a fixed-sized fleet, finding a feasible solution to the VRPTW itself is a non-deterministic polynomial complete (NPC) problem. As a result, research on VRPTW has focused on heuristics [10]. The electric vehicle routing problem (EVRP) is an extension of the VRP that considers the use of EVs in the logistics distribution. Most studies on EVRP have focused on changes in the EV model; however, the changes brought by the new technologies should also be considered.

New technologies make logistics more transparent. Global positioning systems (GPS), sensors, mobile communication, and radio frequency identification (RFID) techniques can be used to record various types of data regarding delivery vehicles. Energy consumption can be forecasted using these technologies, albeit with some inaccuracies, and furthermore some emergent situations can be detected. The method to optimize the vehicle schedule also needs to be improved to ensure that vehicle schedules can be changed in real time.

The main gap in current studies is the lack of consideration of many factors that affect energy consumption and can be monitored when building the EV energy model. In this study, the objective is to develop a new model that considers both the optimal route and new energy management strategies for EVs. Therefore, a new energy management system is proposed for EVs based on the VRPTW model to solve the aforementioned challenges. The proposed system has two functions: first, recording historical energy consumption and forecasting future energy consumption; and second, applying a genetic algorithm (GA) to optimize vehicle scheduling using forecast energy consumption. The proposed system also alerts operators to emergencies in real time, helping operators make timely interventions.

This study is organized as follows: Section 2 briefly reviews previous research on optimization algorithms for vehicle scheduling problems and systems for energy management. In Section 3, the entire framework for the proposed system is introduced, and a method for forecasting the vehicle energy consumption is described. A new problem model is built for EVs using the predicted energy consumption; and a GA is used to prepare the vehicle schedule. Section 4 discusses the simulation, results, and performance of the proposed system. Section 4 presents the research conclusions and provides an outlook for future work.

2. Literature Review

Vehicle schedule management is a classical VRP. Researchers VRPs have studied various VRP. Alba and Dorronsoro solved the classical VRP using a cellular GA combined with a specialized local search method [11]. Tarantilis and Kiranoud developed a generalized route construction algorithm to find the optimal solution for the distribution of perishable products and ready-mixed concrete for construction companies [12]. Hwang developed a GA-TSP model by improving the GA to solve a typical VRPTW [13]. Ho and Haugland presented a tabu search heuristics method for the split delivery vehicle routing problem with time window (SDVRPTW), which considers that more than one vehicle can provide service to a customer [14]. Cheung et al. developed a mathematical model that can be used in monitoring systems for dynamic fleet management, which uses dynamic data such as vehicle locations, traveling time, and incoming customer orders [15]. These methods built a research basis for the EVRP.

Regarding EVRP research, Conrad and Figliozzi were the first to extend the traditional VRP to EVRP. They proposed a model that assumes that EVs in a fleet are allowed to recharge at certain customer locations [16]. Juan et al. proposed the use of metaheuristics and heuristics as the most efficient way to deal with VRPs [17]. Zuo et al. considered a concave, nonlinear charging function as a new energy consumption model for the EVRP [18]. Zhang et al. suggested an EV battery swap station (BSS) location-routing problem with stochastic demands to determine a minimum cost scheme. EVRP with BSSs includes the optimal number and location of BSSs in an efficient route plan based on stochastic customer demands [19]. Keskin et al. presented a two-stage simulation-based heuristic using adaptive large neighborhood searches (ALNSs) for an electric vehicle routing problem with a time window (EVRPTW) that considers whether the waiting time at the stations is longer than expected [20]. Napoli et al. discussed the issue of the production of electricity required for EVs to carry out daily missions [21]. Ferro et al. developed a new mixed-integer programming model for the EVRP and used the CPLEX solver [22]. Xiao et al. investigated an EVRPTW that included the energy and electricity consumption rates (ECR) per unit distance traveled as a function of speed and load; this problem is referred to as EVRPTW-ECR. A mixed-integer linear programming model was developed for the EVRPTW-ECR [23]. Afroditi et al. developed a comprehensive mathematical formulation with multiple constraints owing to capacity limitations, time window restrictions, and the vehicle's predefined charging level to model EVRP [24]. Kancharla and Ramadurai proposed a three-index formulation for an EVRP with nonlinear charging, load-dependent discharging, and an ALNS algorithm to solve the problem with capacitated charging stations [25]. Zhang et al. applied the ant colony algorithm to EVRP to minimize energy consumption [26]. Lin et al. presented a general EVRP and determined an optimal routing strategy that minimizes travel time, energy costs and the number of EVs dispatched. This is the first EVRP model to consider the effect of vehicle load on battery consumption [27]. Soysal et al. proposed a chance-constrained mixed-integer nonlinear programming model and a linear approximation for the pick-up and delivery problem with EVs under the stochastic battery depletion assumption [28]. Raeesi and Zografos introduced an alternative to intra-route recharging of electric commercial vehicles used for freight distribution by utilizing new pertinent technological developments that enable mobile battery swapping. They further proposed a methodology for the exact evaluation of each given solution in the context of EVRPTW [29]. Jie et al. presented a two-echelon capacitated electric vehicle routing problem with battery swapping stations (2E-EVRP-BSS) to determine the delivery strategy that considers battery driving range limitations for deliveries within metropolitan areas most effectively. An integer programming formulation and a hybrid algorithm that combines column generation and adaptive large neighborhood search (CG-ALNS) were proposed to solve this problem [30].

In addition to the aforementioned studies that used different models to calculate specific energy consumption values, other studies have considered energy consumption as an uncertainty factor for the EVRP. For example, Zhang et al. used fuzzy numbers to denote service time, battery energy consumption, and travel time inconsistencies, and applied fuzzy theory to solve the EVRPTW [31]. Pelletier et al. proposed a robust optimization framework to consider inconsistencies in the context of an EVRP. Furthermore, a two-phase heuristic method based on a large neighborhood search was used to solve larger instances of the problem. Several numerical tests were conducted to assess the effectiveness of the proposed methodology [32]. Notably, the energy consumption and traveling time must be considered because of the difficulties involved in predicting the energy consumption.

The main difference between the traditional VRP and EVRP is that the latter considers energy consumption in its model. For energy management, Basso et al. proposed a method for calculating the energy cost coefficients of a road network. These coefficients embed information regarding road topography, vehicle speed, power train efficiency, and the effects of acceleration and braking at traffic lights and intersections. Using this method, an accurate energy consumption estimation can be obtained [33]. Kessler and Bogenberger analyzed the existing energy consumption models [34]. Alqahtani and Hu developed an integrated VR and energy scheduling decision model to adaptively dispatch vehicles to balance temporally and spatially distributed energy requests. This model considers vehicle mobility constraints to maximally exploit the potential of mobile prosumer networks for cost savings and carbon emission reductions [35].

Based on these previous studies, it is difficult to accurately predict energy consumption. Multiple factors exist apart from the vehicle and the traveling distances that affect energy consumption including the weather, road conditions, and driver behavior. The effects of factors such as the number of starts and stops at intersections and traffic lights, and the speed dropping below a certain threshold must also be considered for dynamic traffic information. Consequently, a case-based reasoning (CBR) system is considered to forecast the energy consumption and travel time.

For research on the CBR system, Shen et al. built an approximate CBR model that uses neural network technology to process fuzzy inference with the dualities of fuzzy logic and approximate reasoning [36]. The main characteristic of this system is its ability to solve new problems by using the results of past cases, which is similar to the current models. Sadek et al. proposed a prototype CBR system that can create routes for real-time freeway traffic. The results of the aforementioned study indicated the successful generation of high-quality solutions using case-bases of reasonable sizes in real time [37]. Moreover, it could automatically update the case-bases by modifying the coefficients. For instance, Anthony and Xun successfully dealt with planning problems in development control using a developed CBR system. The system helped the user make decisions for new cases by recycling similar previous cases [38]. Passone et al. incorporated an expert database into a GA that was implemented for the CBR adaptation phase. The proposed system is suitable for numerical modeling applications [39]. Maria and Maite proposed retention and forgetting strategies to add and remove cases, with strategies that automatically update the case-base of a CBR system and maintain it at a certain scale. The results showed that the case-base was effectively maintained by the proposed strategies [40]. Castro et al. developed a fuzzy CBR system to solve the risk problems. The fuzzy algorithm helped the CBR system use the most suitable case in the case base instead of the most similar one [41].

3. Vehicle Management System

3.1. System Architecture

3.1.1. Vehicle Management System Architecture

With the emerging sensor and mobile communication techniques, vehicle status and environmental factors can be immediately determined and transmitted to back-end management systems. However, it is difficult to predict accurate values of energy consumption and travel time through mathematical formulations as many factors can affect the results. Therefore, a CBR system can be developed to obtain a range of values and apply the static method to obtain a conservative result for further optimization of the vehicle schedule.

Figure 1 illustrates the framework of the proposed system. This system has two subsystems: a CBR system and an optimization system. The CBR system is used to estimate the energy consumption and travel time for each delivery task. By combining the static optimization, an amplification parameter is first determined; then the energy consumption and traveling time for the calculation are obtained and output to the optimization system. With this information, the optimization system optimizes the schedule for the vehicles and the company then follows the schedule. The optimization system adjusts the schedule using heuristics if the real-time monitoring system detects abnormal situations.

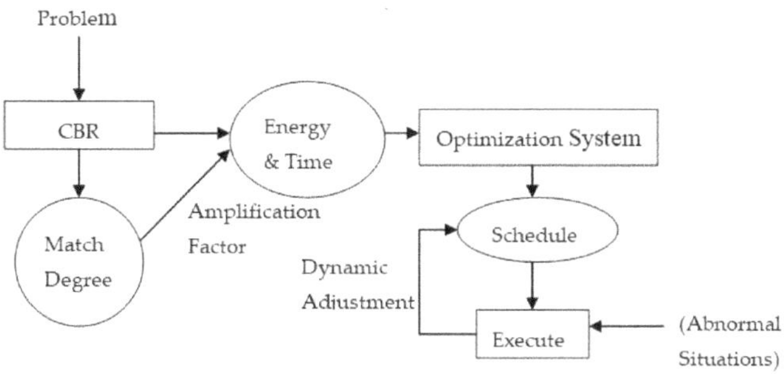

Figure 1. Vehicle management system architecture.

3.1.2. CBR System Architecture

In the proposed vehicle scheduling system, a CBR system is incorporated to forecast the energy consumption and travel time. The system architecture is illustrated in Figure 2. The system is composed of two parts. The first part is the route division. It separates the planned route into several segments. Route segmentation simplifies finding the same route from the case base. If the exact same route does not exist in the database, then fuzzy logic and CBR are used to select the most similar one. The second part of the system is the calculation part. The CBR is applied to calculate the time of each route's small segment. Neural network theory is applied to train the weightings of the CBR; rule-based strategies are then used to update the case base in the final step.

Compared with the traditional CBR system, the proposed system integrates fuzzy logic and neural network techniques. Consequently, the system is much more intelligent. The case base automatically updates itself. If the deviation between the estimated and the actual traveling result is considerable, the weightings will be trained using the neural network and the existing case will be replaced.

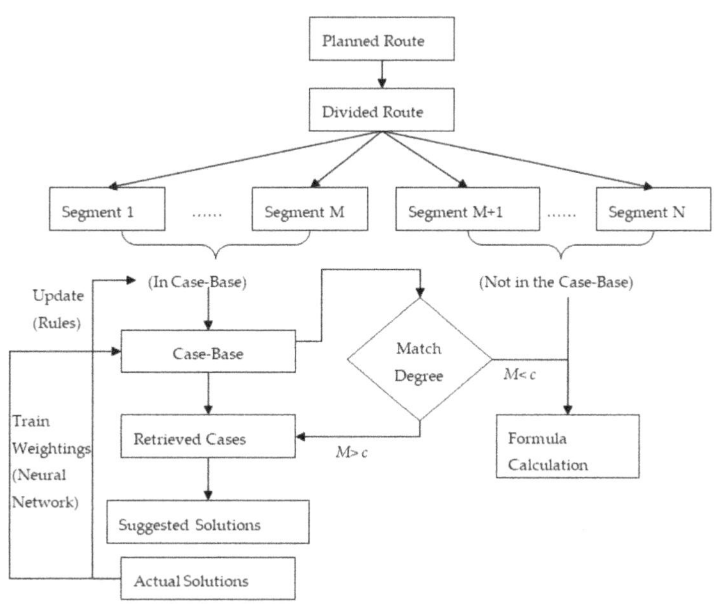

Figure 2. The sub case-based reasoning (CBR) system architecture.

3.2. CBR System

3.2.1. Weighting Design

Weightings are used to calculate the degree to which cases match in the case-base in the CBR system. A database is designed to save the weight coefficients of all factors that affect the process of degeneration. The design of the database for traveling time estimation is shown in Table 1, while the database design for energy consumption is shown in Table 2.

Different data types exhibit different weightings (w_i), with the weightings satisfying the following constraints:

$$\sum_{i=1}^{n} w_i = 1 \tag{1}$$

where $0 \leq w_i \leq 1$ ($i = 1, 2 \ldots \ldots n$).

The vehicles will provide the information listed in Tables 1 and 2 to the back-end system when they have completed their delivery tasks. The system stores the data in the case base and produces a new case identification number for the data.

When the system begins to estimate the traveling time and energy consumption, it first searches in the case base. If the data of the unsolved problem matches the data of a case in the case base, the system records the value of x_i as 1 in the blank space of the match degree. Then, it multiplies the match degree with the weighting of this type of data ($w_i \times x_i$) to produce the result. The system summates all calculation results for all types of data belonging to the case. The sum (M) is obtained as the case degree, which matches the problem that needs to be solved.

$$x_i = \begin{cases} 0, & \text{not match,} \\ 1, & \text{match,} \end{cases} \tag{2}$$

$i = 1, 2 \ldots \ldots n$,

$$M = \sum_{i=1}^{n} w_i x_i, \tag{3}$$

where M indicates the match degree of the case matches to the problem.

The system chooses the data with the highest match degree in the database for all cases of the same route and then uses the case's time or energy consumption as the predicted result for the matching segment after calculating the results. The sum of the times or energies needed for all segments is the time or energy consumption required for the vehicle to arrive at the destination.

Table 1. Weighting coefficient database for traveling time/energy consumption.

Weighting (w_i)	Factor	Match_Degree (x_i)
w_i	Weather	1/0
w_i	Workday	1/0
w_i	Time_Period	1/0
w_i	Vehicle_Type/Battery_Type	1/0
w_i	Driver	1/0
w_i	Products Weight	1/0
......	......	......
w_i	Energy level/Tire Pressure	1/0
w_i	Sum	$\sum w_i x_i$

If the match degree of a case to the problem is less than c (c is a coefficient, which can be adjusted based on real conditions by users), the deviation of the most similar existing route segments under the conditions of the best-matched case and of the unsolved problem will be calculated. The deviation ratio of the distance is the deviation ratio of the route segment of the problem and the most similar case. Then, the result can be calculated using Equation (4):

$$\begin{cases} t = t_s \cdot \frac{L}{L_s} \\ E = E_s \cdot \frac{L}{L_s} \end{cases}, \quad (4)$$

where t is the traveling time for the route segment of the problem, t_s is the traveling time for the most similar case, E is the energy consumption for the route segment of the problem, E_s is the energy consumption for the most similar case, L is the length of the route segment for the problem, and L_s is the length of the route for the most similar case.

Finally, the total energy consumption and the time spent for each segment are the energy and time needed for delivery, respectively.

3.2.2. Case Update

The factors that affect logistics are constantly changing and developing. With the development of new vehicles, changes in the transportation infrastructure, and the amount of traffic in a city, the time needed to travel or the energy consumed between the same starting point and destination under the same conditions will change. Thus, the results of the cases in the database are not applicable to new cases. Consequently, the database should be updated once the deviation becomes more pronounced. The method is as follows: First, the system checks whether the segment in the case-base is the most similar route to the problem; if it is, the case-base stores the actual result of the segment as a new case. If the new case is more similar to the previous case, for which $M = 1$, the new case will replace the previous case. The system trains the weighting of the case when the results are different. If $M \neq 1$, the case base also adds a new case. The case base is updated according to the following rules:

Rule 1: If the segment in the case-base is the same as the actual segment then go to Rule 2.

Rule 2: If the match degree is not less than c, go to Rule 3; otherwise, go to Rule 4.

Rule 3: If $M \neq 1$ go to Rule 4; otherwise, go to Rule 5

Rule 4: Add the case to the case base and train the weightings using a neural network.

Rule 5: If $\frac{ResultNew - ResultOld}{ResultOld} > a$ or $\frac{ResultNew - ResultOld}{ResultOld} < -a$ then add the case to the case base and train the weighting of this segment.

3.2.3. Weight Training

In the proposed CBR system, the result is calculated from the weight of each factor in a case w_i. In some cases, the weights in the database are incorrect. Therefore, it is necessary to apply a neural network to train the weights. The details are as follows:

$X = \{x_1, x_2, \ldots, x_n\}$ is a set of n vectors, where the components of each vector represent the match degree of a case with w_i as the coefficient, the value of which is determined by specific segments. The different segments have different sets of w_i. A single-layer neural network is applied to train the weightings [42].

Step 1: Initialization

Set initial weights w_i and threshold θ as random numbers.

Step 2: Activation

Activate the perceptron by applying inputs $x_i(q)$ and the desired output $Y_d(q)$, which is the actual traveling time. The actual output at iteration $q = 1$ is calculated.

$$Y(q) = step[\sum_{i=1}^{n} x_i(q)w_i(q) - \theta], \quad (5)$$

where n is the number of perception inputs and Equation (5) is a step activation function.

Step 3: Training

Update the weights of the perceptron

$$w_i(q+1) = w_i(q) + \Delta w_i(q), \quad (6)$$

where $\Delta w_i(q)$ is the weight correction at iteration q.

The weight is corrected based on the delta rule:

$$\Delta w_i(q) = \alpha \times x_i(q) \times e(q), \tag{7}$$

$$e(q) = Y_d(q) - Y(q), \tag{8}$$

Step 4: Iteration

Increase p by 1, return to Step 2, and repeat the process until convergence.

Then, w_i can be determined.

3.2.4. Amplification Coefficient

In this study, the uncertainties of the predicted time and energy consumption are considered. Therefore, an amplification coefficient is used to ensure that the arrival time of the product will be within an acceptable time window and to prevent the vehicle from running out of energy while it is in use.

The CBR has two outputs: the estimated time and the match degree. An amplification coefficient A is created by considering these two factors.

$$A = 1 + \frac{c}{M} \tag{9}$$

where c is a coefficient that can be adjusted based on practical situations, experiments, and simulation results. M represents the match degree of the most similar case. Figure 3 shows the relationship between A and M.

Then, the estimated traveling time t is adjusted to t':

$$t' = A \cdot t \tag{10}$$

The estimated energy consumption E is adjusted to E':

$$E' = A \cdot E \tag{11}$$

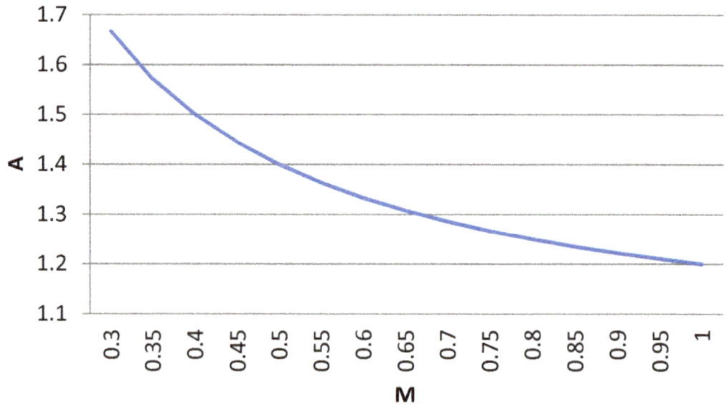

Figure 3. The relationship between the amplification coefficient (A) and the match degree (M).

3.3. *Model of Problem*

The recharging of EVs in a charging station during a delivery is considered neither in the proposed model nor in most other EVRPTWs. Therefore, the vehicle can only be charged in the depot, in the proposed system. There are two reasons for this design. First, in most countries, recharging stations for EVs are not common; hence, it is possible that there are no recharging stations on the delivery route. Second, current batteries can last longer than previous batteries owing to the improved EV designs. For example, Tesla states that its electric trucks can travel 800 km between successive charges. Other commercial electrical trucks can travel as much as 400 km between charges. This distance is normally sufficient for the daily tasks of a delivery truck. In the future, a solution will need to be considered for the charging problem.

The schedule problem can be defined on a direct graph $G = (V, A)$, where A is the set of arcs and $V = \{1,2, \ldots, v\}$ is the set of destination locations. Furthermore, A_m is the set of arcs of vehicle m. For any $i \rightarrow j \in A$, let $t\prime_{ij}$ denote the normal adjusted traveling time from destination i to destination j, let e_{ij} denote the energy consumption from destination i to destination j, and let d_{ij} denote the distance from destination i to destination j. All the vehicles start from the same distribution center, called the depot. Set k as the available vehicles, where vehicle m has the capacity C_m and full energy E_m. Set p_m as the cost of vehicle m running at 1 km. Set f_m as the fee for using vehicle m, which includes the driver salary and depreciation cost of the vehicle. There are n products that must be delivered. In this problem, a vehicle is only allowed to deliver product i in a given time window $[a_i, b_i]$, which means that the destination only handles the consignment after a_i and before b_i. A vehicle is only allowed to arrive at the distribution center before a_i, but the vehicle can wait until the destination becomes available; however, but arrivals after b_i are not allowed. Set s_i as the service time for product i. Set w as the weight of product i. The duration for which the destinations are open is defined as $[a, b]$.

The objective function of the problem is stated as follows:

$$\text{Minimize} \sum_{m \in K} (p_m \sum_{(i,j) \in A_m} d_{ij} + f_m), \tag{12}$$

Each vehicle's schedule is subject to the following constraints:

$$t' + t'_{ij} <= b_j \tag{13}$$

where b_j denotes the latest time for product j. This means that the vehicle must arrive before the upper time of the time window for destination j.

$$t' = \max(t', a_j) + s_j, \tag{14}$$

This formula is used to calculate the ready time that the vehicle can leave the destination j.

$$\sum W_k \leq C_m, \tag{15}$$

where W_k denotes the weight of product k in vehicle m. This means that the total weight of the products must be less than the capacity of vehicle m.

$$\sum_{i,j \in A_m} e_{ij} \leq E_m \tag{16}$$

This means that the total energy consumption by vehicle m must be less than its energy capacity.

A GA is applied to solve this problem. A specific method can be found in reference [43]. In the algorithm, the chromosome string is composed of the serial number of the products, where a gene means a product and its order means an arrangement order. The initial population was randomly selected. A selection factor was set to select the parents using the roulette wheel. It also contains a mutation operation to prevent the population from becoming trapped in local optimization. Subsequently, a new generation was produced. Equation (12) is a fitness function that is used to evaluate the arrangement performance. This process is repeated. After each iteration, the best solution is obtained. When the number of iterations reaches the stopping number or time reaches the stopping time, the vehicle schedule is arranged using the best result.

3.4. Dynamic Adjustment

Three situations are considered for dynamic adjustment: traffic jams, environmental changes, and urgent consignments.

3.4.1. Traffic Congestion and Vehicle Problem

It is difficult to forecast traffic situations and vehicle accidents. If the monitoring system finds a truck stuck in a traffic jam or in need of maintenance, it will check whether the current estimated delivery time will exceed the latest starting time of the delivery or whether the remaining energy is sufficient to complete all deliveries. If the system finds that some products cannot be delivered within their corresponding time windows or that the vehicle's energy will be insufficient, the system will remove the deliveries from the original schedule of the vehicle and reassign them.

3.4.2. Environmental Effects

Some products are sensitive to certain environmental conditions. Changes in factors such as temperature, humidity, and concentration of dangerous gases strongly affect the quality of these products. If there are problems with environmental factors in a vehicle, the consignments in the vehicle are removed from the original schedule by the system and reincorporated into the schedule.

3.4.3. New Urgent Tasks

If there are any urgent products arriving after the static optimization, dynamic optimization is applied to ensure that these products can be incorporated into the original schedule.

The three aforementioned situations, as well as some other situations, can be handled by incorporating deliveries into the original schedule, but the parameters of the problem need to be adjusted. The methods used to handle such situations would be the same. The method shown in Figure 4 is described in detail in reference [43]. Subsequently, a new schedule is generated.

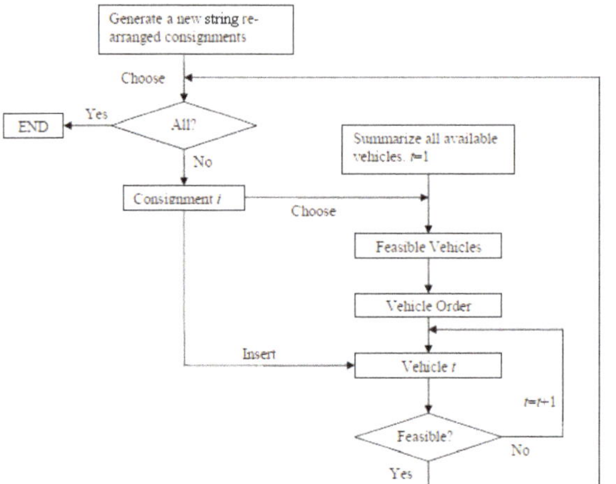

Figure 4. Procedure of dynamic adjustment.

3.5. Simulation

Solomon benchmark problems were employed for the simulation to test the proposed system. The data can be obtained from http://w.cba.neu.edu/~msolomon/problems.htm (Accessed on 7 October 2020). Because of a lack of energy data for EVs in the Solomon database, only time was considered in the simulation. For the CBR system, a normal distribution was used to build the traveling time model. Because the traveling time of each issue was independent, it obeyed a normal distribution. The model was used to simulate the traveling time. The expected value and variance were related to the results of the case archived from the CBR system. The model was built as follows:

The probability density function of the normal distribution is

$$f(x) = \frac{1}{\sqrt{2\pi}\sigma} e^{-\frac{(x-\mu)^2}{2\sigma^2}}, -\infty < x < +\infty, \tag{17}$$

As shown in Equation (17), the normal distribution has two variables: μ and σ. In this model, the estimated time can be set as μ, and

$$\sigma = \frac{1}{c \cdot M}, \tag{18}$$

where c represents a coefficient that can be adjusted, and M represents the match degree of the most similar case.

Based on the 3σ rule, which states that for a normal distribution, nearly all values lie within three standard deviations of the mean, the amplification coefficient A can be set based on Equation (19):

$$A = \frac{\mu + 3\sigma}{\mu}, \tag{19}$$

For this problem, the traveling time is assumed to follow a normal distribution between the two distribution centers and between each distribution center and the depot. The corresponding amplification coefficient is obtained from the data of the CBR system $A_{ij}(i, j \in A; i \neq j)$. Subsequently, $t\prime_{ij}$ can be calculated. With the calculated result, the GA is used for further optimization.

Table 2 shows an example of some cases between the depot and the two distribution centers in the case base. From the data, $\mu = 15.25D = \sigma^2 = 1.337$, and $\sigma = 1.156$ can be obtained. The amplification coefficient can be calculated as follows:$A_{02} = \frac{\mu+3\sigma}{\mu} = 1.227$, $t\prime_{02} = A_{02} \cdot \mu_{02} = 18.718$. Then, $t\prime_{02}$ is used as the new traveling time between the depot and Point 2 for the Solomon benchmark problem to test the optimization method. The traveling times between points are calculated using the same method. The calculated traveling time replaces the original time in the Solomon benchmark problem for optimization. Subsequently, the GA is applied.

Table 2. CBR database of distribution center and second customer.

ID	Weather	Workday/Holiday	Time	Driver	Weight	Travelling Time
001		Workday	8:00	A	10	15.1
002		Workday	9:00	A	10	12.5
003		Workday	8:00	A	10	16.5
004		Workday	8:30	A	10	17.2
005		Workday	8:30	B	8	13.0
006		Holiday	9:00	B	9	15.8
007		Holiday	9:00	B	8	15.0
008		Holiday	8:30	B	10	16.2
009		Workday	8:25	B	9	15.2
010		Workday	8:10	A	10	15.5
011		Workday	8:00	A	10	15.0
012		Holiday	8:10	B	9	15.1
013		Holiday	8:00	B	10	16.0
014		Workday	8:00	A	10	15.2
015		Workday	8:20	A	10	15.5

For the GA, the population size is set as 500, p_c is set as 0.8, p_m is set as 0.03, and the maximum *Gen* is set as 3000. The cost unit $p = 1$, the loading of the vehicle $q = 200$, the penalty for waiting $p_e = 1$, and the penalty for late arrival $p_l = 10$. The running results of R101 in the Solomon benchmarks and the optimal route are shown in Table 3 and Figure 5.

The main objective of the tests is to compare the performance for certain and uncertain times. Table 4 shows the results of the optimized cost for the R101 problem. Then, the real conditions are simulated. After implementing the optimal schedule, a random traveling time is generated following the normal distribution of the cases in the case base. This may cause schedules with a certain time to not satisfy the time requirement. Then, a penalty is added to the cost. The results are listed in Table 5. The results show that the real cost considering uncertain time is less than the cost that only considers a certain time.

Table 3. Operating results for 10 tests.

Test	Result
1	1892.60
2	1895.06
3	1900.24
4	1906.41
5	1892.60
6	1895.06
7	1892.60
8	1902.61
9	1892.60
10	1892.60
Average	1896.78

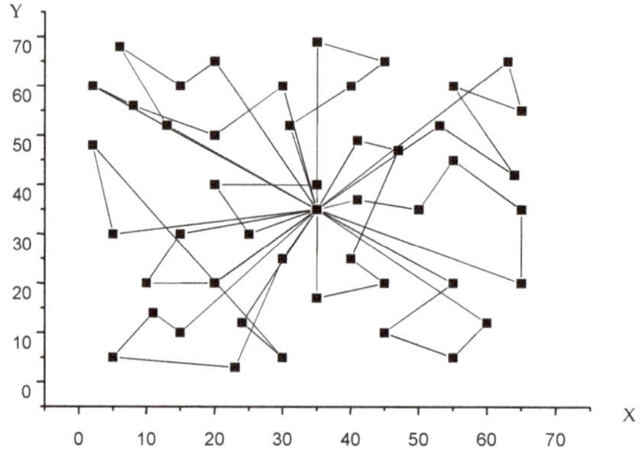

Figure 5. Optimal route.

As mentioned before, the Solomon benchmark has three types of problems. The serial numbers begin with C, R, and RC respectively. For the cases that begin with C, the points can be divided into clusters according to their locations. For the cases that begin with R, the time windows are narrow. For the cases that begin with RC, the locations of the points can be divided into clusters, and the time windows are narrow. Therefore, three groups of ten tests were conducted. From the results shown in Tables 4, 6, and 7, for the original schedule, the cost for a certain time is better than that for an uncertain time. This is because the traveling time for a certain case uses the time for an average number of cases, which is less than the traveling time in uncertain cases, and it can arrange fewer vehicles and incur less cost. However, ideal cases do not occur in reality. Real conditions may cause schedules with a certain time to not satisfy the time requirement, and a penalty will be produced. Therefore, after the simulation, the situation is different. Regardless of the type of problem, the results show that the real cost considering uncertain time is less than the cost that only considers a certain time. This can be observed from Tables 5, 8, and 9.

Table 4. Comparison of scheme costs for R101.

Test	Cost (Certain Time)	Vehicle (Certain Time)	Cost (Uncertain Time)	Vehicle (Uncertain Time)
1	1629.99	12	1892.60	12
2	1640.94	13	1895.06	12
3	1635.31	12	1900.24	13
4	1629.99	12	1906.41	13
5	1631.54	12	1892.60	12
6	1629.99	12	1895.06	12
7	1629.99	12	1892.60	12
8	1631.54	12	1902.61	13
9	1633.12	12	1892.60	12
10	1636.02	12	1892.60	12
Average	1632.84	12.1	1896.78	12.3

Table 5. Simulation results of R101.

Test	Cost (Certain Time)	Cost (Uncertain Time)
1	2598.56	2437.28
2	2690.75	2341.65
3	2571.67	2276.59
4	2498.63	2284.61
5	2601.43	2348.10
6	2571.62	2283.94
7	2489.61	2310.48
8	2701.46	2401.82
9	2613.20	2199.62
10	2894.02	2294.84
Average	2623.09	2317.89

The Solomon benchmark has three types of problems. Simulations are conducted for C101 and RC 101. The results of the initial costs are presented in Tables 6 and 7. Tables 8 and 9 present the data after the implementation of the schedule. The schedules considering uncertain time are better for these two types of problems.

Table 6. Comparison of scheme costs for C101.

Test	Cost (Certain Time)	Vehicle (Certain Time)	Cost (Uncertain Time)	Vehicle (Uncertain Time)
1	363.25	5	458.62	5
2	370.89	6	462.06	6
3	363.25	5	458.62	5
4	369.79	6	459.36	5
5	363.25	5	458.62	5
6	363.25	5	458.62	5
7	365.62	5	462.14	6
8	363.25	5	460.84	6
9	365.62	5	459.36	5
10	368.02	5	458.62	5
Average	365.02	5.2	459.71	5.3

Table 7. Comparison of scheme Costs for RC101.

Test	Cost (Certain Time)	Vehicle (Certain Time)	Cost (Uncertain Time)	Vehicle (Uncertain Time)
1	1023.65	9	1345.29	9
2	1038.34	9	1347.06	9
3	1034.06	9	1345.29	9
4	1023.65	9	1345.29	9
5	1034.06	9	1348.20	9
6	1023.65	9	1350.62	10
7	1023.65	9	1345.29	9
8	1040.12	10	1345.29	9
9	1034.06	9	1347.06	9
10	1023.65	9	1353.07	11
Average	1029.89	9.1	1347.25	5.3

Table 8. Simulation results of C101.

Test	Cost (Certain Time)	Cost (Uncertain Time)
1	583.95	495.17
2	593.41	509.46
3	602.74	521.03
4	630.15	559.46
5	573.24	496.86
6	584.36	539.16
7	591.03	529.43
8	604.53	560.14
9	640.13	499.62
10	596.70	520.94
Average	600.02	523.13

Table 9. Simulation results of RC101.

Test	Cost (Certain Time)	Cost (Uncertain Time)
1	1476.32	1302.68
2	1452.04	1446.92
3	1464.65	1386.34
4	1623.86	1450.46
5	1500.96	1409.62
6	1546.21	1395.26
7	1689.01.	1345.10
8	1489.62	1406.98
9	1600.45	1400.23
10	1584.23	1384.63
Average	1390.74	1275.82

Finally, the conclusions are summarized. Considering the uncertainty of the factors, including energy consumption and travel time, will be more suitable to reality. The final result is the most cost-saving solution.

4. Conclusions

In summary, a new EV schedule management system was presented. The system consists of two parts.

(1) The forecasting system obtains a set of fuzzy data, in which the CBR is applied to forecast the time and energy consumption. The proposed CBR system includes a case-base design, weight training, and case updates. For weight training, an ANN is incorporated into the algorithm design. Additionally, considering the development of cities, case base

updates are also included in the system design. The main innovation of the system is that an amplification coefficient that complies with the match degree is proposed, which will also be used in the second part of the system. An amplification coefficient is generated to guarantee that the vehicle can complete the tasks in the allotted time without running out of energy.

(2) The optimization process with a GA and hybrid heuristic method is applied to optimize the vehicle schedule. In this process, the amplification coefficient was considered; if there are any abnormal situations, dynamic adjustment can be performed.

Finally, a simulation was conducted to prove that the consideration of uncertain times and energy could effectively reduce delivery loss and cost. The Solomon benchmark problem was used to test the system. The results showed that, in the beginning, the cost of the proposed system was higher than that of a traditional system. However, the proposed system could effectively decrease the occurrence of default issues, which would affect the reputation of a company.

This study built a new energy management and vehicle routing model for EVs. In the model, the energy consumption and travelling time can be forecasted by CBR system and adjusted for further processed. With the result a more reasonable and reliable routing can be made for EVs. With the help of the model, companies can manage EVs' schedule more effective in logistics.

In future work, the system would be deployed in a logistics company to collect real data to perfect the proposed model and CBR system. For EVs, the factors that affect battery life are limited to the theoretical part (at least for now). However, in the real world, there may be some other factors that have not been considered previously, such as road situations and driver behaviors.

Author Contributions: L.W. and Z.W. conceived the ideas and designed the framework of the system; L.W. contributed to the case-based reasoning system design and the simulation to test the system; Z.W. contributed to the optimization system design; L.W. prepared the draft; C.C. supervised the work and contributed to the manuscript writing and editing. all authors commented the manuscript. All authors have read and agreed to the published version of the manuscript.

Funding: This research was funded by the National Natural Science Foundation of China under Grant 71502029 and Grant 61703087.

Informed Consent Statement: Not applicable.

Data Availability Statement: The data presented in this study are openly available in http://w.cba.neu.edu/~msolomon/problems.htm (Accessed on 7 October 2020).

Acknowledgments: The authors would like to acknowledge the financial support from the China Scholarship Council (CSSA) and technical support from Michigan State University. We would like to thank the anonymous reviewers very much, as their comments and suggestions were very helpful in this paper's revision.

Conflicts of Interest: The authors declare no conflict of interest.

Abbreviations

EV	Electric vehicle
VRP	vehicle routing problem
VRPTW	vehicle routing problem with time windows
NPC	non-deterministic polynomial complete
EVRP	electric vehicle routing problem
GPS	Global positioning systems
RFID	radio frequency identification
GA	genetic algorithm

TSP	Traveling salesman problem
BSS	battery swap station
SDVRPTW	Split Delivery Vehicle Routing Problem with Time Window
ALNS	adaptive large neighborhood search
EVRPTW	electric vehicle routing problem with a time window
ECR	electricity consumption rate
2E-EVRP-BSS	two-echelon capacitated electric vehicle routing problem with battery swapping station
CG-ALNS	column generation and adaptive large neighborhood search
CBR	case-based reasoning

References

1. Ritchie, H. Fossil Fuels. Published Online at OurWorldInData.org. Available online: https://ourworldindata.org/fossil-fuels (accessed on 10 January 2020).
2. Mirhedayatian, S.M.; Yan, S. A framework to evaluate policy options for supporting electric vehicles in urban freight transport. *Transp. Res. Part D Transp. Environ.* **2018**, *58*, 22–38. [CrossRef]
3. Foley, B.; Degirmenci, K.; Yigitcanlar, T. Factors Affecting Electric Vehicle Uptake: Insights from a Descriptive Analysis in Australia. *Urban Sci.* **2020**, *4*, 57. [CrossRef]
4. Meisenzahl, M. *Amazon just Revealed Its First Electric Delivery van of a Planned 100,000-Strong EV Fleet—See How It Was Designed*; Business Insider: New York, NY, USA, 2020.
5. Munoz-Villamizar, A.; Montoya-Torres, J.R.; Faulin, J. Impact of the use of electric vehicles in collaborative urban transport networks: A case study. *Transp. Res. Part D Transp. Environ.* **2017**, *50*, 40–54. [CrossRef]
6. Tomaszewska, A.; Chu, Z.; Feng, X.; O'Kane, S.; Liu, X.; Chen, J.; Ji, C.; Endler, E.; Li, R.; Liu, L.; et al. Lithium-ion battery fast charging: A review. *eTransportation* **2019**, *1*, 100011. [CrossRef]
7. Schneider, M.; Stenger, A.; Goeke, D. The electric vehicle-routing problem with time windows and recharging stations. *Transp. Sci.* **2014**, *48*, 500–520. [CrossRef]
8. Toth, P.; Vigo, D. *The Vehicle Routing Problem*; Society for Industrial and Applied Mathematics: Philadelphia, PA, USA, 2002.
9. Desrochers, M.; Desrosiers, J.; Solomon, M. A new optimization algorithm for the vehicle routing problem with time windows. *Oper. Res.* **1992**, *40*, 342–354. [CrossRef]
10. Arnold, F.; Sörensen, K. What makes a VRP solution good? the generation of problem-specific knowledge for heuristics. *Comput. Oper. Res.* **2019**, *106*, 280–288. [CrossRef]
11. Alba, E.; Dorronsoro, B. Computing nine new best-so-far solutions for capacitated VRP with a cellular genetic algorithm. *Inf. Process. Lett.* **2006**, *98*, 225–230. [CrossRef]
12. Tarantilis, C.D.; Kiranoudis, C.T. A flexible adaptive memory-based algorithm for real-life transportation operations: Two case studies from dairy and construction sector. *Eur. J. Oper. Res.* **2007**, *179*, 806–822. [CrossRef]
13. Hwang, H.S. An improved model for vehicle routing problem with time constraint based on genetic algorithm. *Comput. Ind. Eng.* **2002**, *42*, 361–369. [CrossRef]
14. Ho, S.C.; Haugland, D.A. tabu search heuristic for the vehicle routing problem with time windows and split deliveries. *Comput. Oper. Res.* **2004**, *31*, 1947–1964. [CrossRef]
15. Cheung, B.K.S.; Choy, K.L.; Li, C.L.; Shi, W.Z.; Tang, J. Dynamic routing model and solution methods for fleet management with mobile technologies. *Int. J. Prod. Econ.* **2008**, *113*, 694–705. [CrossRef]
16. Conrad, R.G.; Figliozzi, M.A. The recharging vehicle routing problem. In Proceedings of the IIE Annual Conference, Reno, Nevada, 21–25 May 2011.
17. Juan, A.; Mendez, C.; Faulin, J.; Armas, J.D.; Grasman, S. Electric Vehicles in Logistics and Transportation: A Survey on Emerging Environmental, Strategic, and Operational Challenges. *Energies* **2016**, *9*, 86. [CrossRef]
18. Zuo, X.; Xiao, Y.; You, M.; Kaku, I.; Xu, Y. A new formulation of the electric vehicle routing problem with time windows considering concave nonlinear charging function. *J. Clean. Prod.* **2019**, *236*, 117687. [CrossRef]
19. Zhang, S.; Chen, M.; Zhang, W. A novel location-routing problem in electric vehicle transportation with stochastic demands. *J. Clean. Prod.* **2019**, *221*, 567–581. [CrossRef]
20. Keskin, M.; Çatay, B.; Laporte, G. A simulation-based heuristic for the electric vehicle routing problem with time windows and stochastic waiting times at recharging stations. *Comput. Oper. Res.* **2021**, *125*, 105060. [CrossRef]
21. Napoli, G.; Micari, S.; Dispenza, G.; Andaloro, L.; Polimeni, A. Freight distribution with electric vehicles: A case study in Sicily. RES, infrastructures and vehicle routing. *Transp. Eng.* **2021**, *2021*, 100047. [CrossRef]
22. Ferro, G.; Paolucci, M.; Robba, M. An optimization model for electrical vehicles routing with time of use energy pricing and partial recharging. *IFAC-PapersOnLine* **2018**, *51*, 212–217. [CrossRef]
23. Xiao, Y.; Zuo, X.; Kaku, I.; Zhou, S.; Pan, X. Development of energy consumption optimization model for the electric vehicle routing problem with time windows. *J. Clean. Prod.* **2019**, *225*, 647–663. [CrossRef]
24. Afroditi, A.; Boile, M.; Theofanis, S.; Sdoukopoulos, E.; Margaritis, D. Electric vehicle routing problem with industry constraints: Trends and insights for future research. *Transp. Res. Procedia* **2014**, *3*, 452–459. [CrossRef]

25. Kancharla, S.R.; Ramadurai, G. Electric vehicle routing problem with non-linear charging and load-dependent discharging. *Expert Syst. Appl.* **2020**, *160*, 113714. [CrossRef]
26. Zhang, S.; Gajpal, Y.; Appadoo, S.; Abdulkader, M. Electric vehicle routing problem with recharging stations for minimizing energy consumption. *Int. J. Prod. Econ.* **2018**, *203*, 404–413. [CrossRef]
27. Lin, J.; Zhou, W.; Wolfson, O. Electric vehicle routing problem. *Transp. Res. Procedia* **2016**, *12* (Suppl. SC), 508–521. [CrossRef]
28. Soysal, M.; Çimen, M.; Belba, S. Pickup and delivery with electric vehicles under stochastic battery depletion. *Comput. Ind. Eng.* **2020**, *146*, 106512. [CrossRef]
29. Raeesi, R.; Zografos, K.G. The electric vehicle routing problem with time windows and synchronised mobile battery swapping. *Transp. Res. Part B Methodol.* **2020**, *140*, 101–129. [CrossRef]
30. Jie, W.; Yang, J.; Zhang, M.; Huang, Y. The two-echelon capacitated electric vehicle routing problem with battery swapping stations: Formulation and efficient methodology. *Eur. J. Oper. Res.* **2019**, *272*, 879–904. [CrossRef]
31. Zhang, S.; Chen, M.; Zhang, W.; Zhuang, X. Fuzzy optimization model for electric vehicle routing problem with time windows and recharging stations. *Expert Syst. Appl.* **2020**, *145*, 113123. [CrossRef]
32. Pelletier, S.; Jabali, O.; Laporte, G. The electric vehicle routing problem with energy consumption uncertainty. *Transp. Res. Part B Methodol.* **2019**, *126*, 225–255. [CrossRef]
33. Basso, R.; Kulcsár, B.; Egardt, B.; Lindroth, P.; Sanchez-Diaz, I. Energy consumption estimation integrated into the electric vehicle routing problem. *Transp. Res. Part D Transp. Environ.* **2019**, *69*, 141–167. [CrossRef]
34. Kessler, L.; Bogenberger, K. Dynamic traffic information for electric vehicles as a basis for energy-efficient routing. *Transp. Res. Procedia* **2019**, *37*, 457–464. [CrossRef]
35. Alqahtani, M.; Hu, M. Integrated energy scheduling and routing for a network of mobile prosumers. *Energy* **2020**, *200*, 117451. [CrossRef]
36. Shen, Z.L.; Lui, H.C.; Ding, L.Y. Approximate Case-Based Reasoning on Neural Networks. *Int. J. Approx. Reason.* **1994**, *10*, 75–98. [CrossRef]
37. Sadek, A.W.; Smith, B.L.; Demetsky, M.J. A prototype case-based reasoning system for real-time freeway traffic routing. *Transp. Res. Part C Emerg. Technol.* **2001**, *9*, 353–380. [CrossRef]
38. Yeh, A.G.O.; Shi, X. Case-based reasoning (CBR) in development control. *JAG* **2001**, *3*, 238–251. [CrossRef]
39. Passone, S.; Chung, P.W.H.; Nassehi, V. Incorporating domain-specific knowledge into a genetic algorithm to implement case-based reasoning adaptation. *Knowl. Based Syst.* **2006**, *19*, 192–201. [CrossRef]
40. Maria, S.; Maite, L.S. Adaptive case-based reasoning using retention and forgetting strategies. *Knowl. Based Syst.* **2011**, *24*, 230–247.
41. Castro, J.L.; Navarro, M.; Sanchez, J.M.; Zurita, J.M. Introducing attribute risk for retrieval in case-based reasoning. *Knowl. Based Syst.* **2011**, *24*, 257–268. [CrossRef]
42. Negnevitsky, M. *Artificial Intelligence: A Guide to Intelligent Systems*, 2nd ed.; UTAS: Sandy Bay, Australia, 2005.
43. Wang, L.; Ting, S.L.; Ip, W.H. Design of a Radio Frequency Identification (RFID) based Monitoring and Vehicle Management System. In Proceedings of the 2014 International Conference on Wireless, Shenzhen, China, 16–17 November 2014.

MDPI
St. Alban-Anlage 66
4052 Basel
Switzerland
Tel. +41 61 683 77 34
Fax +41 61 302 89 18
www.mdpi.com

Energies Editorial Office
E-mail: energies@mdpi.com
www.mdpi.com/journal/energies

www.ingramcontent.com/pod-product-compliance
Ingram Content Group UK Ltd.
Pitfield, Milton Keynes, MK11 3LW, UK
UKHW060123010826
14089UKWH00006B/126